AF411178

# Impact of Pre-Mortem Factors on Meat Quality

# Impact of Pre-Mortem Factors on Meat Quality

Editor

**Gen Kaneko**

MDPI • Basel • Beijing • Wuhan • Barcelona • Belgrade • Manchester • Tokyo • Cluj • Tianjin

*Editor*
Gen Kaneko
School of Arts and Sciences
University of Houston - Victoria
Victoria, TX
United States

*Editorial Office*
MDPI
St. Alban-Anlage 66
4052 Basel, Switzerland

This is a reprint of articles from the Special Issue published online in the open access journal *Foods* (ISSN 2304-8158) (available at: www.mdpi.com/journal/foods/special_issues/Factors_on_Meat_Quality).

For citation purposes, cite each article independently as indicated on the article page online and as indicated below:

LastName, A.A.; LastName, B.B.; LastName, C.C. Article Title. *Journal Name* **Year**, *Volume Number*, Page Range.

**ISBN 978-3-0365-2815-1 (Hbk)**
**ISBN 978-3-0365-2814-4 (PDF)**

# Contents

# About the Editor

**Gen Kaneko**

Dr. Gen Kaneko is an evolutionary biologist interested in comparative aspects of energy metabolism. He received his PhD from the University of Tokyo and has worked at the University of Tokyo, Yale University, and the University of Houston - Victoria. His expertise ranges from biochemistry to bioinformatics. Major findings by Dr. Kaneko and his colleagues include the following: (1) inheritance of calorie restriction-induced longevity in a rotifer; (2) characterization of adipocytes in fish skeletal muscle; and (3) clarification of the complex molecular evolution of the metazoan 70 kDa heat shock proteins (HSP70s). He is currently expanding the study by extensive use of nuclear magnetic resonance (NMR) spectroscopy and bioinformatics.

*Editorial*

# Impact of Pre-Mortem Factors on Meat Quality: An Update

**Gen Kaneko** 

School of Arts & Sciences, University of Houston-Victoria, Victoria, TX 77901, USA; kanekog@uhv.edu



**Citation:** Kaneko, G. Impact of Pre-Mortem Factors on Meat Quality: An Update. *Foods* **2021**, *10*, 2749. https://doi.org/10.3390/foods10112749

Received: 18 October 2021
Accepted: 8 November 2021
Published: 10 November 2021

**Publisher's Note:** MDPI stays neutral with regard to jurisdictional claims in published maps and institutional affiliations.



Meat quality is closely associated with the chemical composition of skeletal muscle and is therefore influenced by the pre-mortem metabolic state of skeletal muscle tissue. Muscle metabolism is affected by various pre-mortem factors such as diet, age, genetic background, and environmental temperature. The importance of muscle metabolism has been increasingly recognized as an intermediate element that links meat quality and pre-mortem factors (i.e., growth conditions) in meat science [1–3].

This special issue of Foods, "Impact of Pre-Mortem Factors on Meat Quality" (ISSN 2304-8158), aims to compile the recent literature with a focus on pre-mortem factors, muscle metabolism, and meat quality. It includes nine research articles about various types of meat (beef [4], lamb [5], pork [6], chicken [7–9], goat [10,11] and fish [12]) as well as one review article about beef quality [13]. These articles, while their aims are different, provide an accurate representation of the current frontier of meat science and the direction in which it is heading. This editorial article is written to introduce three aspects of food science, highlighted by articles in this special issue.

The first keyword is sustainability. A sustainable food system is a system that "delivers food security and nutrition for all in such a way that the economic, social and environmental bases to generate food security and nutrition for future generations are not compromised" [14]. The practical concept of sustainability in the food industry includes the maximum utilization of known materials, the identification of new alternative foods, ensuring economic stability for producers, and food safety/security. In this regard, most food science studies should be of sustainable value, but studies on indigenous chickens [7], algae or insect supplementation in chicken diets [8], and milk replacers [10,11] clearly offer sustainable solutions to current problems in meat science. The continued characterization of novel pre-mortem factors that influence meat quality contributes to optimizing the growth conditions of animals.

The second aspect is the emergence of the omics approach, which has allowed us to understand the overall changes in muscle metabolism induced by pre-mortem factors. Metabolomics is probably the most common omics technology in meat science because meat samples are often not "fresh" enough to be analyzed by proteomics or transcriptomics. In their novel and highly relevant study, Biondi et al. (2019) analyzed the effect of diet on a microbiome using lamb meat [5]. Tuell et al. (2020) demonstrated that metabolomics is sensitive enough to predict the effect of photoperiod, which seems indirect compared to other factors such as diets and temperature, on oxidative stability in broiler fillet [9]. The omics technologies are not yet applicable on-site, mainly due to their high cost but will be introduced to assess meat quality [15].

The last aspect is the application of advanced computational procedures. In particular, the accumulation of data from omics studies has facilitated the use of computational procedures in meat science, and in fact, some articles in this special issue are closely related to the omics "big data" approaches from the same group [16–18]. Research articles in this special issue analyze factors that affect meat quality via mathematical modeling [4], machine learning [12], and correlation-clustering analysis [6]. One review article summarizes the effect of diet and genetics on beef quality characteristics [13]. Along with the prevalence of omics approaches, greater computational efforts will be required in meat science.

Due to the inclusion of rigorous and well-researched articles, this special issue offers a valuable contribution to meat science, highlighting the potential role of omics and computational analysis to further advance meat science in a sustainable manner. A possible addition to current research efforts would be the integrated experimental design and writing style, which facilitate the use of the systematic approach. Systematic reviews have become an important method to provide evidence-based interventions in medicine, but their application has been, and is still, uncommon in meat science. This is partly because of the diverse nature of muscle metabolism—results acquired from a specific species cannot be directly compared with those from other species. However, that is indeed the reason why data from a single study should be effectively utilized. A concerted effort is necessary to move the field forward, providing promising protein sources for humans in the present era of food security concerns.

**Funding:** This research received no external funding.

**Institutional Review Board Statement:** Not applicable.

**Informed Consent Statement:** Not applicable.

**Data Availability Statement:** Not applicable.

**Conflicts of Interest:** The authors declare no conflict of interest.

## References

1. Matarneh, S.K.; Silva, S.L.; Gerrard, D.E. New insights in muscle biology that alter meat quality. *Annu. Rev. Anim. Biosci.* **2021**, *9*, 355–377. [CrossRef] [PubMed]
2. Gonzalez-Rivas, P.A.; Chauhan, S.S.; Ha, M.; Fegan, N.; Dunshea, F.R.; Warner, R.D. Effects of heat stress on animal physiology, metabolism, and meat quality: A review. *Meat Sci.* **2020**, *162*, 108025. [CrossRef] [PubMed]
3. Kaneko, G.; Ushio, H.; Ji, H. Application of magnetic resonance technologies in aquatic biology and seafood science. *Fish. Sci.* **2019**, *85*, 1–17. [CrossRef]
4. Ellies-Oury, M.-P.; Hocquette, J.-F.; Chriki, S.; Conanec, A.; Farmer, L.; Chavent, M.; Saracco, J. Various statistical approaches to assess and predict carcass and meat quality traits. *Foods* **2020**, *9*, 525. [CrossRef] [PubMed]
5. Biondi, L.; Randazzo, C.L.; Russo, N.; Pino, A.; Natalello, A.; Van Hoorde, K.; Caggia, C. Dietary supplementation of tannin-extracts to lambs: Effects on meat fatty acids composition and stability and on microbial characteristics. *Foods* **2019**, *8*, 469. [CrossRef] [PubMed]
6. Needham, T.; Gous, R.M.; Lambrechts, H.; Pieterse, E.; Hoffman, L.C. Combined effect of dietary protein, ractopamine, and immunocastration on boar taint compounds, and using testicle parameters as an indicator of success. *Foods* **2020**, *9*, 1665. [CrossRef] [PubMed]
7. Dalle Zotte, A.; Gleeson, E.; Franco, D.; Cullere, M.; Lorenzo, J.M. Proximate composition, amino acid profile, and oxidative stability of slow-growing indigenous chickens compared with commercial broiler chickens. *Foods* **2020**, *9*, 546. [CrossRef] [PubMed]
8. Gkarane, V.; Ciulu, M.; Altmann, B.A.; Schmitt, A.O.; Mörlein, D. The effect of algae or insect supplementation as alternative protein sources on the volatile profile of chicken meat. *Foods* **2020**, *9*, 1235. [CrossRef] [PubMed]
9. Tuell, J.R.; Park, J.-Y.; Wang, W.; Cooper, B.; Sobreira, T.; Cheng, H.-W.; Kim, Y.H.B. Effects of photoperiod regime on meat quality, oxidative stability, and metabolites of postmortem broiler fillet (*M. Pectoralis major*) muscles. *Foods* **2020**, *9*, 215. [CrossRef] [PubMed]
10. Ripoll, G.; Alcalde, M.J.; Argüello, A.; Córdoba, M.d.G.; Panea, B. Effect of rearing system on the straight and branched fatty acids of goat milk and meat of suckling kids. *Foods* **2020**, *9*, 471. [CrossRef] [PubMed]
11. Ripoll, G.; Alcalde, M.J.; Córdoba, M.G.; Casquete, R.; Argüello, A.; Ruiz-Moyano, S.; Panea, B. Influence of the use of milk replacers and pH on the texture profiles of raw and cooked meat of suckling kids. *Foods* **2019**, *8*, 589. [CrossRef] [PubMed]
12. Fu, B.; Kaneko, G.; Xie, J.; Li, Z.; Tian, J.; Gong, W.; Zhang, K.; Xia, Y.; Yu, E.; Wang, G. Value-added carp products: Multi-class evaluation of crisp grass carp by machine learning-based analysis of blood indexes. *Foods* **2020**, *9*, 1615. [CrossRef] [PubMed]
13. Mwangi, F.W.; Charmley, E.; Gardiner, C.P.; Malau-Aduli, B.S.; Kinobe, R.T.; Malau-Aduli, A.E. Diet and genetics influence beef cattle performance and meat quality characteristics. *Foods* **2019**, *8*, 648. [CrossRef] [PubMed]
14. Nguyen, H. *Sustainable Food Systems Concept and Framework*; Food and Agriculture Organization of the United Nations: Rome, Italy, 2018.
15. Prakash, B.; Singh, P.P.; Kumar, A.; Gupta, V. Prospects of omics technologies and bioinformatics approaches in food science. In *Functional and Preservative Properties of Phytochemicals*; Elsevier: Amsterdam, The Netherlands, 2020; pp. 317–340.
16. Chen, L.; Liu, J.; Kaneko, G.; Xie, J.; Wang, G.; Yu, D.; Li, Z.; Ma, L.; Qi, D.; Tian, J. Quantitative phosphoproteomic analysis of soft and firm grass carp muscle. *Food Chem.* **2020**, *303*, 125367. [CrossRef] [PubMed]

17. Yu, E.; Fu, B.; Wang, G.; Li, Z.; Ye, D.; Jiang, Y.; Ji, H.; Wang, X.; Yu, D.; Ehsan, H.; et al. Proteomic and metabolomic basis for improved textural quality in crisp grass carp (*Ctenopharyngodon idellus* C. et V) fed with a natural dietary pro-oxidant. *Food Chem.* **2020**, *325*, 126906. [CrossRef] [PubMed]
18. Hocquette, J.-F.; Ellies-Oury, M.-P.; Legrand, I.; Pethick, D.; Gardner, G.; Wierzbicki, J.; Polkinghorne, R.J.; Hocquette, J.-F.; Polkinghorne, R. Research in beef tenderness and palatability in the era of big data. *Meat Muscle Biol.* **2020**, *4*, 4. [CrossRef]

 *foods*

*Article*

# Various Statistical Approaches to Assess and Predict Carcass and Meat Quality Traits

**Marie-Pierre Ellies-Oury** [1,2,3,*], **Jean-François Hocquette** [2,3], **Sghaier Chriki** [4],
**Alexandre Conanec** [1,2,3,5] , **Linda Farmer** [6], **Marie Chavent** [5] and **Jérôme Saracco** [5]

1. Bordeaux Science Agro, 1 cours du Général de Gaulle, CS 40201, 33175 Gradignan, France; alexandre.conanec@agro-bordeaux.fr
2. INRAE, UMR1213 Herbivores, 63122 Saint Genès Champanelle, France; jean-francois.hocquette@inrae.fr
3. Clermont Université, VetAgro Sup, UMR1213 Herbivores, BP 10448, 63000 Clermont-Ferrand, France
4. Isara Agro School for Life, 23 rue Jean Baldassini, 69364 Lyon CEDEX 07, France; schriki@isara.fr
5. Université de Bordeaux, UMR5251, INRIA, 33400 Talence, France; marie.chavent@math.u-bordeaux.fr (M.C.); jerome.saracco@math.u-bordeaux.fr (J.S.)
6. Agri-Food and Biosciences Institute, 18a Newforge Lane, Belfast BT9 5PX, UK; Linda.Farmer@afbini.gov.uk
* Correspondence: marie-pierre.ellies@inrae.fr; Tel.: +33-5-57353870

Received: 25 March 2020; Accepted: 8 April 2020; Published: 22 April 2020

**Abstract:** The beef industry is organized around different stakeholders, each with their own expectations, sometimes antagonistic. This article first outlines these differing perspectives. Then, various optimization models that might integrate all these expectations are described. The final goal is to define practices that could increase value for animal production, carcasses and meat whilst simultaneously meeting the main expectations of the beef industry. Different models previously developed worldwide are proposed here. Two new computational methodologies that allow the simultaneous selection of the best regression models and the most interesting covariates to predict carcass and/or meat quality are developed. Then, a method of variable clustering is explained that is accurate in evaluating the interrelationships between different parameters of interest. Finally, some principles for the management of quality trade-offs are presented and the Meat Standards Australia model is discussed. The "Pareto front" is an interesting approach to deal jointly with the different sets of expectations and to propose a method that could optimize all expectations together.

**Keywords:** optimization; meat quality; trade-off; meat standards Australia; carcass; bovine

---

## 1. Introduction

If a global rise in meat consumption is predicted worldwide, it can be seen that this increase does not concern all meat types [1]. For example, if we focus on the European Union, we can forecast an increase of consumption of poultry and pork meat (collectively referred to as white meat) with a simultaneous decrease of beef (referred to as red meat) consumption. Beef consumption in Europe has decreased in the past twenty years (−14%: from 17 kg of carcass equivalent in 2007 to 15 kg of carcass equivalent in 2018). In the same period, pork consumption has only decreased by 3% and poultry consumption has increased by 19% [2].

In the future, European beef consumption is expected to decline or stabilize, with a per capita consumption of 10.2 kg in 2028 (vs. 10.3 kg currently) [1]. In particular, consumption of fresh meat is expected to decrease, although this is likely to be offset by the increased use of meat products as ingredients in processed food [3].

In France, according to a recent survey made on 625 French consumers [4] the major reasons for the decline in beef consumption reflect the reasons previously identified in the literature [5,6].

Indeed, 37% of respondents associated the decline of beef consumption with too high variability in sensory quality, although also with a high price of beef. Mastering sensory quality is therefore a priority issue for the industry in order to stem the fall in consumption. Moreover, 35% of respondents were concerned about the possible health risks associated with beef consumption, combined with its suspected low nutritional value (i.e., high saturated fat level). Mastering nutritional quality is thus also a crucial issue for beef industry. Furthermore, the current exposure of meat production systems to the public has raised new social questions about environmental issues, human health, animal welfare and, indeed, whether animals should be slaughtered to produce human food [7].

Beyond consumers, it is also important to consider the whole meat supply chain, which includes all operators from farm to consumption. Thus, upstream of the supply chain, farmers are nowadays dependent on a fluctuating market where they have little control over prices. Good control of production costs is therefore their main lever to ensure a suitable margin. In cattle farming, as in most livestock farming, animal feed is the first item of expenditure (50% to 60% on average). It is therefore in the interests of farmers to select and breed efficient animals, that is to say, animals that are able to convert efficiently distributed feed into sales products. The individual efficiency of animals is therefore a key parameter for operators in the sector, particularly upstream operators [8].

Slaughterers and processors also have to overcome their own constraints in order to ensure the sustainability of their activity. Their main concern is the market structure and consumers' demands (in relation to their expectations and consumption habits). The quality of carcasses is thus an important parameter for the meat sector, insofar as it determines the payment of the farmer, the remuneration of the intermediate link and the assurance of an optimized meat quality [8].

However, it is difficult to reach the expectations of all the stakeholders at the same time. So, in order to manage these conflicting requirements and the trade-offs needed, it is necessary to know precisely retailers' and consumers' expectations and to know how to assess carcass and meat quality according to these expectations.

With this in mind, the present work will review various methodological approaches that could allow the simultaneous control of expectations that are not always positively correlated, with the final aim being to better manage the trade-offs between different measures of quality in the beef chain sector. The objective here is therefore to propose methods and tools that can help in the evaluation and prediction of different types of quality. These methods will be illustrated for some carcass traits and beef tenderness.

## 2. What Are the Expectations Concerning Carcass and Meat Quality?

### 2.1. Carcass Quality Expectations

In Europe, the EUROP carcass classification system is based on global indicators, including the category of animals determined by gender (including steers), age, conformation, fat scores of carcasses, and hot carcass weight. A carcass is, however, a complex and heterogeneous entity, which, within the same EUROP classification, can comprise varying proportions of muscle and/or varying proportions of muscles with a higher or lower commercial value. The global characterization of carcasses by the EUROP system does not take this complexity into account. Indeed, the European grid is the simplest system in the world to grade carcasses. Unlike more complex systems as in Asia and North America (USDA), it does not take into account marbling, color or other traits recorded in the chiller [9].

Nevertheless, all grading systems (except the Australian system) are focused on carcasses, rather than on meat. The only grading system focused on meat is the Meat Standards Australia (MSA) grading scheme, which grades meat not carcasses. In addition, it includes traits measured not only in the slaughterhouse and in the chiller as other grading schemes do, but also traits recorded pre-slaughter and post-chiller (Table 1).

**Table 1.** Major beef carcass classification systems implemented throughout the world (adapted from [9]).

<table>
<tr><td>Country</td><td>Europe</td><td>S. Africa</td><td>Canada</td><td>Japan</td><td>S. Korea</td><td>USA</td><td>Australia</td></tr>
<tr><td>Scheme</td><td>EUROP</td><td>S. Africa</td><td>Canada</td><td>JMGA</td><td>Korea</td><td>USDA</td><td>MSA</td></tr>
<tr><td>Grading unit</td><td colspan="6">Carcass</td><td>Cut</td></tr>
<tr><td>Pre slaughter factors</td><td colspan="6"></td><td>HGP implants & Bos Indicus</td></tr>
<tr><td rowspan="3">Slaughter-floor</td><td colspan="6">Carcass weight and sex</td><td rowspan="2">Electrical stimulation</td></tr>
<tr><td>Conformation</td><td>Dentition</td><td rowspan="2">Conformation</td><td colspan="3" rowspan="2"></td></tr>
<tr><td>Fat cover</td><td>ribfat</td><td>Hang</td></tr>
<tr><td rowspan="9">Chiller</td><td colspan="2" rowspan="9"></td><td colspan="5">Marbling score</td></tr>
<tr><td colspan="5">Meat color</td></tr>
<tr><td colspan="3">Fat color and fat thickness</td><td colspan="2">Ossification score</td></tr>
<tr><td rowspan="6">Texture</td><td colspan="3">Eye muscle area</td><td>Fat thickness</td></tr>
<tr><td>Meat brightness</td><td>Texture</td><td>Meat texture</td><td>Hump height</td></tr>
<tr><td>Fat luster</td><td>Firmness</td><td>Rib fat</td><td>Ultimate pH</td></tr>
<tr><td>Fat texture</td><td>Lean maturity</td><td>Kidney fat</td><td rowspan="3"></td></tr>
<tr><td>Fat firmness</td><td rowspan="2"></td><td>Perirenal fat</td></tr>
<tr><td>Rib thickness</td><td></td></tr>
<tr><td rowspan="2">Post chiller</td><td colspan="6" rowspan="2"></td><td>Ageing time</td></tr>
<tr><td>Cooking method</td></tr>
</table>

In order to improve the assessment of carcass quality obtained by the EUROP system, Monteils et al. (2017) [10] conducted a study based on a literature survey associated with a hierarchical structure according to the interests of different stakeholders in the meat chain. This work allowed the authors to propose a set of additional indicators taking into account their frequency of citations in the literature as well as their complementarity with the "historical" indicators of the EUROP grid [10]. These authors propose to complete the current set of EUROP indicators (based on carcass weight and sex, conformation and fat cover) with five other ones, namely: hindquarter weight, meat color, retail cut yield, rib eye area and marbling score.

In 2018 and 2019, a survey was conducted in various French slaughterhouses to determine, from the point of view of operators, which are the most important carcass quality indicators to consider (unpublished data). This work was carried out with 13 organizations marketing meat after slaughter, representative of the diversity of operators: four producer organizations or cooperatives, four slaughterhouses, three butchers and two breeders engaged in direct sales. In this study, we processed and analyzed data using an interface of R (R Core Team 2018), named IRaMuTeQ. Based on R software and python language, IRaMuTeQ extracted information from texts using descriptive statistics [11]. This survey did not intend to be exhaustive but rather to provide elements to validate the indicators proposed by [10] within the 1st French agricultural region: "Nouvelle-Aquitaine". The objectives of the survey questionnaire were to:

- determine what are the expectations of the operators in the sector (slaughterers, butchers, direct sales farmers, cooperatives, etc.) regarding carcasses, according to their customers and market requirements

- determine what constitutes an optimal quality carcass for different breeds and categories of animals according to the various stakeholders
- establish minimum quality thresholds to be reached for each of the specifications or each of the customer types
- highlight the criteria for assessing carcass quality

Outputs of this study were first the expectations of the various stakeholders in the meat sector in the region of "Nouvelle-Aquitaine" in terms of carcass quality for suckler cows. The main expectations were fairly homogeneous regardless of the outlet of the carcasses (supermarket shelves, cutting, parts, etc.) (Figure 1). Expectations were mainly oriented towards the muscular development of the carcass through performance such as yields (83% of citations) and conformation indicators (75% of citations). Operators specified that a minimum conformation was required for carcasses, which must also be as homogeneous as possible (in terms of weight, fat cover, etc.) in order to facilitate the processing of the carcass and preparation of meat cuts.

**Figure 1.** Main expectations expressed by the 13 operators in the sector surveyed. (The size of the expectations was proportional to its percentage of citation. For instance, yield was recorded in 83% of citations; minimal conformation 75%; carcass homogeneity and meat color: 67%; marbling: 58%; high live weight and fatness score 3 or 4: 50%; the other expectations were recorded in less than 50% of citations).

Marbling attracts the attention of stakeholders (58% of citations), since the consumer is becoming more and more interested in it. This descriptor is quite important, although the fattening state is rarely mentioned as a determining criterion. The operators indicate that the same fattening state can hide carcasses with a very different fat and marbling development. What is important for the stakeholders is to have carcasses of high quality (in terms of distribution of forequarter and hindquarter muscles, suitability for storage and maturation, etc.) but also to have ad hoc cutting of carcasses to allow marketing of small portions, suitable for self-service marketing. Meat color and tenderness (evaluated through handling and appreciation of the "meat grain" previously defined by [12] are also determinant descriptors (67% and 50% of citations respectively), especially for consumer satisfaction.

It is assumed that the addition of the five new indicators (hindquarter weight, meat color, retail cut yield, rib eye area and marbling score) proposed by [10] would complement the EUROP carcass classification in a beneficial way, permitting an improved meeting of the expectations of operators in the sector, and thus the expectations of final consumers.

As indicated earlier, additional indicators could improve assessment of carcass quality obtained with the EUROP system. However, the proposal to add new indicators to the current carcass rating parameters is also partly due to the fact that the EUROP rating does not in any way predict the potential eating quality [13] or nutritional quality.

*2.2. Meat Quality Expectations*

Meat eating quality refers to the characteristics of the product itself and includes especially sensory traits (e.g., tenderness, flavor, juiciness, overall liking), and healthiness, reviewed by [14,15]. Consumers' expectations are thus of different kinds, but above all are very numerous. Indeed, a recent survey on 625 individuals indicates that the main reasons for the decline in beef consumption are: the too high price of beef, the possible health risks, the environmental impact of farming, the development of new consumption behaviors (less quantity and more quality), the lack of consistency in tenderness and taste, animal welfare concerns or the impact of health scandals [4]. Thus, many experts have made a distinction between intrinsic and extrinsic quality attributes of meat. The first refers to the product itself and includes, for instance, safety and health aspects, as well as sensory properties. The latter refers to traits more or less associated with the product, namely production system characteristics (including animal welfare, environmental aspects, and social considerations, for instance), as well as marketing variables. Each quality trait (intrinsic or extrinsic) is itself the aggregation of sub-criteria [16]. This gives rise to two questions: how to measure all these traits and how to aggregate them. Hocquette et al. [15] have suggested some ways to combine different quality criteria based on the existing literature. In the past, this aggregation was conducted by experts such as butchers who used to provide advice to consumers. However, this method is not exhaustive and also not consistent across butchers or meat experts. Another simple way is to define minimum thresholds: for instance, "this meat should contain a minimum of fat, a minimum of Poly Unsaturated Fatty Acids (PUFA) or of any type of vitamin". This method is easy to understand and to implement but is a rough evaluation and requires routine measurements of the components of interest. A ranking system could also be defined to classify meat samples from the best (rank 1) to the worst (rank n), with a summation of the ranks with different traits. However, this is only a "relative" judgment, comparing alternatives among themselves, and not an "absolute" assessment. The best way is to convert quality traits into value scores (e.g., quantitative information on a common scale) which are then compounded, as done in the Australian system MSA (see below).

## 3. Modulation and Prediction of Quality Traits

*3.1. Regression Models*

We recently developed a new computational methodology that simultaneously selects the best regression model and the most interesting covariates [17] to predict carcass and/or meat quality. With such a method, we might predict one parameter using many variables. For instance, it could be possible to predict beef tenderness by various breeding factors and/or by animal performance.

In the modvarsel R-package, different models were tested (the linear regression, the PCR and the Slice Inverse regression, but also the random forest) but other ones (Support Vector Machine—[SVM], Ridge, Partial Least Squares—[PLSR]) could easily be practically implemented by the user. For each model, a number of selected variables has been reported. More precisely, this R package was used to select the proteins that could be considered predictive of meat tenderness among a pool of 21 candidate proteins assayed in *semitendinosus* muscle from 71 young bulls of the European ProSafeBeef project (Figure 2). The occurrence of each variable was calculated, leading to a ranking of variables according to their importance (Figure 3). By using the preselected settings, an algorithm proposes an optimal number of factors (in this case, an optimal number of proteins) to predict the variable of interest (in this case, tenderness), but it is also possible for the user to select the optimal number himself.

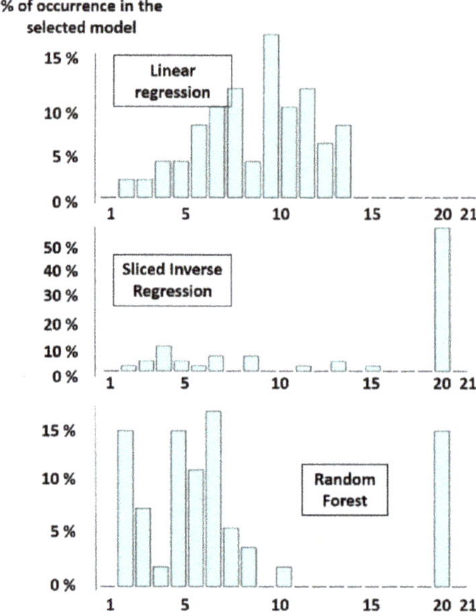



**Figure 2.** Number of selected variables for three different models (in this example, 21 factors were used to predict the variable of interest, which is tenderness) (adapted from [17]). In this example, 17% of the linear regression (LR) models use 10 variables and all of the LR models use less than 14 variables. On the contrary, 60% of the slice inverse regression (SIR) models use 20 variables out of 21 to predict the parameter of interest. About 15% of the Random Forest (RF) models use 20 variables out of 21 to predict tenderness, whereas, more than 80% of the RF models use between two and eight variables to predict this parameter.

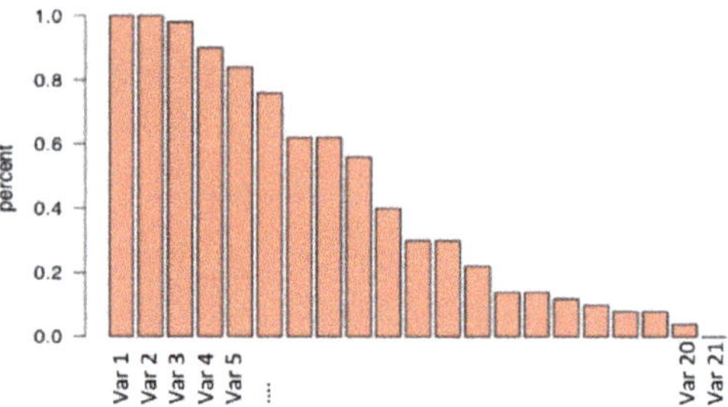



**Figure 3.** Occurrence of each variable in the selected models (adapted from [17])). In this example, the variables 1 and 2 are selected in 100% of models. The variable 21 is selected in less than 1% of the models and is therefore not very informative and not necessary in the model for predicting the parameter of interest (which is tenderness in this example).

As a further development of the modvarsel R-package, we developed another statistical approach to select variables both in the group of co-variables and in the output parts which contain a group of variables to predict [17]. This method, called data-driven sparse partial least squares implemented in the ddsPLS R-package, may allow the prediction of several variables (for example, the tenderness scores of different muscles) by the same pool of factors (for example, breeding characteristics and/or animal performances). The use of a multi-block model allows for highlighting significant links between the tenderness and some co-variables. This approach made it possible to select and combine, respectively, three and four proteins capable of predicting the tenderness of the Triceps brachii and Gluteobiceps muscles. This confirms the interest and relevance of the method to accurately predict meat tenderness, as it appears that the combination of several variables (individually poorly correlated with tenderness) can provide a relevant prediction of tenderness.

### 3.2. Interrelations between the Various Quality Traits

Beyond the predictive models that allow prediction of a parameter of interest, such as tenderness, from a certain number of variables, it is important to determine how the different parameters of interest (such as carcass weight, carcass fatness or conformation, tenderness, flavor liking, juiciness, etc.) interact with each other. Indeed, knowing the interrelationships between these parameters is essential to propose breeding practices that allow the production of carcasses and meat with optimized qualities.

To determine how the different parameters of interest interact with each other, we recently proposed a methodological approach that could explore how to establish the links between different data sets, by using a variable clustering method [18] instead of the standard individual clustering as is usually done by principal component analysis [19].

This approach allows:

(1) clarifying the interactions among different parameters of interest (for instance: animal performances, nutritional value, meat quality traits), and

(2) assessing how to simultaneously control different parameters of interest that are not always positively correlated.

To test this method, data from 71 young bulls of the European ProSafeBeef project were used [14]. For each animal, 97 variables were collected and organized in three sets of data, characterizing animal efficiency and performance, nutritional value and sensory quality [20].

A clustering of variables was conducted using the R-package ClustOfVar. This is a dimension reduction method that can be a helpful tool to select variables. Indeed, each synthetic variable from ClustOfVar is a linear combination of a subset of original variables (whereas the principal components in principal component analysis are a linear combination of all the original variables). The clustering of variables allows the establishment of a total of 15 synthetic indexes (five per set of data). Then, a second clustering, realized on these 15 synthetic indexes, establishes the proximities between the three data sets.

The ClustOfVar approach used in this paper provided homogeneous groups of variables defined by a squared Pearson correlation [21]. This method, quite new in animal science, has already delivered obvious results in a number of areas such as the automobile industry or tourist cruise ships industry [21–24].

Our previous work [18] gave some new information to manage the trade-offs between different data sets (animal efficiency, nutritional value and meat quality traits in our case). This study aimed to replace the usual data mining analysis with a variable clustering approach. This method appears to be an effective tool to integrate various concepts in an optimized management of breeding factors and breeding practices.

After having analyzed the relationships between the different parameters of interest, there is a need to set up a method to manage the trade-offs between conflicting parameters to help the beef cattle industry to design specifications, taking into account the possible interactions between these different parameters of interest.

### 3.3. Trade-Off Management

Trade-offs are needed when a decision maker is in a situation when improving a criterion automatically implies a decrease in the score of another criterion. Such a definition implies that there exists a dependence between the two criteria.

In meat, such a situation often exists, for example in the case of tenderness and lipid content in a meat cut. There is a dependence since it was shown that higher lipid content induces a higher score for the evaluation of tenderness [25]. Therefore, if a meat brand wants to market a new product having a high tenderness and a low intramuscular fat content, the decision maker would have to make a trade-off between these two criteria.

However, there are other cases where improving a criterion does not automatically imply decreasing or increasing the other criterion. This means that the correlation between the two criteria is not strong ($|r| \ll 1$). Then, before choosing the trade-off between the two criteria, an optimal set has to be found. This set is also called the Pareto Front (PF) and is the solution of the multi-objective optimization (MOO) problem:

$$\min_{x}\big(f_1(x), \dots, f_p(x)\big) \tag{1}$$

where $x$ is the decision vector holding q practice management parameters (for example, breed, feed intake, etc.), $f_j$ is a link function between the decision space and the j$^e$ objective (for example, tenderness grade or lipids content), j∈{1, … ,p} where p is the number of objective functions (Figure 4). The "min" operand is defined according the Pareto optimality based on the principle of domination, described in [26].

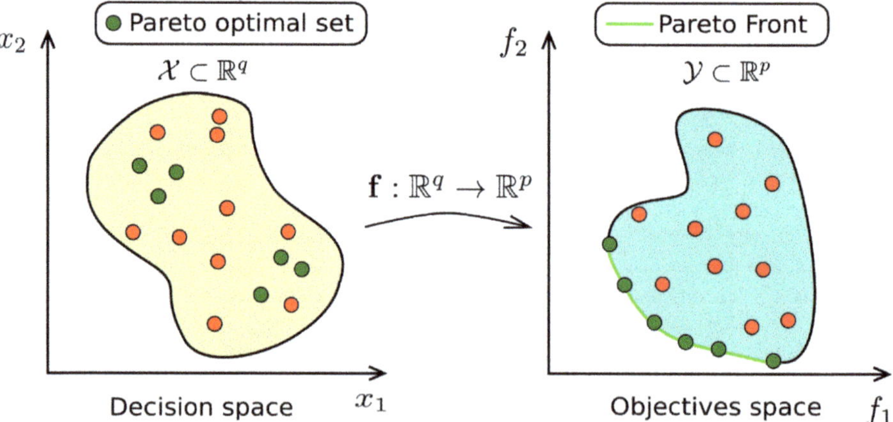


**Figure 4.** Scheme of the Pareto front and its optimal set (adapted from [26]).

This stage was absolutely objective in the sense that no choices were made in the process of optimization. In contrast, the second stage seeking to perform the trade-off is subjective since the preferences of the decision maker are integrated in the process. Therefore, different final solutions for a given problem can occur if a decision maker integrates differing sets of preferences. Numerous techniques to deal with multicriteria decisions exist [27], varying in complexity but giving a hierarchical order in most of the cases to choose in a relative way the best trade-off amongst all the possible ones contained in the Pareto Front.

In the meat field, an approach using weight aggregation was used by [28] to find a trade-off between nutritional and sensory quality. This study reported that weight setting has a high influence on the final result and must be managed carefully. The study also observed that the model used was not taking into account all the complexity of the reality since compensation between criteria can result in selecting poor overall trade-off.

The underlying difficulty of the described process is to dispose of trusty link functions (the $f_j$ used later in the optimization search). Indeed, as discussed in the previous section, there is no analytical formula to model the links between animal practices and meat qualities. However, most of the time surrogate models can approximate them:

$$y_j = f_j(x) + \varepsilon_j \tag{2}$$

where $y_j$ is the variable of interest to model (tenderness, for example) and $\varepsilon_j$ is a random error term. The chosen statistical model (such as parametric, semiparametric or nonparametric model, with specific assumptions on $\varepsilon_j$) will depend on each $y_j$ to be modeled. Consequently, the optimization search with a determinist approach (assuming that $\varepsilon_j$ is negligible) can lead to sub-optimal solutions. Fortunately, there exist more complex methods to deal with this uncertainty, like stochastic and robust optimization methods [29–31]. Our own approach to dealing with this uncertainty is promising but still in a validation stage.

### 3.4. Modelling Approaches Combining Different Quality Indicators including Their Interactions

The main pitfall concerning the management of trade-offs lies in the use of the threshold method. Indeed, whereas trade-off management allows for removing unsatisfactory beef samples based on the chosen thresholds, it can also discard a significant proportion of "good" samples. Indeed, in the area of eating quality (focused on tenderness, juiciness, flavor liking and overall liking), certain thresholds are often used. For instance, by using thresholds such as pH > 6, age > 18 months (for young bulls) or 30 months (for other bovines), fat score < 3, conformation < 0+, ageing time < 14 days (young bulls) or <7 days (other bovines), unsatisfactory grilled striploins will be removed from the market which is good, but 42% of striploins assessed as good quality by consumers would be discarded as well, which is wasteful (Figure 5; reviewed by [32]). A comparison of different eating quality systems showed that those which were based on such thresholds for whole animals/carcasses discarded a higher proportion of good quality meat compared to the Meat Standards Australia (MSA), which estimated quality on a cut-by-cut basis (Farmer, personal communication). This demonstrates the need for more complex quality prediction methods. The global modelling approaches which have been set up to solve this problem aim to provide breeders with some decision-making keys to adapt farm management in order to optimize carcass and meat quality. The "downstream management" consists of predicting eating quality based on a consideration of consumer responses and the use of these to determine the importance of production parameters rather than putting the emphasis on production and carcass conformation.

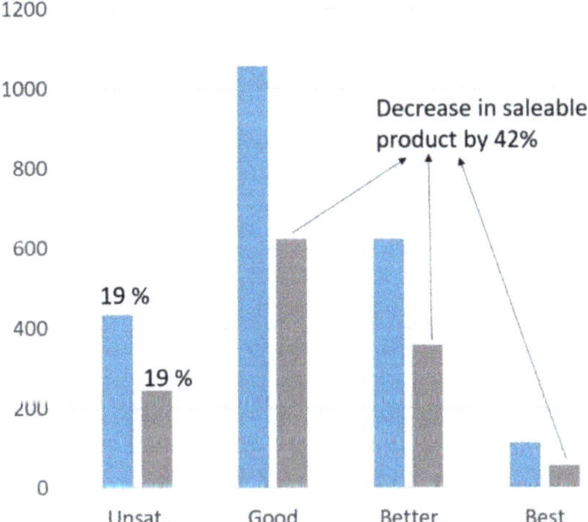


**Figure 5.** Number of samples in each quality category for grilles striploins (blue) and after applying threshold criteria (grey). The % failure is displayed above the fall column [32]. Threshold criteria are pH (>6), age (>30 months for females and steers, >18 months for young bulls), conformation (<0+), fat score (<3), aging length (<7 days for females and steers; <14 days for young bulls).

The MSA grading scheme (which has been developed to predict sensory quality of beef) is based on innovative statistical approaches using scores from the direct assessment of tenderness, juiciness, flavor and overall liking of cooked meat on a 0–100 scale by untrained consumers [33,34].

The first principle of the MSA system is to work with untrained consumers rather than expert panelists because they represent the "normal consumer/customer (who is not expert)" who buys meat without any training. The second principle is to combine these four traits to give a global quality score called "MQ4". Statistical analysis, which is crucial in the modeling approach, has defined the best weighting coefficients which are roughly 30% for tenderness, flavor, and overall liking and 10% for juiciness. In addition, discriminant analysis was used to match these consumer sensory scores with the quality ranking of meat given by consumers. Consumers are also asked to class meat as unsatisfactory, good every day (3 *), better than every day (4 *) and premium (5 *). Furthermore, the values of the global MQ4 score defining the limits between each quality class are precisely calculated for each data set and are regularly refined: they are about 46 (between unsatisfactory and 3 *), 64 (between 3 * and 4 *) and 76 (between 4 * and 5 *) on a scale of 0 to 100.

The MSA model has been tested successfully in different countries (reviewed by Bonny et al., (2018) [35]) including France [36–38], South Korea [39], Northern Ireland [40–43], the USA [44,45], Japan [46], Ireland [47], South Africa [48], New Zealand [49], and Poland [50,51]. The general conclusion is that the MSA methodology is relevant in all these countries, indicating that the MSA model is likely to be generally applicable. However, the relative weighting coefficients for tenderness, flavor liking, juiciness, and overall liking in the optimal calculation of the MQ4 score vary slightly between countries, and the optimal limits between quality classes can be refined for each country or for each group of consumers.

For quality prediction, a mathematical model has been built to predict the eating quality of beef for each "muscle × cooking method" combination. This model was constructed from a large database of consumer tests using a standard protocol. A dozen parameters having a statistically significant effect on eating quality, such as traits characterizing animals, pre-slaughter and slaughter conditions, meat, and post-mortem events are considered in the model as well as the interactions between them.

In practice, the slaughterhouse is the backbone of the system. A specific grader, who is accredited after training and who is periodically trained, grades the carcasses for marbling, fat and meat color, ossification, which is an indicator of age and more precisely of physiological maturity. Temperature and pH are also recorded. The MSA model then predicts the overall quality score of MQ4 on a scale of 0 to 100, as well the quality class for each piece of meat associated with a cooking method and a specific ageing time. All factors related to the animal, to its carcass and its meat are recorded and included in the model for prediction. So far, marbling, ageing time and the carcass hanging method were not taken into account in France. However, a French private company (Beauvallet/CV Plainemaison) has launched a new premium beef breed called OR ROUGE based on these traits.

Research on the application of the substantial data gained using MSA protocols in Europe and elsewhere will continue within the activities of the International Meat Research 3G Foundation. This Foundation was launched in 2017 under the auspices of the United Nations Economic Commission for Europe (UNECE) [52,53]. The Meat3G foundation is likely to develop a new MSA-like model to predict eating quality of beef across countries and based on data gathered in different countries. The standard protocols of carcass grading based on the MSA protocols have been approved by the United Nations Economic Commission for Europe (UNECE).

As previously detailed, many models were recently developed in order to predict each quality trait and to evaluate the possible trade-offs that could be accepted in order to satisfy all the operators of the beef chain at the same time. In order to summarize our subject, it is possible to group together in Table 2 the main objectives, advantages and disadvantages of the statistical methods developed in this paper.

**Table 2.** Objectives, advantages and disadvantages of the statistical approaches developed in the present paper.

| | Objective | Advantages | Disadvantages |
| --- | --- | --- | --- |
| Regression model | Estimation of model to explain a single parameter by many covariates. | Easy model interpretability thanks to a parametric modeling. Easy prediction method. | Linear model. Single parameter modeling. Single block of covariates. Need of a sample size greater than the number of covariates. |
| modvarsel R-package | Regression model benchmark and variable selection | Wide choice between several parametric, semi-parametric or non-parametric regression models. Ranking of variables according to their importance allowing simple selection of variables. Easy prediction method. Easy to use. | Computational burden. Single parameter modeling. Single block of covariates. |
| ddsPLS R-package | Modeling and selection of variables to predict and of traits to be predicted | Prediction of several parameters by the same pool of factors. Multi-block approach: various blocks of covariates and one block of parameters to explain. Adapted for a small sample size much lower than the number of covariates. | Linear model. Only numerical covariates and response blocks. Interpretation of the outputs slightly more technical. |
| ClustOfVar R-package | Approach providing a clustering of variables based on their correlations | Identification of interactions/links allowing dimensional reduction of variables-via the scores (synthetic variables) associated with each cluster. Easy interpretability of the scores. Method adapted to quantitative and qualitative variables. Hierarchical clustering or not. Easy to use. | Possible correlation between the cluster scores. Only linear correlations (or correlation ratios) taken into account. |
| Trade-off management | Decision-making methodology for a compromise between different quality objectives. | Integration of priority preference of the decision maker. Easy to use. | Need of a big amount of data to be accurate. Discard of unsatisfactory but also relevant samples |
| Meat Standards Australia (MSA) | Decision-making methodology based on the combination of different sensory quality traits | Inclusion in the model of different variables and of their interactions. Easy interpretability of the scores. Continuous improvement of the model. Method already implemented in the Australian beef industry with success. | Need of a big amount of data to be accurate. |

## 4. Conclusions

Consumer satisfaction when eating beef is a complex response based on subjective and emotional assessments. Safety and healthiness are very important in addition to taste and convenience for consumers, but some other parameters, such as yield and conformation, are really important for breeders. A variety of modelling approaches have been tested to assess and predict carcass and meat quality traits. Amongst these, the "Pareto front" approach proves to be an interesting method to optimize all stakeholder expectations. The MSA model has been proven to deliver improved eating quality when tested in numerous countries around the world. The first step will be to look for the set of non-dominant solutions (i.e., possible compromises from which the decision-maker has to choose). It is, therefore, now necessary to draw up a list of quality parameters sought to determine their limits and the minimum/maximum acceptable values for each parameter. It will also be necessary to study the existence of possible combinations between the different expectations. Then, the list of criteria to be optimized will have to be drawn up and prioritized by experts in order to know how to satisfy the expectations of the different stakeholders of the sector. These different aspects are currently in progress.

**Author Contributions:** Conceptualization: M.-P.E.-O., J.-F.H.; methodology: M.-P.E.-O., S.C.; J.-F.H., M.C., J.S.; software: M.-P.E.-O., A.C., M.C., J.S.; validation: M.-P.E.-O., S.C., A.C., L.F., J.-F.H., M.C., J.S.; writing—original draft preparation: M.-P.E.-O., S.C., A.C., L.F., J.-F.H., M.C., J.S.; writing—review and editing: M.-P.E.-O., S.C., A.C., L.F., J.-F.H., M.C., J.S. All authors have read and agreed to the published version of the manuscript.

**Funding:** This research received no external funding.

**Acknowledgments:** The authors would like to thank the students of Bordeaux Sciences Agro who contributed to the study on the expectations of operators in the meat sector: F Clerc, F Farison, E Janodet, A Malgoire, C Maniero, Y Moulard, E Perceau and O Valais.

**Conflicts of Interest:** The authors declare no conflict of interest.

## References

1. Food and Agriculture Organization of the United Nations. OECD-FAO Agricultural Outlook 2019–2028. Available online: http://www.fao.org/3/ca4076en/ca4076en.pdf (accessed on 25 March 2020).
2. FranceAgriMer. *Les Marchés des Produits Laitiers, Carnés et Avicoles. Bilan 2018, Perspectives 2019;* FranceAgriMer: Lyon, France, 2019.
3. Hocquette, J.-F.; Ellies-Oury, M.-P.; Lherm, M.; Pineau, C.; Deblitz, C.; Farmer, L. Current situation and future prospects for beef production in Europe—a review. *Asian Australas. J. Anim. Sci.* **2018**, *31*, 1017. [CrossRef] [PubMed]
4. Ellies-Oury, M.-P.; Lee, A.; Jacob, H.; Hocquette, J.-F. Meat consumption–what French consumers feel about the quality of beef? *Ital. J. Anim. Sci.* **2019**, *18*, 646–656. [CrossRef]
5. Henchion, M.; McCarthy, M.; Resconi, V.C.; Troy, D. Meat consumption: Trends and quality matters. *Meat Sci.* **2014**, *98*, 561–568. [CrossRef] [PubMed]
6. Piazza, J.; Ruby, M.B.; Loughnan, S.; Luong, M.; Kulik, J.; Watkins, H.M.; Seigerman, M. Rationalizing meat consumption. *4Ns. Appet.* **2015**, *91*, 114–128. [CrossRef]
7. Henchion, M.; De Backer, C.J.S.; Hudders, L. Ethical and sustainable aspects of meat production; consumer perceptions and system credibility. In *New Aspects of Meat Quality*; Elsevier: Amsterdam, The Netherlands, 2017; pp. 649–666.
8. Dockès, A.-C.; Magdelaine, P.; Daridan, D.; Guillaumin, A.; Rémondet, M.; Selmi, A.; Gilbert, H.; Mignon-Grasteau, S.; Phocas, F. Attentes en matière d'élevage des acteurs de la sélection animale, des filières de l'agroalimentaire et des associations. *Prod. Anim.* **2011**, *24*, 285. [CrossRef]
9. Polkinghorne, R.J.; Thompson, J.M. Meat standards and grading: A world view. *Meat Sci.* **2010**, *86*, 227–235. [CrossRef]
10. Monteils, V.; Sibra, C.; Ellies-Oury, M.-P.; Botreau, R.; De la Torre, A.; Laurent, C. A set of indicators to better characterize beef carcasses at the slaughterhouse level in addition to the EUROP system. *Livest. Sci.* **2017**, *202*, 44–51. [CrossRef]

11. Chaves, M.M.N.; dos Santos, A.P.R.; dos Santosa, N.P.; Larocca, L.M. Use of the software IRAMUTEQ in qualitative research: An experience report. In *Computer Supported Qualitative Research*; Springer: Berlin, Germany, 2017; pp. 39–48.

12. Ellies-Oury, M.P.; Durand, Y.; Delamarche, F.; Jouanno, M.; Lambert, J.; Micol, D.; Dumont, R. Relationships between the assessment of "grain of meat" and meat tenderness of Charolais cattle. *Meat Sci.* **2013**, *93*, 397–404. [CrossRef]

13. Bonny, S.P.F.; Pethick, D.W.; Legrand, I.; Wierzbicki, J.; Allen, P.; Farmer, L.J.; Polkinghorne, R.J.; Hocquette, J.-F.; Gardner, G.E. European conformation and fat scores have no relationship with eating quality. *Animal* **2016**, *10*, 996–1006. [CrossRef]

14. Hocquette, J.-F.; Van Wezemael, L.; Chriki, S.; Legrand, I.; Verbeke, W.; Farmer, L.; Scollan, N.D.; Polkinghorne, R.; Rødbotten, R.; Allen, P.; et al. Modelling of beef sensory quality for a better prediction of palatability. *Meat Sci.* **2014**, *97*, 316–322. [CrossRef]

15. Hocquette, J.-F.; Botreau, R.; Picard, B.; Jacquet, A.; Pethick, D.W.; Scollan, N.D. Opportunities for predicting and manipulating beef quality. *Meat Sci.* **2012**, *92*, 197–209. [CrossRef] [PubMed]

16. Grunert, K.G.; Bredahl, L.; Brunsø, K. Consumer perception of meat quality and implications for product development in the meat sector—A review. *Meat Sci.* **2004**, *66*, 259–272. [CrossRef]

17. Ellies-Oury, M.P.; Chavent, M.; Conanec, A.; Bonnet, M.; Picard, B.; Saracco, J. Statistical model choice including variable selection based on variable importance: A relevant way for biomarkers selection to predict meat tenderness. *Sci. Rep.* **2019**, *9*, 10014. [CrossRef] [PubMed]

18. Ellies-Oury, M.-P.; Cantalapiedra-Hijar, G.; Durand, D.; Gruffat, D.; Listrat, A.; Micol, D.; Ortigues-Marty, I.; Hocquette, J.-F.; Chavent, M.; Saracco, J.; et al. An innovative approach combining Animal Performances, nutritional value and sensory quality of meat. *Meat Sci.* **2016**, *122*, 163–172. [CrossRef]

19. Chriki, S.; Gardner, G.E.; Jurie, C.; Picard, B.; Micol, D.; Brun, J.-P.; Journaux, L.; Hocquette, J.-F. Cluster analysis application identifies muscle characteristics of importance for beef tenderness. *BMC Biochem.* **2012**, *13*, 29. [CrossRef] [PubMed]

20. Gagaoua, M.; Micol, D.; Picard, B.; Terlouw, C.E.; Moloney, A.P.; Juin, H.; Meteau, K.; Scollan, N.; Richardson, I.; Hocquette, J.-F. Inter-laboratory assessment by trained panelists from France and the United Kingdom of beef cooked at two different end-point temperatures. *Meat Sci.* **2016**, *122*, 90–96. [CrossRef]

21. Chavent, M.; Kuentz, V.; Liquet, B.; Saracco, L. Clustofvar: An r package for the clustering of variables. *J. Stat. Softw.* **2012**, *50*, 1–16. [CrossRef]

22. Simonet, V.K.; Lyser, S.; Candau, J.; Deuffic, P.; Chavent, M.; Saracco, J. Une approche par classification de variables pour la typologie d'observations: Le cas d'une enquête agriculture et environnement. *J. Société Française Stat.* **2013**, *154*. No. 2.

23. Plasse, M.; Niang, N.; Saporta, G.; Villeminot, A.; Leblond, L. Combined use of association rules mining and clustering methods to find relevant links between binary rare attributes in a large data set. *Comput. Stat. Data Anal.* **2007**, *52*, 596–613. [CrossRef]

24. Brida, J.G.; Scuderi, R.; Seijas, M.N. Segmenting cruise passengers visiting Uruguay: A factor–cluster analysis. *Int. J. Tour. Res.* **2014**, *16*, 209–222. [CrossRef]

25. Corbin, C.H.; O'Quinn, T.G.; Garmyn, A.J.; Legako, J.F.; Hunt, M.R.; Dinh, T.T.N.; Rathmann, R.J.; Brooks, J.C.; Miller, M.F. Sensory evaluation of tender beef strip loin steaks of varying marbling levels and quality treatments. *Meat Sci.* **2015**, *100*, 24–31. [CrossRef]

26. Jaimes, A.L.; Martınez, S.Z.; Coello, C.A.C. An introduction to multiobjective optimization techniques. *Optim. Polym. Process.* **2009**, 29–57.

27. Mardani, A.; Jusoh, A.; Nor, K.; Khalifah, Z.; Zakwan, N.; Valipour, A. Multiple criteria decision-making techniques and their applications–a review of the literature from 2000 to 2014. *Econ. Res. Ekon. Istraživanja* **2015**, *28*, 516–571. [CrossRef]

28. Conanec, A.; Picard, B.; Durand, D.; Cantalapiedra-Hijar, G.; Chavent, M.; Denoyelle, C.; Gruffat, D.; Normand, J.; Saracco, J.; Ellies-Oury, M.-P. New approach studying interactions regarding trade-off between beef performances and meat qualities. *Foods* **2019**, *8*, 197. [CrossRef] [PubMed]

29. Chapman, J.L.; Lu, L.; Anderson-Cook, C.M. Incorporating response variability and estimation uncertainty into Pareto front optimization. *Comput. Ind. Eng.* **2014**, *76*, 253–267. [CrossRef]

30. Lee, K.-H.; Park, G.-J. Robust optimization considering tolerances of design variables. *Comput. Struct.* **2001**, *79*, 77–86. [CrossRef]

31. Mattson, C.A.; Messac, A. Pareto frontier based concept selection under uncertainty, with visualization. *Optim. Eng.* **2005**, *6*, 85–115. [CrossRef]

32. Farmer, L.; Bowe, R.; Troy, D.; Bonny, S.; Birnie, J.; Dell'Orto, V.; Polkinghorne, R.; Wierzbicki, J.; De Roest, K.; Scollan, N.; et al. Compte-rendu du congrès intitulé "Qualité durable de la viande bovine en Europe". *Rev. Française Rech. Viandes Prod. Carnés AB CORP Int.* **2016**, *32*, 1–10.

33. Polkinghorne, R.; Thompson, J.M.; Watson, R.; Gee, A.; Porter, M. Evolution of the Meat Standards Australia (MSA) beef grading system. *Aust. J. Exp. Agric.* **2008**, *48*, 1351–1359. [CrossRef]

34. Watson, R.; Gee, A.; Polkinghorne, R.; Porter, M. Consumer assessment of eating quality–development of protocols for Meat Standards Australia (MSA) testing. *Aust. J. Exp. Agric.* **2008**, *48*, 1360–1367. [CrossRef]

35. Bonny, S.P.F.; Hocquette, J.-F.; Pethick, D.W.; Legrand, I.; Wierzbicki, J.; Allen, P.; Farmer, L.J.; Polkinghorne, R.J.; Gardner, G.E. The variability of the eating quality of beef can be reduced by predicting consumer satisfaction. *Animal* **2018**, *12*, 2434–2442. [CrossRef]

36. Hocquette, J.F.; Legrand, I.; Jurie, C.; Pethick, D.W.; Micol, D. Perception in France of the Australian system for the prediction of beef quality (Meat Standards Australia) with perspectives for the European beef sector. *Anim. Prod. Sci.* **2011**, *51*, 30–36. [CrossRef]

37. Legrand, I.; Hocquette, J.F.; Polkinghorne, R.J.; Wierzbicki, J. Comment prédire la qualité de la viande bovine en Europe en s'inspirant du système australien MSA. *Innov. Agron.* **2017**, *55*, 171–182.

38. Legrand, I.; Hocquette, J.-F.; Polkinghorne, R.J.; Pethick, D.W. Prediction of beef eating quality in France using the Meat Standards Australia system. *Animal* **2013**, *7*, 524–529. [CrossRef] [PubMed]

39. Thompson, J.M.; Polkinghorne, R.; Hwang, I.H.; Gee, A.M.; Cho, S.H.; Park, B.Y.; Lee, J.M. Beef quality grades as determined by Korean and Australian consumers. *Aust. J. Exp. Agric.* **2008**, *48*, 1380–1386. [CrossRef]

40. Chong, F.S.; Farmer, L.J.; Hagan, T.D.J.; Speers, J.S.; Sanderson, D.W.; Devlin, D.J.; Tollerton, I.J.; Gordon, A.W.; Methven, L.; Moloney, A.P. Regional, socioeconomic and behavioural-impacts on consumer acceptability of beef in Northern Ireland, Republic of Ireland and Great Britain. *Meat Sci.* **2019**, *154*, 86–95. [CrossRef]

41. Farmer, L.J.; Devlin, D.J.; Gault, N.F.S.; Gordon, A.W.; Moss, B.W.; Polkinghorne, R.J.; Thompson, J.M.; Tolland, E.L.C.; Tollerton, I.J.; Watson, R. Adaptation of Meat Standards Australia quality system for Northern Irish beef. *Adv. Anim. Biosci.* **2010**, *1*, 127. [CrossRef]

42. Farmer, L.J.; Devlin, D.J.; Gault, N.F.S.; Gee, A.; Gordon, A.W.; Moss, B.W.; Polkinghorne, R.; Thompson, J.; Tolland, E.L.C.; Tollerton, I.J. Effect of type and extent of cooking on the eating quality of Northern Ireland beef. In Proceedings of the 55th International Congress of Meat Science and Technology, Copenhagen, Denmark, 16–21 August 2009; pp. 7–33.

43. Farmer, L.J.; Devlin, D.J.; Gault, N.F.S.; Gordon, A.W.; Moss, B.W.; Polkinghorne, R.J.; Thompson, J.M.; Tolland, E.L.C.; Tollerton, I.J. Prediction of eating quality using the Meat Standards Australia system for Northern Ireland beef and consumers. In Proceedings of the International Congress on Meat Science and Technology, Copenhagen, Denmark, 16–21 August 2009; pp. 16–21.

44. O'Quinn, T.G.; Legako, J.F.; Brooks, J.C.; Miller, M.F. Evaluation of the contribution of tenderness, juiciness, and flavor to the overall consumer beef eating experience. *Transl. Anim. Sci.* **2018**, *2*, 26–36. [CrossRef]

45. Polkinghorne, R. Targeting the consumer demand for beef in Australia, Japan, Korea, Ireland, and the United States. In *Proceedings of the 60th Annual Reciprocal Meat Conference, Brookings, SD*; American Meat Science Association: Champaign, IL, USA, 2007; pp. 27–33.

46. Polkinghorne, R.J.; Nishimura, T.; Neath, K.E.; Watson, R. Japanese consumer categorisation of beef into quality grades, based on Meat Standards Australia methodology. *Anim. Sci. J.* **2011**, *82*, 325–333. [CrossRef]

47. McCarthy, S.N.; Henchion, M.; White, A.; Brandon, K.; Allen, P. Evaluation of beef eating quality by Irish consumers. *Meat Sci.* **2007**, *132*, 118–124. [CrossRef]

48. Strydom, P.; Burrow, H.; Polkinghorne, R.; Thompson, J. Do demographic and beef eating preferences impact on South African consumers' willingness to pay (WTP) for graded beef? *Meat Sci.* **2019**, *150*, 122–130. [CrossRef]

49. Garmyn, A.J.; Polkinghorne, R.J.; Brooks, J.C.; Miller, M.F. Consumer assessment of New Zealand forage finished beef compared to US grain fed beef. *Meat Muscle Biol.* **2019**, *3*, 22–32. [CrossRef]

50. Guzek, D.; Głąbska, D.; Gutkowska, K.; Wierzbicki, J.; Woźniak, A.; Wierzbicka, A. Analysis of the factors creating consumer attributes of roasted beef steaks. *Anim. Sci. J.* **2015**, *86*, 333–339. [CrossRef] [PubMed]

51.  Pogorzelski, G.; Woźniak, K.; Polkinghorne, R.; Póltorak, A.; Wierzbicka, A. Polish consumer categorisation of grilled beef at 6 mm and 25 mm thickness into quality grades, based on Meat Standards Australia methodology. *Meat Sci.* **2020**, *161*, 107953. [CrossRef] [PubMed]
52.  Pethick, D.; Crowley, M.; Polkinghorne, R.; Webster, J.; Hocquette, J.-F.; McGilchrist, P.; Osborne, T.; McCamley, I.; Maguire, T.; Inglis, M. Comment les professionnels de la viande en Australie ont valorisé les résultats de R&D. *Viandes Prod. Carnés* **2019**, *35*, 1–2.
53.  Pethick, D.W.; Mcgilchrist, P.; Polkinghorne, R.; Warner, R.; Tarr, G.; Garmyn, A.; Thompson, J.; Hocquette, J.-F. Travaux de recherche internationaux sur la qualité sensorielle de la viande ovine et bovine. *Rev. Française Rech. Viandes Prod. Carnés AB CORP Int.* **2018**, *32*, 1–2.

*Review*

# Diet and Genetics Influence Beef Cattle Performance and Meat Quality Characteristics

Felista W. Mwangi [1], Edward Charmley [2], Christopher P. Gardiner [1], Bunmi S. Malau-Aduli [3], Robert T. Kinobe [1] and Aduli E. O. Malau-Aduli [1,*]

[1]  Animal Genetics and Nutrition, Veterinary Sciences Discipline, College of Public Health, Medical and Veterinary Sciences, Division of Tropical Health and Medicine, James Cook University, Townsville, QLD 4811, Australia; felista.mwangi@my.jcu.edu.au (F.W.M.); christopher.gardiner@jcu.edu.au (C.P.G.); robert.kinobe@jcu.edu.au (R.T.K.)
[2]  CSIRO Agriculture and Food, Private Mail Bag Aitkenvale, Australian Tropical Sciences and Innovation Precinct, James Cook University, Townsville, QLD 4811, Australia; Ed.Charmley@csiro.au
[3]  College of Medicine and Dentistry, Division of Tropical Health and Medicine, James Cook University, Townsville, QLD 4811, Australia; bunmi.malauaduli@jcu.edu.au
*   Correspondence: aduli.malauaduli@jcu.edu.au; Tel.: +61-747-815-339

Received: 14 November 2019; Accepted: 27 November 2019; Published: 6 December 2019

**Abstract:** A comprehensive review of the impact of tropical pasture grazing, nutritional supplementation during feedlot finishing and fat metabolism-related genes on beef cattle performance and meat-eating traits is presented. Grazing beef cattle on low quality tropical forages with less than 5.6% crude protein, 10% soluble starches and 55% digestibility experience liveweight loss. However, backgrounding beef cattle on high quality leguminous forages and feedlot finishing on high-energy diets increase meat flavour, tenderness and juiciness due to improved intramuscular fat deposition and enhanced mono- and polyunsaturated fatty acids. This paper also reviews the roles of stearoyl-CoA desaturase, fatty acid binding protein 4 and fatty acid synthase genes and correlations with meat traits. The review argues that backgrounding of beef cattle on *Desmanthus*, an environmentally well-adapted and vigorous tropical legume that can persistently survive under harsh tropical and subtropical conditions, has the potential to improve animal performance. It also identifies existing knowledge gaps and research opportunities in nutrition-genetics interactions aimed at a greater understanding of grazing nutrition, feedlot finishing performance, and carcass traits of northern Australian tropical beef cattle to enable red meat industry players to work on marbling, juiciness, tenderness and overall meat-eating characteristics.

**Keywords:** diet; genetics; meat quality characteristics; tropical beef cattle; stearoyl-CoA desaturase; fatty acid binding protein 4; fatty acid synthase; *Desmanthus* legumes; supplementation; growth performance

---

## 1. Introduction

Beef plays a significant role in global human nutrition. It is the third most consumed meat in the world after poultry and pork at 6.4, 14.0 and 12.2 kg per capita, respectively [1]. Beef consumption continues to rise in line with growth and increase in population and household incomes. By 2027, it is estimated that beef consumption will be 8% and 21% higher in the developed and developing countries, respectively, compared to the 2015–2017 average [2]. Beef is a nutrient-dense food that provides health-beneficial macro- and micro-nutrients for humans. A 100 g serving of beef provides more than the 25% recommended dietary intake (RDI) of protein, niacin, vitamin $B_6$, vitamin $B_{12}$, zinc and selenium, and more than 10% RDI of phosphorus, iron and riboflavin. Beef protein is of certain characteristics and contains all the essential amino acids [3] and provides antioxidants such as carnosine and anserine [4,5].

On the world stage, Australia is among the major global meat industry players in terms of beef production and exports. In 2018, Australia was ranked seventh in world beef production and third in beef exports after Brazil and United States at 2.1, 1.6, and 1.5 million tons of carcass weight equivalents (CWE), respectively [6]. The beef cattle industry contributes significantly to the Australian economy, accounting for 20% ($12.1 billion) of the 2016–2017 total gross value of farm production and 22% of the total value of export income. The Australian beef cattle population is currently 23 million head and occupies about half of Australian farms and 75% of the total agricultural land mass [7]. Half of the national beef cattle herd is in northern Australia with 43% in Queensland and 16% in Western Australia and Northern Territory [8]. Queensland alone accounted for 1.1% of the global beef herd in 2017 and 8% of world beef exports in 2016 [9]. The major breeds in northern Australia are Brahman, Santa Gertrudis and Droughtmaster; all bred for tick resistance and heat tolerance, but their meat is comparatively different to temperate breeds [10]. To increase productivity and meat characteristics, these cattle breeds are sometimes crossed with *Bos taurus* to maintain at least 5/8 *Bos indicus* genetic composition to ensure adequate heat tolerance and tick resistance [9,10]. Several composite breeds consisting of half *Bos taurus* and *Bos indicus* such as Belmont Red, NAPCO Composite and AACO have been developed by crossing the Brahman with British, European and African breeds [11]. However, the challenge of low pasture quality and quantity remains a major limitation to beef production [9,12] and this is where the use of nutritional supplementation with *Desmanthus*, an environmentally well-adapted and vigorous tropical legume that can persistently survive under harsh tropical conditions, has the potential to drive animal performance, and improve meat characteristics.

Meat characteristics is the culmination of the acceptability of a meat product in relations to its colour, intramuscular fat content, healthy fatty acid (FA) composition, tenderness, juiciness, flavour and aroma. Increasingly, market demands for meat products with healthy nutritional attributes and overall sensory characteristics are key factors strongly influencing willingness-to-pay decisions of beef consumers. Wolcott et al. [13] reported the findings of a broad consumer taste panel assessment of beef from cattle of various genetic, nutritional and environmental backgrounds in Australia, which demonstrated a measurable and negative impact of *Bos indicus* content on meat characteristics traits of tenderness, marbling and juiciness. FA composition of ruminant muscle tissues is essential to meat-eating characteristics due to its influence on flavour and tenderness, and published results suggest it is controlled by genetic factors such as genes responsible for lipids synthesis and metabolism [14]. Omega-3 long-chain polyunsaturated fatty acids (n-3 LC-PUFA) are beneficial in improving brain and retinal development, maternal and offspring health, cognitive function and psychological status in humans [15]. Delta-5 ($\Delta5$), $\Delta6$ and $\Delta9$ desaturases are crucial enzymes in polyunsaturated fatty acids (PUFA) metabolism, and their activity can be influenced by several factors like dietary fatty acids and type of biological tissue [16]. Fatty acid binding proteins (FABPs) are conserved intracellular lipid-binding proteins that bind FA and other lipids reversibly. Fatty acid binding protein 4 (FABP4) is among the nine identified tissue-specific cytoplasmic FABPs [17]. FABP4 gene is expressed in the adipose tissue and plays an essential role in lipid metabolism and homeostasis. It interacts with peroxisome proliferator-activated receptors, binds to hormone-sensitive lipase [18] and is an essential candidate gene affecting intramuscular fat deposition. However, FABP4's association with fatness traits in cattle varies from one study to another [19]. For example, FABP4 gene polymorphisms were significantly associated with backfat thickness [20], marbling and carcass weight [21] in Korean Hanwoo cattle. In contrast, FABP4 was associated with palmitoleic acid only in Japanese Black cattle [22]. Other genetic determinants of meat characteristics traits include the stearoyl-coA desaturase (SCD) and fatty acid synthase (FASN) genes.

The primary aim of this review was to explore the published literature reporting the effects of nutritional grazing, dietary supplementation and roles of SCD, FABP4 and FASN genes on beef cattle performance and subsequent carcass and meat characteristics. The review also identifies current knowledge gaps that could underpin future research in nutrition-genetics interactions aimed at a greater understanding of grazing nutrition, feedlot finishing performance and carcass traits with a focus on tropical northern Australian beef cattle and the effect on marbling, juiciness, tenderness and overall meat-eating characteristics.

## 2. Tropical Northern Australian Pastures and Beef Production

Beef production in northern Australia is heavily dependent on extensive tropical pasture grazing systems [23] of mainly native pastures dominated by C4 grasses [24]. In the northern rangelands, pastures are mainly unimproved with limited use of exotic pasture species in some regions of Queensland [25]. Black Speargrass (*Heteropogon contortus*) and *Aristida-Bothriochloa* grasslands dominate the more productive areas of eastern Queensland. In northern and western Queensland, Northern Territory and Western Australia Mitchell grass (Astrebla spp.), perennial tallgrass and shortgrass grass species, and spinifex (Triodia spp.) dominate [25,26]. *Stylosanthes* legumes are widely sown across northern Australia's light textured soils to improve pasture nutritive value [27]. On the contrary, there was lack of suitable legume pasture for clay soils until recently [28]. Clay soil typify much of northern Australia pasture land [29]. For example, Vertosol (cracking clay) soils occupy 28% of Queensland's total area, and are associated with grasslands, eucalypt woodland and brigalow/gidgee forests [30]. The most predominant pastures in these clay soils are *Asterbla* spp (Mitchell grasses) and *Iseilema* spp (Flinders grasses) with few sown pastures such as *Cenchrus ciliaris* (Buffel grass) in the Brigalow belt [31]. With the exception of young leaves and seeds, native pastures are of relatively low nutritional value at the end of summer growing season. During winter, growth is limited by temperature [32] and most native pastures are susceptible to frost leading to rapid decline in nutrient value [33,34].

The pastures are highly seasonal with growth occurring in the wet season (November to April), and ceases in 4 to 7 months of the year when conditions are too dry and/or too cold [35,36]. During the transition period from rainy to dry season, pastures decline in leaf to stem ratio caused by over 50% loss in leaf mass, crude protein content drops below 8% and the proportion of dead material increases, thus rendering the pastures less nutritionally beneficial and less palatable to cattle [37,38]. In addition, pastures deteriorate after several years of grazing due to nitrogen run-down stress [29]. The resulting poor nutrition leads to poor reproductive performance, slow growth rate, loss of body condition, increased susceptibility to parasites and diseases, increased turn-off age [35,36,39] and increased enteric methane emissions [40].

Forage quality is determined by nutrient concentration, intake, nutrient availability, and partitioning of metabolized products within animals [41]. Low quality forages contain less than 10% soluble sugars and starches, crude protein is below 8% and digestibility less than 55%. Utilization of these forages is limited by low intake due to physical fill limits and slow digestion as a result of high cell wall content and minimal nutrients available to support an efficient rumen microbial growth [38,41]. A study summary of data from 11 studies depicted a linear relationship between forage crude protein content and liveweight gain in cattle. Forage CP below 5.6% resulted in weight loss of up to 6 g/kg MBW (metabolic body weight), but above 5.6% resulted in 5–27 g/kg MBW gain daily [42].

### 2.1. Beef Cattle Responses to Under-Nutrition

Beef cattle use their evolutionary adaptation mechanisms which are either short (days), medium (weeks) or long-term, to cope with periods of under-nutrition. Short-term adaptations are in response to diurnal feeding frequency or daily changes in feed intake; mid-term changes appear within weeks of change; while long-term adaptations necessitate that the animals get into a new equilibrium involving different nutritional and physiological changes [43].

#### 2.1.1. Decrease in Liveweight

Decrease in liveweight (LW) after short-term underfeeding takes place due to gut fill variation amounting to 4–5 kg LW/kg decrease in DM intake. For instance, digesta in the reticulo-rumen of a fed ruminant animal weighs up to 15% of body weight [44]. Medium-term under nutrition leads to organ and tissue mass variation. Liver and digesta-free gastrointestinal track reduction of more than 50%

was reported after three weeks of restricted dietary access to maintain body weight [45]. Mid- and long-term weight losses are due to decreases in portal and hepatic blood flows as well as mobilization of fat, muscle and bone tissues in the reverse order of how they were deposited. The latest maturing tissues are fairly more sensitive due to physiological priority [43,44].

### 2.1.2. Metabolic and Body Composition Changes

Storage triglycerides in the adipose tissue are hydrolysed to release FA which are oxidized directly to energy and broken down into ketone bodies (aceto-acetate, hydroxy-butyrate and acetones) in the liver [44]. The liver also incorporates FA into lipoproteins and triacylglycerols in the blood. Ketone bodies, lipoproteins and triacylglycerols act as sources of energy in peripheral tissues (Figure 1) [44]. Mobilized FA from the adipose tissue results in elevation of blood non-esterified fatty acids (NEFA). The liver removes 10% of NEFA from the blood during each cycle pass and converts half of all NEFA into ketone bodies. Hence, an increase in NEFA results in an increase in blood ketone bodies [40].

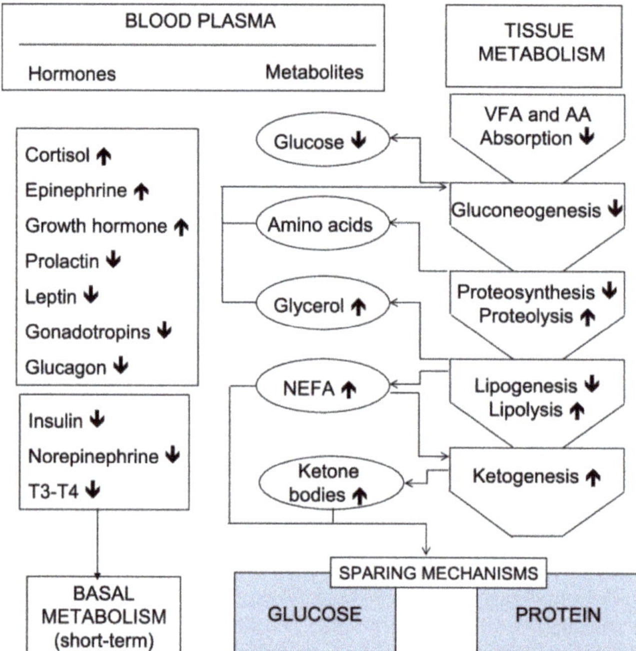


**Figure 1.** Metabolic and endocrine adaptations to undernutrition in the ruminant. NEFA: non-esterified fatty acids, VFA: volatile fatty-acids, AA: Amino-acids, T3-T4: thyroid hormones. Bold upward and downward pointing arrows indicate increase or decrease in tissue metabolism, blood plasma hormone or metabolite levels, respectively [43].

In periods of undernutrition, gluconeogenesis from propionate decreases due to a decrease in propionate availability. This is partially compensated by gluconeogenesis from amino acid (AA) proteolysis, glycerol lipolysis and lactate recycling. These metabolic changes are controlled by teleophoretic hormones such as insulin, glucagon and norepinephrine and together with decreased splanchnic tissues mass and variation in body composition, result in reduced energy expenditure. Mid-term experiments (several weeks) showed that portal-drained viscera, liver and skeletal muscles contributed to changes in energy expenditure of 17–61%, 14–44% and 5–7%, respectively [43,46]. These responses reduced growth rate and resultant beef since carcass fat content is a major factor that defines meat characteristics parameters like texture and taste [47,48].

## 2.2. Nutritional Supplementation to Improve Beef Cattle Performance on Low Quality Pastures

### 2.2.1. Feed Supplements During Grazing

Numerous studies (Table 1) indicate that high feed conversion efficiencies and medium to high levels of production can be achieved by ruminants fed poor-quality tropical forages that are adequately supplemented with critical nutrients [49]. Metabolizable energy utilization efficiency of a forage can exceed that of grain-based diets when supplemented appropriately. The supplements optimize availability of nutrients for rumen fermentative digestion and utilization of nutrients that are products of fermentation [38]. Batista and colleagues [46] observed that supplements with a high proportion of rumen degradable protein, favour nitrogen recycling and promote increased microbial protein synthesis in beef cattle. Supplementation of the drinking water of steers fed Pangola grass (*Digitaria eriantha*) hay with Spirulina was found to increase ammonia-N concentration, propionate and branched- chain fatty acids in the rumen fluid. However, this study did not observe any positive effect of Spirulina supplementation on steer liveweight gains [50]. Non-protein N supplements are also supplied together with molasses to provide readily available energy for rumen microbes to synthesize microbial protein [42].

**Table 1.** Impact of supplements on beef cattle performance.

| Pasture | Supplement | Outcome | Reference |
| --- | --- | --- | --- |
| *Urochloa decumbens* | Corn, Corn gluten, Soybean, Urea, | ADG up to 0.75 kg | [51] |
| *Urochloa decumbens* hay | Pure casein, urea and ammonia | Increase NDF digestion | [52] |
| *Urochloa decumbens* hay | Urea, ammonium sulphate and albumin | Increased DMI and NDF digestion | [53] |
| *Urochloa brizantha* | Cottonseed meal, corn and urea | ADG of up to 0.3 kg | [54] |
| *Urochloa brizantha* | Soybean meal, urea and grain sorghum | ADG of up to 0.3 kg | [55] |

ADG: average daily gain, NDF: neutral detergent fibre, DMI: dry matter intake.

Supplements are reported to stimulate feed intake and liveweight gain (Table 1) to achieve up to one kg daily [56]. Supplementing cattle with urea together with molasses or other readily available energy sources at 2.8% N increases forage intake and prevents liveweight loss [42]. However, the cost of supplementation during grazing in an extensive grazing system is a limiting factor, hence it is mainly used for weaners and the breeding herd or for whole herd survival [56,57].

### 2.2.2. Augmenting Pastures with Legumes

Incorporating a highly digestible forage into low digestible pastures supplies vitamin A, essential minerals, ammonia, peptides and amino acids. It also provides a highly colonized fibre source to 'seed' bacteria on the less-digestible fibre, thus improving total digestibility [38,58].

*Nutritional benefits:* Many studies have recognized the potential of legume pastures to improve beef cattle production in the tropics [36,39,59]. Legumes are rich in protein compared to tropical grasses due to the different biochemical pathways of carbon fixation during photosynthesis [60]. The protein therefrom can avail a renewable protein source to cattle grazing low quality grass pastures at a low cost [61]. Although the growth rate of cattle under grass only, or grass-legume pasture combination is similar early in the growing season, as the season progresses, legume-grass pasture-fed cattle gain more weight than grass-only fed cattle [62]. This is due to slower nitrogen decline in legumes as compared to all-grass pastures that leads to higher nutrition value of mature or dry legume pastures [36]. The introduction of legumes on grass-based pastures improves animal energy and protein intake, feed conversion and rumen function, as well as increases mineral and vitamin availabilities [39,63]. Steers grazed on *Leucaena leucocephala* and *Urochloa brizantha* pastures had higher weight gains compared to those on *Urochloa brizantha* only pasture [64]

*Effect on rumen lipid and protein degradation:* Some pasture legumes are known to produce tannins [65]. Moderate levels of tannins (<50 g/kg DM) reduce protein degradation in the rumen without depressing rumen fibre digestion or voluntary feed intake. At 20–40 g/kg DM, tannins may bind to the dietary proteins during mastication, thereby shielding them from microbial degradation [66,67]. This increases outflow of dietary protein to the duodenum and protein digestion and absorption in the small intestines [68,69].

Tannins may inhibit or slow down lipid biohydrogenation (BH) in the rumen [70]. Dietary herbage lipid composition is made up of membrane lipids: glycolipids and phospholipids, while seed lipids are polar lipids, mainly triacylglycerides [71]. After ingestion, dietary triglycerides and phospholipids are hydrolysed into glycerol, FA and small amounts of mono- and diglycerides by microbial lipases in the rumen. Glycerol undergoes rapid fermentation to yield propionic acid as the major product [72], while unsaturated FA are hydrogenated into saturated trans FA by microbes [44,71]. Biohydrogenation involves isomerization of cis-12 double bonds in unsaturated FA to a trans-11 isomer, followed by hydrogenation of the cis-9 bond in linoleic acid by microbial reductase into saturated FA (Figure 2) [71].

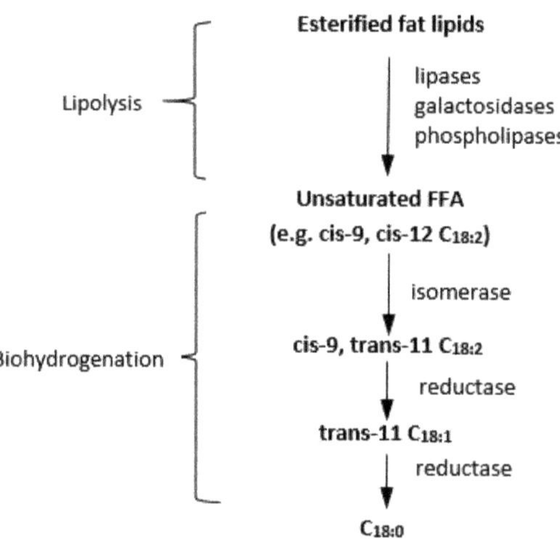


**Figure 2.** Key steps in lipolysis and biohydrogenation to convert esterified fatty acids to saturated fatty acids in the rumen. FFA: free fatty acids [71].

In an in vitro study, tannin extract from *Schinopsis lorentzii* reduced biohydrogenation of α-linolenic acid (ALA) from 43% to only 13% in flaxseed diet after 24 h incubation [73]. Tannin-containing forage (Sainfoin) had no effect on ALA BH, but diets containing tannin extracts (7.9% of dietary DM) reduced BH by 20% in vitro [74]. Incubating hay and concentrate diet in vitro at 1.0mg/ml of cow buffered ruminal fluid increased vaccenic acid and reduced stearic acid concentration by 23 and 16%, respectively, compared to the control. Tannin was extracted from *Ceratonia siliqua*, *Acacia cyanophylla* and *Schinopsis lorentzii* [75]. Feeding lambs with *Cistus ladanifer* at 200g/kg DM reduced complete rumen BH by 36% in lambs, but had no effect when fed at 50g/kg [76]. The inclusion of Sainfoin in a Timothy grass silage diet of lambs increased the accumulation of ALA in rumen digesta [77]. Protecting unsaturated FA from biohydrogenation in the rumen increases the levels of unsaturated FA absorbed through the intestines into the blood stream. Plasma n-3 PUFA of steers infused sunflower oil into the duodenum was two-fold compared to control [78]. These studies were conducted in controlled environments and some used tannin extracts and/or included oil supplements in the feeds [74,76].

*Effect on parasite control:* Tannin-containing diets impose higher parasite tolerance in different species of grazing animals [66,79]. Tannins exert their anti-parasitic effect by decreasing the viability of larvae, thus interfering with egg hatching and/or improved immunity as a result of improved protein nutrition from reduced rumen protein degradation [80]. In vitro and in vivo studies in small ruminants have reported significant effects of tannin extracts and tannin-containing legumes on faecal egg count, development of eggs to larvae and decreased larvae motility [81,82]. Naturally-infected lambs grazing chicory (*Cichorium intybus*) and birdsfoot trefoil were reported to have fewer helminth parasites than lambs grazing ryegrass or white clover. Birdsfoot trefoil grazing reduced fecal egg count significantly compared to all the other forages [83]. Red deer calves infected with deer-origin gastrointestinal nematodes and lungworm (*Dictyocaulus* sp.) larvae for five weeks were allocated into either lucerne (0.1% condensed tannins; CT), birdsfoot trefoil (1.9% CT) or sulla (*Hedysarum coronarium* L.) (3.5% CT) and slaughtered after seven weeks. Abomasal nematode burdens had significant negative linear relationships with dietary CT concentration, although no substantial differences were observed in faecal egg counts [84].

2.2.3. Use of Legumes in Northern Australia

For the past six decades, the northern Australian beef industry has used tropical pasture legumes [85]. These legumes have been recognized as the best long-term alternative to increase grass pasture productivity. However, adoption level still remains low [86]. For a long time, most attention was directed at lighter textured soil pastures [87] and only little emphasis was placed on legumes adapted to dark clay soils, resulting in genotypes that were either not sufficiently productive or persistent [88–90]. However, attention has been directed to pastures of heavier textured soils of central and southern Queensland in recent years, leading to the development of more suitable perennial legume species and varieties such as *Stylosanthes scabrana* (Caatinga stylo), *Clitoria ternatea* (Butterfly pea), *Macroptilium bracteatum* (Burgundy bean) and *Desmanthus*. Inspite of this, only limited published literature on animal growth and performance from these legumes exists. Stylosanthes, Butterfly pea and *Desmanthus*-grass pastures were observed to improve weight gains compared to grass only pastures numerically throughout the year [59]. Steers grazed on Butterfly pea-grass and Caatinga stylo-grass pastures had no difference in weight gain compared to grass only pasture in the first year of establishment. However, 31 and 68 kg/ha difference, respectively, was observed over a five-year period [91].

*Desmanthus* has gained attention in recent years due to its palatability, high protein content, non-toxic characteristics, anti-methanogenesis demonstrated in vitro [92] and its ability to establish and persist well in clay soils [88–90]. Two varieties of *Desmanthus*, *D. virgatus and D. leptophyllus*, were found in old trial sites after 25 years of establishment on black cracking clay soil surviving droughts, floods, frost and commercial grazing [93]. *Desmanthus* is highly nutritious with at least 14% CP in whole plant and 22% in leaves [39,94]. *Desmanthus* is, therefore, a good legume choice for the clay soils of northern Australia [95,96]. Since 2012, 35,000 ha of *Desmanthus* have been established in northern Australia [97] and necessitates studies to be carried out to determine its effect on beef cattle performance and meat characteristics. A short term 90-day study, reported up to 40kg higher liveweight in steers grazed on *Desmanthus*–Buffel grass pasture than Buffel grass only [98]. Supplementing Rhodes grass fed goats with *Desmanthus* increased dry matter intake, liveweight gain, loin-eye muscle area and hot carcass weight significantly, compared to urea and cottonseed meal supplements [99]. Similarly, a 10-week study reported that *Desmanthus*–Mulato grass diet increased liveweight gain in goats significantly compared to those on Mulato grass only [100]. In contrast, growing goats demonstrated poor acceptance of *Desmanthus bicornutus* compared to Leucaena, alfalfa, and lablab, resulting in lower weight gains [39]. Gardiner (pers com) however has observed *D. bicornutus* in pasture to be very palatable to cattle and cv JCU-4 fed in metabolic chambers was well accepted.

*2.3. Feedlot Finishing of Tropical Pasture-Backgrounded Cattle*

Feedlot finishing is an important phase in the beef supply chain of pasture backgrounded beef cattle. In 2017, 50% of Queensland beef herds were finished in the feedlot with grains high in sugar and fat [9]. Lot-feeding helps to finish cattle during periods of limited pasture availability. This allows beef products to meet certain yardsticks of a wide range of markets, marketing of more even products, reduce farm-stocking pressure during the dry season and help to plan for the marketing season [101,102]. A comprehensive review by Drouillard and Kuhl [103] reported that diet characteristics during backgrounding affects cattle performance in the feedlot. Cattle grazed on poor pastures that restricted growth moderately led to complete compensatory growth during lot-feeding, while those backgrounded on poor pastures resulting to weight loss failed to achieve compensatory growth. Cattle grazed on endophyte-infected fescue were compared to those grazed on endophyte-infected fescue-clover mix and endophyte-free fescue. Cattle grazing on endophyte-infected fescue-clover mixture consistently performed best during grazing and finishing [104,105]. A meta-analysis of 20 dry-lot and 12 grazing studies showed that cattle fed high energy diets during backgrounding had lower final body weights than those grazing or fed on restricted energy. However, this study did not analyse the effect of dietary protein content [106].

## 3. Meat characteristics

The ultimate goal of the beef cattle industry is to provide consumers with beef that is safe and of high eating characteristics. The major determinant of meat characteristics is eating characteristics, influenced mainly by intramuscular fat content, low fat melting point, tenderness, juiciness and flavour [107]. Carcass fat deposition and meat FA composition both play important roles in eating characteristics variation [108–110]. Consumption of beef fat can help in the transport and absorption of fat-soluble vitamins and exerts positive effect on immune response [111].

Beef fat is primarily categorized into three; subcutaneous, intermuscular and intramuscular fat [112]. Saturated FA (SFA) whose levels are high in ruminant meat due to hydrogenation of dietary unsaturated FA in the rumen, are associated with health risks such as coronary heart disease [113,114], although this association remains controversial [115,116]. Monounsaturated FA (MUFA) is reported to be associated with lower mortality rate [117] although other studies did not find any association [115]. Increasing the level of n-3 polyunsaturated FA (PUFA) in the human diet is important to overcome the imbalance resulting from high consumption of plant oils rich in linoleic acid [118]. Long-chain n-3 and n-6 PUFA improve growth, brain and retinal development, maternal and offspring health, cognitive function and psychological status in humans [15]. Also, conjugated linoleic acid (CLA) and n-3 FA confer anti-inflammatory effects [119]. Recommendations for various dietary fat fractions are 15–35%, <10%, <2.5–9%, <2–3% and <1% of total energy intake from total fat, SFA, n-6 PUFA, n-3 PUFA and trans fatty acids, respectively [120,121] and ratios of PUFA:SFA at 0·45 and n-6:n-3 PUFA below 4 [122]. As a result, there is increasing focus on studies aimed at elevating the levels of beneficial n-3 LC-PUFA and reducing saturated fatty acids in beef, especially in intramuscular fat (IMF), commonly referred to as marbling, since it cannot be trimmed out [3,113].

Marbling is associated with carcass fatness. A positive correlation between total carcass fat content and subcutaneous fat thickness with marbling has been observed [123,124]. For instance, a genetic correlation of 0.91 between marbling score and muscle lipid content was reported [125]. Increase in subcutaneous fat thickness from below 0.19 mm to over 1.40 mm transitioned marbling score from 'devoid' to 'abundant' [126]. An increase in carcass fat content and subcutaneous fat thickness from 187 to 217 g/kg and 6.6 to 8.3 mm, respectively, increased marbling score from 2.2 to 2.6 in steers [127], while an increase in carcass fatness influenced FA composition and PUFA:SFA ratio [128]. Marbling fat consists of more unsaturated FA compared to other fats in beef, hence a higher PUFA:SFA ratio. It also contains more oleic acid and less stearic acid [3,129].

### 3.1. Effect of Intramuscular Fat on Beef-Eating Characteristics

#### 3.1.1. Tenderness

Variation in tenderness is attributed to animal age, pre- and post-slaughter carcass handling, post-mortem pH decline, genetic make-up and carcass composition, mainly marbling [108,130,131]. Subcutaneous and intermuscular fats provide insulation for muscles to prevent cold shortening. Muscles cool at a slower rate and rigour is attained at higher temperatures. Leaner lamb carcasses with lower marbling scores and less subcutaneous fat thickness were reported to be tougher than those with more fat [132,133]. Similarly, Jeremiah [126] reported a higher tenderness score for steers and heifers with higher subcutaneous fat thickness and marbling as scored by both trained and untrained panel of consumers. High marbling score as in Kobe beef that can exceed <200 mg/g fresh meat, cause dilution of fibrous proteins by soft fat, thus lowering the bulk density that may reduce resistance to shearing. Marbled fat cell expansion forces muscle bundles apart to result in opened up muscle structure [109,134]. Marbling fat concentration values above 30 mg/g muscle are suggested to result in optimum tenderness [48].

#### 3.1.2. Flavour

Animal nutrition status, diet, sex, breed and genetic make-up are factors that influence meat flavour [135]. Meaty flavour of cooked meat develops from a complex interaction of precursors from the fat and lean components of meat. Products of Maillard reactions between carbohydrates and proteins, such as pyrazines and thiazoles, and lipid degradation of aldehydes, alcohols and ketones, are the most important determinants of flavour [136,137]. Hence, meat composition plays an important role in flavour, which could explain the increase in flavour intensity with age in meat animals [108]. A trained panel reported higher flavour scores for beef from carcasses with higher subcutaneous fat thickness than those with minimal fat [126]. FA composition of the fat also plays a significant role in meat flavour. Linolenic acid was found to be positively correlated with milky-oily and sour flavour in beef, while oleic acid was negatively correlated [138]. Oleic acid is considered to be of major effect on the flavour of cooked beef [139]. FA oxidative degradation to form alkyl radicals occurs faster in PUFA than MUFA, while linolenic acid derivatives, eicosapentaenoic acid (EPA) and docosahexaenoic acid (DHA), are highly susceptible to oxidation giving rise to aldehydes [109].

Fats act as storage sites for skatoles and indoles, two compounds that play a significant role in meat flavour, but moreso in sheep than cattle. They are produced in the rumen through microbial deamination and decarboxylation of tryptophan. When they exceed the liver's metabolism capacity, they are deposited in body fats, thus contributing to pastoral flavours in ruminant meat. At low levels, skatoles contribute to desirable odours and flavours, but at high levels, they produce a nauseating faecal odour [110,111]. Finishing grass-fed cattle with concentrate diet for at least 54 days reduces pastoral flavour significantly [112].

#### 3.1.3. Juiciness

Meat juiciness is the initial impression of moisture released on the meat surface during chewing and the degree of induced salivation [124]. Meat juiciness relies on water and fat contents, hence factors influencing water holding capacity and fat content of meat may influence juiciness [140]. Marbled fat provides lubrication between muscle fibres and increases the perception of juiciness by stimulating salivation while chewing [134]. Fat prevents drying out of meat during cooking [141]. However, some studies did not find any positive correlation between beef subcutaneous fat and marbling with juiciness [126].

### 3.2. Factors Influencing Beef Intramuscular Fat Content and Fatty Acid Composition

In beef, the lipid fraction generally contributes between 4–15% of carcass weight on fresh basis [142]. Out of these, four fifths are SFA, mainly composed of palmitic, stearic and oleic acid. The remaining fifth comprises 30 different FA [143]. In the intramuscular lipid fraction, values of 2–30% in the *Longissimus dorsi* muscle (LDM) has been reported [144].

Lipid fraction and FA composition are influenced by three major factors; namely, age of the animal, diet and breed. A study of fat content and FA profile in three breeds of cattle reported an increase in IMF and saturated FA percentage and a decrease in unsaturated FA of LDM with age [145]. In contrast, Jersey and Limousin cattle showed decrease in total SFA and increase in MUFA [146], while phospholipids showed a decrease in palmitate, stearate and oleate, but an increase in PUFA with age [147]. Age had no effect on Japanese Black cattle FA composition [148]. Beef cattle producers target early turn-off age of steers and heifers, usually below 28 months [149,150]. Details on mechanisms in which age influences beef fatty acids composition are not given as this is beyond the intended scope of this review.

Composition of backgrounding and finishing diets influence beef fatty acids profile. Differences are reported between cattle fed pasture versus concentrate diets [114], pastures containing varying plant secondary metabolites [151], and diets supplemented with oils [152], vitamins and minerals [153,154].

### 3.2.1. Pasture Versus Concentrate Diets

Beef from pasture raised cattle contains higher levels of n-3 and MUFA compared to concentrate-fed cattle [114,155]. Tume [156] suggested that the effect is mainly attributable to individual ingredients in the diet and their combinations. Plants are the primary sources of n-3 PUFA due to their unique ability to synthesize ALA, which comprises at least half of the FA content of forages. ALA forms the building block of n-3 essential FA, and its elongation and desaturation results in the synthesis of EPA, DHA and docosahexaenoic acid (DPA) [113]. Moreover, biohydrogenation of unsaturated FA in the rumen is followed by microbial FA synthesis from dietary long-chain FA and de novo synthesis, amounting to 10–15% of bacterial dry mass that influences the FA profile of absorbed lipids [71,157]. Feeding Angus crossbred steers on forages and pasture only, increased muscle rumenic acid by four-folds compared to high-grain diet [158]. Continental crossbred steers fed on grass pasture had the highest intramuscular PUFA content and increasing dietary concentrate supplement led to a linear increase in SFA, increased n-6:n-3 PUFA ratio as well as a decrease in PUFA:SFA ratio [114]. Angus-cross steers backgrounded on pasture only were finished on corn-silage concentrate or pasture. Pasture finished steers had 61%, 21% and 22% less total fat, oleic acid and total MUFA compared to the concentrate group. Individual (linolenic acid, EPA, DPA, and DHA acids) and total n-3 FA concentrations and the ratio of n-6 to n-3 fatty acids were greater in forage than concentrate finished steers [159]. Grass-fed German Holstein and Simmental bulls had higher total PUFA and lower n-6-n3 ratio than concentrate fed bulls, but total SFA was similar [155]. Similarly, grass-fed Hereford steers had higher PUFA, lower MUFA and n6:n3 ratio than concentrate-fed steers, while SFA was not affected [160]. Fat deposition relies on consumption of surplus net energy [161], hence, grain feeding increases carcass total fat content due to high energy levels [124].

Time on feed also plays a significant role in beef FA composition. Steers raised on native range stocker operation were divided into eight groups, finished on a high concentrate diet and slaughtered serially at 28 days intervals from zero (control) to 196 days on a finishing diet. Carcass marbling score, subcutaneous fat thickness, LDM MUFA and total lipid percentage increased, while PUFA decreased with increase in days on concentrate diet. Differences in these parameters were observed after 112 days on the diet after which a plateau was reached [162]. These results should be interpreted with caution as age also affects carcass fat content and composition as well as marbling [163–165].

Some studies have reported the effect of pasture species on beef FA composition, while some reported no difference. For instance, alfalfa-finished steers had higher concentrations of linoleic acid and ALA than those finished on pearl millet and a combination of white clover, blue grass, orchard grass and tall fescue, but forage species did not affect total lipid content of the LDM in a 40-day study [159]. In a four-month study where steers grazed on tall fescue only, or combined with red clover or alfalfa, there was no effect of pasture on meat FA content [152]. Similarly, rib eye rolls from steers finished on tall fescue and meadow brome or birdsfoot trefoil for four months had similar marbling scores, n-6-n3 ratios and total SFA, MUFA and PUFA, but EPA was higher in birdsfoot trefoil finished steers [166]. The effect of pasture type on FA composition was reported in a 90-day study where lambs with access to shrubs produced meat with higher percentage of ALA, n-3, n-6, total PUFA and lower MUFA and n-6:n-3 ratio than those on grass only, but total SFA was similar [167]. The effect of different forage species may be due to plant secondary metabolites. Cattle grazing botanically diverse pastures with different plant secondary metabolites had higher intramuscular n-3 and total PUFA compared to cattle grazing predominantly ryegrass pastures with similar pasture FA profile [168]. Red clover reduces ruminal biohydrogenation of PUFA, possibly due to the protective effects of the polyphenol oxidase enzyme [169]. As discussed earlier, dietary tannin may inhibit or minimize rumen biohydrogenation of unsaturated FA and increase the level of n-3 PUFA in the blood circulation. LDM of lambs fed Sulla (*Hedysarum coronarium* L.) containing 1.8% condensed tannins had 24% more ALA compared to lambs fed Sulla and drenched with polyethylene glycol, a compound that binds and inactivates tannins [151]. *Desmanthus* contains up to 4.5% condensed tannins [92], hence grazing cattle on *Desmanthus* pastures may increase n-3 PUFA in beef.

### 3.2.2. Oil Supplements

Dietary supplementation with n-3 LC-PUFA-rich oils has been shown to increase PUFA in the meat of ruminants [152] because at high concentrations, rumen microorganisms cannot hydrogenate these oils to any significant extent [170] and oil supplements also enhance *de novo* FA synthesis from their dietary precursors [118]. Steers supplemented with fish oil doubled the EPA and DHA contents in muscle phospholipids, while those supplemented with linseed increased the levels of ALA in muscle phospholipids from 9.5 to 19 mg/100 g and enhanced EPA synthesis from 10 to 15 mg/100 g in muscle with no effect on feed intake [118]. Lorenzen and colleagues [123] reported over 80% increase in CLA in beef from soybean oil supplementation compared to the control during finishing of steers. Soybean oil supplement increased CLA in the adipose tissue of steers [154]. Fish oil supplement increased n-3 LC-PUFA, including linolenic acid, EPA and DHA concentrations in the LDM of bulls and steers [153] and slightly increased the total FA in supplemented steers compared to the control [118]. Diet-protected fish oil and free fish oil increased total muscle EPA and DHA from 13 to 19 mg/100 g and 3 to 12 mg/100 g, respectively [171]. Effect of oil on beef FA composition is not unique to pure oil supplements. British x Continental crossbred steers were fed grass hay or red clover silage only or supplemented with either sunflower-seed or flaxseed concentrates to provide 5.4% oil in diet DM basis. Sunflower-seed or flaxseed supplements increased vaccenic, rumenic and n−6 FA in the *Longissimus thoracis* muscle significantly. ALA was over two-fold in flaxseed compared to sunflower-seed supplemented steers [142].

### 3.2.3. Micronutrients

*Vitamin A:* Vitamin A or β-carotene deficiency results in elevated IMF content. Angus steers were fed low β-carotene and vitamin A cereal-based ration for 308 days with or without Vit A supplementation before slaughter. Supplemented steers scored 19% less marbling and the LDM IMF content was 35% lower than the control [172]. Supplementing Japanese Black cattle with vit A after 15 months of age reduced marbling score significantly. A correlation of −0.38 was observed between marbling and serum vit A just before slaughter [173]. Effect of vit A is proposed to be due to its derivative retinoic acid that restricts hyperplasia and/or by regulating the growth hormone gene resulting in a decrease in fat deposition [125]. Trans-retinoic acid, a metabolite of retinol, subdues differentiation of preadipocytes by suppressing the expression of peroxisome proliferator-activated receptor gamma (PPARγ) gene [174,175].

*Vitamin C:* Domestic animals normally do not receive dietary Vitamin C supplementation due to their ability to synthesize the vitamin in the liver [176]. However, plasma Vitamin C levels in beef cattle drop below the normal 2.4–4.7 mg/L range during the late fattening period, showing that Vitamin C plays an important role in adipogenesis [177]. Supplemented Japanese Black cattle receiving high-concentrate diets with Vitamin C during the late [178] or from middle fattening stage produced fatter carcasses with higher marbling scores than the control [177]. Increased lipogenesis is as a result of the positive effect of Vitamin C on adipocyte differentiation [179].

*Vitamin D and Calcium:* $1\alpha,25$-dihydroxyvitamin $D_3$, the biologically active form of Vitamin D, inhibits the differentiation of preadipocytes through direct suppression of PPARγ protein [174]. Since $1\alpha,25$-dihydroxyvitamin $D_3$ is critical for calcium homeostasis, low dietary intake of calcium leads to increased plasma $1\alpha,25$-dihydroxyvitamin $D_3$ that suppresses adipocyte differentiation and reduces marbling [177]. Feedlot cattle with low plasma $1\alpha,25$-dihydroxyvitamin $D_3$ levels had higher marbling scores than those with higher levels [180]. In contrast, high marbling scores were reported in Hanwoo steers finished on low calcium diets leading to high levels of plasma $1\alpha,25$-dihydroxyvitamin $D_3$ than in steers finished on high calcium diets [181]. $1\alpha,25$-dihydroxyvitamin $D_3$ may exert two contrasting functions on adipogenesis; inhibit adipocyte differentiation and promote fat accumulation in adipocytes, depending on the animals' stage of growth [177].

### 3.2.4. Cattle Breed

Studies have reported variation in beef fat deposition and FA composition due to genetic differences between cattle. Beef breeds differ from milk breeds and such differences are well documented. Generally, meat type breeds are able to deposit more fat than milk breeds. FA synthesis was reported to be higher in beef cattle subcutaneous and perirenal adipose tissues than in same tissues from dairy cattle of similar age and weight [182]. Differences between Jersey and Limousin [146,147], Japanese Black and Holstein [183] have been reported. German Holstein bulls had higher SFA and total PUFA compared to German Simmental bulls on similar diets, but breed had no effect on n-3 FA [184]. Nuernberg and colleagues [155] reported higher n-3, n-6, n-6:n-3 ratio and total PUFA in German Simmental beef bulls than in German Holstein bulls, but SFA levels were similar. CLA isomer cis-9, trans-11 concentration was higher in the LDM of German Holstein than German Simmental bulls. In a 24-month study, Galloway, White–Blue Belgian and German Holstein bulls were fed on the same diet and slaughtered at different ages between zero and 24 months. Carcass subcutaneous fat and LDM intramuscular fat were highest in German Holstein and least in White-Blue Belgian. At birth, stearic acid, oleic acid and n-3 FA were highest in the LDM of Galloway, while total unsaturated FA, PUFA and n-6 FA were highest in White-Blue Belgian. A similar profile was observed at 18 months of age except in n-3 FA that were similar in Galloway and WBB, but lower in German Holstein [145]. Comparing the FA profile of Japanese Black and Holstein steers subcutaneous neutral lipids showed lower myristic acid, palmitic acid, stearic acid and total SFA, but higher oleic acid, total MUFA and MUFA:SFA ratio in the Japanese Black than Holstein steers. However, intramuscular phospholipid FA profile was not affected except for palmitic acid [183]. Simmental and Red Angus steers at similar back fat finished level of 10 mm were compared for LDM fat profile. Total lipids, myristoleic acid, palmitoleic acid, vaccenic acid, along with n-6:n-3 ratio, were greater while EPA and total n-3 PUFA were lower in Simmental than Red Angus steers. Time on grain diet was a confounding effect in this study as the Angus spent 70 days less on the grain diet and were slaughtered 73 days younger than the Simental steers [185].

Sires may influence the FA content of their offspring. Japanese Black Wagyu cattle sired by different bulls were reported to have significantly different SFA and MUFA contents [186]. Heritabilities of FA and other carcass traits were reported to range from 14 to 36% in crossbred cattle [187]. These breed variances are probably due to differences in the activities of enzymes influencing gene expression and/or enzyme function [113]. The activity of Δ9-desaturase enzyme to convert palmitic to palmitoleic acid was observed to be greater in Simmental than Red Angus lipids [185].

### 3.3. Genes that Influence Carcass Fat Content and Fatty Acid Profiles

Several genes are reported to be responsible for variation in fat content and FA composition in beef. The genes encode for cocaine- and amphetamine-regulated transcript [188], leptin [189], diacylglycerol O-acyltransferase, the growth hormone 1 [190] sterol regulatory element-binding protein 1 [191], fatty acid synthase, stearoyl-CoA desaturase and fatty acid binding protein 4 [112,192–194]. This review will focus on fatty acid synthase, stearoyl-CoA desaturase and fatty acid binding protein 4 genes.

### 3.3.1. Stearoyl-CoA Desaturase (SCD)

SCD gene encodes for Δ9 desaturase enzyme and introduces a single double bond in SFA to convert them to MUFA. For instance, the enzyme desaturates stearic acid to oleic acid and trans-vaccenic acid. High concentration of oleic acid in beef is associated with soft fat and overall palatability in Wagyu and Hanwoo cattle. As a result, high activity of Δ9 desaturase enzyme is associated with soft fat in beef [144,195]. SCD catalytic activity is about twice higher in bovine marbled muscle tissue than in the subcutaneous adipose tissue. This agrees with higher MUFA levels observed in the muscle than subcutaneous adipose tissue [196–198]. SCD gene expression and activity is reported to increase after weaning [199] and preceding lipid filling in preadipocytes. Similarly, a gradual increase in de novo FA biosynthesis is observed after weaning, indicating that SCD activity is required for lipogenic activity in the subcutaneous adipose tissue to develop [144]. In another study, subcutaneous adipose tissue samples were collected from carcasses of pasture and feedlot cattle fed for 100, 200 and 300 days. Pasture-fed cattle adipose tissue had lower total SFA and higher total UFA than in feedlot cattle. Δ9 desaturase activity was much higher in pasture-fed than feedlot cattle [47].

SCD gene expression varies between and within breeds. Full-length bovine SCD cDNA from 20 Japanese Black steers was compared. Two types of SCD genes with single nucleotide polymorphisms (SNPs) in the open reading frame where valine (V) was replaced by alanine (A) were observed. The two SCD genes were genotyped and classified in 1003 Japanese Black carcasses into VV, VA and AA genotypes. Comparison of FA composition from the carcasses showed that SCD type A gene was associated with higher percentage of MUFA with 0.8% effect and lower IMF melting point. They concluded that SCD is one of the causes of genotype variation [200]. In contrast, SCD (878C>T) SNP was observed to have no association with FA profile in upper sirloin cuts of Aberdeen Angus and Blonde d'Aquitaine cattle [112].

### 3.3.2. Fatty Acid Synthase (FASN)

FASN gene is abundantly expressed in the adipose tissue and encodes for fatty acid synthase, an enzyme that regulates the biosynthesis of long chain fatty acids. The enzyme plays a central role in de novo lipogenesis by catalysing all the reaction steps to convert acetyl-CoA and malonyl-CoA to palmitate. Association of FASN expression or polymorphisms with fat metabolism and obesity traits in cattle has been reported [188,201]. Analysing polymorphisms in thioesterase domain of FASN gene, which regulates the termination of FA synthesis, in Hanwoo cattle showed a significant association between g.17924G > A SNP genotypes with palmitic and oleic acid concentrations. For instance, GG genotype had 3.2% and 2.8% higher oleic acid concentration than the AA and AG genotypes, respectively. However, they did not observe any significant association between g.17924A > G genotypes and other examined FA such as myristic, stearic and linoleic acids [191]. GG genotype of g.17924A > G SNP was reported to result in higher UFA and fairly lower amounts of SFA than the AG and AA genotypes in other studies [192,202]. Another study was carried out to determine exonic SNPs in the gene encoding FASN with FA composition in Korean cattle. It was found that all the SNPs (g.12870 T > C, g.13126 T > C, g.15532 C > A, g.16907 T > C and g.17924 G > A), were associated with varying FA compositions and marbling. Genotypes CC, TT, AA, TT, and GG were associated with higher MUFA and lower SFA [203].

Some studies reported no relationship between FASN gene with fat thickness and marbling score. However, a significant relationship of the fat with DNA-protein kinase, known to play a role in transcriptional activation of FASN, was reported [204,205].

### 3.3.3. Fatty Acid Binding Protein 4 (FABP4)

FABP4 is a gene highly expressed in the adipose tissue and encodes for fatty acid binding protein 4 that belongs to a group of FABPs. These binding proteins play a significant role in absorption, transport and metabolism of FA, and glucose homeostasis by interacting with peroxisome proliferator-activated receptors [18,206]. SNP 7516G > C of FABP4 was analyzed for association with IMF profile of upper sirloin cuts in Aberdeen Angus and Blonde d'Aquitaine cattle. CC genotype in Angus cattle was 52% and 64% lower in Myristoleic acid, and 33% and 35% lower in linoleic acid than CG and GG, respectively. Blonde cattle CC genotype had higher arachidonic acid and EPA, but lower oleic acid and total SFA than the CG. The GG genotype was observed in only one bull [112]. g.7516G > C polymorphisms were analyzed for association with marbling score and subcutaneous fat depth in Wagyu x Limousin crosses. A positive relationship between CC genotype and lower marbling and fat depth was observed. GC genotype had the highest scores while GG was in-between [18]. FABP4 SNPs were also reported to have an association with back fat thickness in Korean Native cattle [20].

## 4. Conclusions and Future Research

Beef is a nutrient-dense food and remains an important dietary component in global human nutrition. In northern Australia, the beef industry contributes immensely to the economy as it accounts for over half of Australia's beef exports. It relies heavily on native pastures that are highly seasonal and of low quality resulting in weight loss during the dry season and a high turn-off age. Although supplementing beef cattle with protein and energy diets improves weight gain, cost is limiting, hence supplementation is not an economical option in extensive grazing systems. However, nutrient-dense diets are used to finish most northern Australian beef cattle herds to produce a more even product that meets certain yardsticks of a wide range of markets. Legumes are known to improve pasture and livestock production at a lower cost. However, most legumes do not survive or persist in clay soils prevalent in northern Australia. In recent years, several persistent pasture legumes in clay soils have been developed and trialed. Of specific interest is *Desmanthus*, a highly palatable, high protein content, non-toxic tropical legume with potential to reduce enteric methane emissions. Few available studies indicate that *Desmanthus* can be used to improve pasture quality and subsequent beef cattle productivity in northern Australia.

- However, only limited peer-reviewed published literature is available on the effect of *Desmanthus* on beef cattle growth and performance. These studies were either conducted indoors or in small sized paddocks (except one in 250 ha paddock) which may not be replicated in normal commercial farm settings. Hence, there is need to conduct more studies under commercial farm settings to determine the suitability of grass–*Desmanthus* pastures in northern Australian beef cattle production system.
- Tannin-containing pastures at 20–40 g/kg DM are reported to increase polyunsaturated fatty acids in meat by reducing rumen biohydrogenation of unsaturated fatty acids. There is need to study the effect of *Desmanthus*, a tannin-containing legume, on performance and meat characteristics of grazing cattle.

Most of beef cattle in northern Australia are *Bos indicus* due to their ability to tolerate ticks, heat and poor-quality pasture. However, meat characteristics from these cattle is low due to low marbling. These breeds are crossed with *Bos taurus* to improve growth rate and meat characteristics of several composite breeds such as Belmont Red and NAPCO Composite. It is irrefutable that genetic make-up plays a significant role in beef fat content and FA profile.

- Several genes such as *SCD*, *FASN* and *FABP4* are reported to influence carcass fat traits in Korean and Japanese cattle as well as Australian temperate breeds such as Angus and Limousin. There is need to investigate the effect of these genes in northern Australian composite breeds.
- In addition, studies are required to determine finishing performance and carcass traits of northern Australian beef composite breeds backgrounded on newly introduced legume pastures, such as *Desmanthus*, to enable industry players to exploit them for greater economic gains.

**Author Contributions:** Conceptualization, A.E.O.M.-A, E.C., C.P.G., B.S.M.-A., R.T.K and F.W.M.; methodology, A.E.O.M.-A., E.C., C.P.G., B.S.M.-A., R.T.K. and F.W.M.; software, A.E.O.M.-A.; validation, A.E.O.M.-A., E.C.,

C.P.G., R.T.K. and B.S.M.-A.; formal analysis, F.W.M.; investigation, F.W.M.; resources, A.E.O.M.-A., E.C., C.P.G., R.T.K. and B.S.M.-A.; data curation, writing—original draft preparation, F.W.M.; writing—reviewing and editing, A.E.O.M.-A., E.C., C.P.G., R.T.K. and B.S.M.-A.; supervision, A.E.O.M.-A., E.C., C.P.G., R.T.K. and B.S.M.-A.; project administration, A.E.O.M.-A., C.P.G.; funding acquisition, A.E.O.M.-A., C.P.G. and E.C.

**Funding:** This research was funded by the Cooperative Research Centre for Developing Northern Australia (CRC-NA) Projects [grant number CRC P-58599] from the Australian Government's Department of Industry, Innovation and Science, and a PhD scholarship funded by CRC-NA and the College of Public Health, Medical and Veterinary Sciences, James Cook University, Queensland, Australia, awarded to the first named author.

**Acknowledgments:** The authors gratefully acknowledge James Cook University's (JCU) College of Public Health, Medical and Veterinary Sciences, the Cooperative Research Centre for Developing Northern Australia (CRC-DNA), Meat and Livestock Australia (MLA) and the Commonwealth Scientific and Industrial Research Organisation (CSIRO)-JCU-Agrimix Joint Research Project.

**Conflicts of Interest:** The authors declare no conflict of interest. The funders had no role in the design of the study; collection, analyses, or interpretation of data; in the writing of the manuscript, or in the decision to publish the results.

## References


1. OECD Meat Consumption (Indicator). Available online: https://data.oecd.org/agroutput/meat-consumption.htm (accessed on 15 May 2019).
2. OECD-FAO Agricultural Outlook 2018–2027. Available online: https://www.oecd-ilibrary.org/agriculture-and-food/oecd-fao-agricultural-outlook-2018-2027_agr_outlook-2018-en (accessed on 15 May 2019).
3. Troy, D.J.; Tiwari, B.K.; Joo, S. Health implications of beef intramuscular fat consumption. *Korean J. Food Sci. Anim. Resour.* **2016**, *36*, 577–582. [CrossRef]
4. Asp, M.L.; Richardson, J.R.; Collene, A.L.; Droll, K.R.; Belury, M.A. Dietary protein and beef consumption predict for markers of muscle mass and nutrition status in older adults. *J. Nutr. Health Aging* **2012**, *16*, 784–790. [CrossRef]
5. Williams, P. Nutritional composition of red meat. *J. Nutr. Diet.* **2007**, *64*, S113–S119. [CrossRef]
6. FAO Overview of Global Meat Market Developments in 2018. Available online: http://www.fao.org/3/ca3880en/ca3880en.pdf (accessed on 10 June 2019).
7. Australian Bureau of Statistics Value of Agricultural Commodities Produced in Australia, 2017–2018. Available online: https://www.abs.gov.au/ausstats/abs@.nsf/0/58529ACD49B5ECE0CA2577A000154456?Opendocument (accessed on 12 August 2019).
8. Meat and Livestock Australia State of the Industry Report 2018: The Australian Red Meat and Livestock Industry. Available online: http://rmac.com.au/wp-content/uploads/2018/09/SOTI18.pdf (accessed on 29 April 2019).
9. Department of Agriculture and Fisheries The Queensland Beef Supply Chain. Available online: https://publications.qld.gov.au/dataset/23f4f979.../2-qld-beef-supply-chain.pdf%0A%0A (accessed on 29 April 2019).
10. Greenwood, P.L.; Gardner, G.E.; Ferguson, D.M. Current situation and future prospects for the Australian beef industry—A review. *Asian Australas. J. Anim. Sci.* **2018**, *31*, 992–1006. [CrossRef]
11. Johnston, M.G.; Jeyaruban, D.J. Estimated additive and non-additive breed effects and genetic parameters for ultrasound scanned traits of a multi-breed beef population in tropical Australia. In Proceedings of the 10th World Congress of Genetics Applied to Livestock Production, Vancouver, BC, Canada, 17–22 August 2014.
12. Bell, A.W.; Charmley, E.; Hunter, R.A.; Archer, J.A. The Australasian beef industries—Challenges and opportunities in the 21st century. *Anim. Front.* **2011**, *1*, 10–19. [CrossRef]
13. Wolcott, M.L.; Johnston, D.J.; Barwick, S.A.; Iker, C.L.; Thompson, J.M.; Burrow, H.M. Genetics of meat quality and carcass traits and the impact of tenderstretching in two tropical beef genotypes. *Anim. Prod. Sci.* **2009**, *49*, 383–398. [CrossRef]
14. Mannen, H. Identification and utilization of genes associated with beef qualities. *Anim. Sci. J.* **2011**, *82*, 1–7. [CrossRef]
15. Mapiye, C.; Vahmani, P.; Mlambo, V.; Muchenje, V.; Dzama, K.; Hoffman, L.C.; Dugan, M.E.R. The trans-octadecenoic fatty acid profile of beef: Implications for global food and nutrition security. *Food Res. Int.* **2015**, *76*, 992–1000. [CrossRef]
16. Zietemann, V.; Kröger, J.; Enzenbach, C.; Jansen, E.; Fritsche, A.; Weikert, C.; Boeing, H.; Schulze, M.B. Genetic variation of the FADS1 FADS2 gene cluster and n-6 PUFA composition in erythrocyte membranes. *Br. J. Nutr.* **2010**, *101*, 1748–1759. [CrossRef]

17. Ordovas, J.M. Identification of a functional polymorphism at the adipose fatty acid binding protein gene (FABP4) and demonstration of its association with cardiovascular disease: A path to follow. *Nutr. Rev.* **2007**, *65*, 130–134. [CrossRef]

18. Michal, J.J.; Zhang, Z.W.; Gaskins, C.T.; Jiang, Z. The bovine fatty acid binding protein 4 gene is significantly associated with marbling and subcutaneous fat depth in Wagyu x Limousin F2 crosses. *Anim. Genet.* **2006**, *37*, 400–402. [CrossRef]

19. Barendse, W.; Bunch, R.J.; Thomas, M.B.; Harrison, B.E. A splice site single nucleotide polymorphism of the fatty acid binding protein 4 gene appears to be associated with intramuscular fat deposition in longissimus muscle in Australian cattle. *Anim. Genet.* **2009**, *40*, 770–773. [CrossRef]

20. Cho, S.A.; Park, T.S.; Yoon, D.H.; Cheong, H.S.; Namgoong, S.; Park, B.L.; Lee, H.W.; Han, C.S.; Kim, E.M.; Cheong, I.C.; et al. Identification of genetic polymorphisms in FABP3 and FABP4 and putative association with back fat thickness in Korean native cattle. *BMB Rep.* **2008**, *41*, 29–34. [CrossRef] [PubMed]

21. Lee, S.H.; van der Werf, J.H.J.; Lee, S.H.; Park, E.W.; Oh, S.J.; Gibson, J.P.; Thompson, J.M. Genetic polymorphisms of the bovine fatty acid binding protein 4 gene are significantly associated with marbling and carcass weight in Hanwoo (Korean Cattle ). *Anim. Genet.* **2010**, *41*, 442–444. [PubMed]

22. Hoashi, S.; Hinenoya, T.; Tanaka, A.; Ohsaki, H.; Sasazaki, S.; Taniguchi, M.; Oyama, K.; Mukai, F.; Mannen, H. Association between fatty acid compositions and genotypes of FABP4 and LXR-alpha in Japanese Black cattle. *BMC Genet.* **2008**, *9*, 84. [CrossRef] [PubMed]

23. Bezerra, R.L.; Ferreira, R.R.; Edvan, L.R.; Neto, G.S.; da Silva, L.A.; de Araújo, J.M. Protein supplementation is vital for beef cattle fed with tropical pasture. In *Grasses as Food and Feed*; Tadele, Z., Ed.; IntechOpen: London, UK, 2018; pp. 97–107.

24. Hattersley, P.W. The distribution of C3 and C4 grasses in Australia in relation to climate. *Oecologia* **1983**, *57*, 113–128. [CrossRef]

25. Hunt, L.P.; Mcivor, J.G.; Grice, A.C.; Bray, S.G. Principles and guidelines for managing cattle grazing in the grazing lands of northern Australia: Stocking rates, pasture resting, prescribed fire, paddock size and water points—A review. *Rangel. J.* **2014**, *36*, 105–119. [CrossRef]

26. Tothill, J.; Gillies, C. *The Pasture Lands of Northern Australia: Their Condition, Productivity and Sustainability*; Occasional; Tropical Grasslands Society of Australia: St Lucia, QLD, Australia, 1992; ISBN 0959094849.

27. Cameron, D.F.; Edye, L.A.; Chakraborty, S.; Manners, J.M.; Liu, C.J.; Date, R.A.; Boland, R.M. An integrated program to improve anthracnose resistance in Stylosanthes—A review. In Proceedings of the Proc 8th Australian Agronomy Conference, Toowoomba, Australia, 30 January–2 February 1996; pp. 112–115.

28. Pengelly, B.C.; Conway, M.J. Pastures on cropping soils: Which tropical pasture legume to use? *Trop. Grassl.* **2000**, *34*, 162–168.

29. Robertson, F.A.; Myers, R.J.K.; Saffigna, P.G. Nitrogen cycling in brigalow clay soils under pasture and cropping. *Soil Res.* **1997**, *35*, 1323. [CrossRef]

30. Soil Science Australia State Soils. Available online: https://www.soilscienceaustralia.org.au/about/about-soil/state-soil (accessed on 15 May 2019).

31. Gleeson, T.; Martin, P.; Misfud, C. Northern Australian Beef Industry: Assessment of Risks and Opportunities. Available online: https://docplayer.net/25202349-Northern-australian-beef-industry-assessment-of-risks-and-opportunities.html (accessed on 29 April 2019).

32. Ivory, D.; Whiteman, P. Effect of Temperature on Growth of Five Subtropical Grasses. II. Effect of Low Night Temperature. *Funct. Plant Biol.* **2006**, *5*, 149. [CrossRef]

33. McMeniman, N.P.; Beale, I.F.; Murphy, G.M. Nutritional evaluation of south-west Queensland pastures. I. The botanical and nutrient content of diets selected by sheep grazing on Mitchell grass and mulga/grassland associations. *Aust. J. Agric. Res.* **1986**, *37*, 289–302. [CrossRef]

34. Wilson, J.R.; Mannetje, L. Senescence, digestibility and carbohydrate content of Buffel grass and Green Panic leaves in swards. *Aust. J. Agric. Res.* **1979**, *29*, 503–516. [CrossRef]

35. Charmley, E.; Stephens, M.L.; Kennedy, P.M. Predicting livestock productivity and methane emissions in northern Australia: Development of a bio-economic modelling approach. *Aust. J. Exp. Agric.* **2008**, *48*, 109–113. [CrossRef]

36. McCown, R.L. The climatic potential for beef cattle production in tropical Australia: Part I-Simulating the annual cycle of liveweight change. *Agric. Syst.* **1981**, *6*, 303–317. [CrossRef]

37. Brandão, R.K.C.; de Carvalho, G.; Silva, R.; Dias, D.; Mendes, F.; Lins, T.; Pereira, M.; Guimarães, J.; Tosto, M.; Rufino, L.; et al. Correlation between production performance and feeding behavior of steers on pasture during the rainy-dry transition period. *Trop. Anim. Health Prod.* **2018**, *50*, 105–111. [CrossRef] [PubMed]

38. Leng, R.A. Factors affecting the utilization of 'poor-quality' forages by ruminants particularly under tropical conditions. *Nutr. Res. Rev.* **1990**, *3*, 277–303. [CrossRef]

39. Kanani, J.; Lukefahr, S.D.; Stanko, R.L. Evaluation of tropical forage legumes (Medicago sativa, Dolichos lablab, Leucaena leucocephala and Desmanthus bicornutus) for growing goats. *Small Rumin. Res.* **2006**, *65*, 1–7. [CrossRef]

40. Johnson, D.E.; Johnson, K.A. Methane emissions from cattle. *J. Anim. Sci.* **1995**, *73*, 2483–2492. [CrossRef]

41. Buxton, D.R.; Mertens, D.R.; Moore, K.J. Forage quality for ruminants: Plant and animal considerations. *Prof. Anim. Sci.* **1995**, *11*, 121–131. [CrossRef]

42. Bowman, J.G.P.; Sowell, B.F.; Paterson, J.A. Liquid supplementation for ruminants fed low-quality forage diets: A review. *Anim. Feed Sci. Technol.* **1995**, *55*, 105–138. [CrossRef]

43. Chilliard, Y.; Bocquier, F.; Doreau, M. Digestive and metabolic adaptations of ruminants to undernutrition, and consequences on reproduction. *Reprod. Nutr. Dev.* **1998**, *38*, 131–152. [CrossRef] [PubMed]

44. Baldwin, R.L. Digestion and metabolism of ruminants. *Bioscience* **1984**, *34*, 244–249. [CrossRef]

45. Burrin, D.G.; Ferrell, C.L.; Britton, R.A.; Bauer, M. Level of nutrition and visceral organ size and metabolic activity in sheep. *Br. J. Nutr.* **1990**, *64*, 439–448. [CrossRef] [PubMed]

46. Ortigues, I.; Doreau, M. Responses of the splanchnic tissues of ruminants to changes in intake: Absorption of digestion end products, tissue mass, metabolic activity and implications to whole animal energy metabolism. *Ann. Zootech.* **1995**, *44*, 321–346. [CrossRef]

47. Yang, A.; Larsen, T.W.; Smith, S.B.; Tume, R.K. Δ9 Desaturase activity in bovine subcutaneous adipose tissue of different fatty acid composition. *Lipids* **1999**, *34*, 971–978. [CrossRef] [PubMed]

48. Smith, G.C.; Cross, H.R.; Carpenter, Z.L.; Murphy, C.E.; Savell, J.W.; Abraham, H.C.; Davis, G.W. Relationship of USDA maturity groups to palatability of cooked beef. *J. Food Sci.* **1982**, *47*, 1100–1107. [CrossRef]

49. Preston, T.R.; Leng, R.A. *Matching Ruminant Production Systems with Available Resources in the Tropics and Sub-Tropics*; Penambul Books: Armidale, Australia, 1987.

50. Panjaitan, T.; Quigley, S.P.; McLennan, S.R.; Poppi, D.P. Effect of the concentration of Spirulina (Spirulina platensis) algae in the drinking water on water intake by cattle and the proportion of algae bypassing the rumen. *Anim. Prod. Sci.* **2010**, *50*, 405–409. [CrossRef]

51. Valente, E.E.L.; Paulino, M.F.; Detmann, E.; Valadares Filho, S.D.C.; Cardenas, J.E.G.; Dias, I.F.T. Requirement of energy and protein of beef cattle on tropical pasture. *Acta Sci. Anim. Sci.* **2013**, *35*, 417–424. [CrossRef]

52. Batista, E.D.; Detmann, E.; Titgemeyer, E.C.; Valadares Filho, S.C.; Valadares, R.F.D.; Prates, L.L.; Rennó, L.N.; Paulino, M.F. Effects of varying ruminally undegradable protein supplementation on forage digestion, nitrogen metabolism, and urea kinetics in Nellore cattle fed low-quality tropical forage. *J. Anim. Sci.* **2016**, *94*, 201–216. [CrossRef]

53. Lazzarini, I.; Detmann, E.; Sampaio, C.B.; Paulino, M.F.; Valadares Filho, S.D.; Souza, M.A.; Oliveira, F.A. Intake and digestibility in cattle fed low-quality tropical forage and supplemented with nitrogenous compounds. *Rev. Bras. Zootec.* **2009**, *38*, 2021–2030. [CrossRef]

54. Fernandes, R.M.; de Almeida, C.M.; Carvalho, B.C.; Neto, J.A.; Mota, V.A.; de Resende, F.D.; Siqueira, G.R. Effect of supplementation of beef cattle with different protein levels and degradation rates during transition from the dry to rainy season. *Trop. Anim. Health Prod.* **2016**, *48*, 95–101. [CrossRef] [PubMed]

55. Neves, D.S.; Silva, R.R.; da Silva, F.F.; Santos, L.V.; Abreu Filho, G.; de Souza, S.O.; Santos, M.D.; Rocha, W.J.; da Silva, A.P.; de Melo Lisboa, M.; et al. Increasing levels of supplementation for crossbred steers on pasture during the dry period of the year. *Trop. Anim. Health Prod.* **2018**, *50*, 1411–1416. [CrossRef] [PubMed]

56. Poppi, D.P.; McLennan, S.R. Nutritional research to meet future challenges. *Anim. Prod. Sci.* **2010**, *50*, 329. [CrossRef]

57. Department of Agriculture and Fisheries Supplementation Feeding Considerations. Available online: https://www.daf.qld.gov.au/business-priorities/agriculture/disaster-recovery/drought/managing/supplementation-feeding-considerations (accessed on 24 August 2019)

58. Krebs, G.; Leng, R.A.; Nolan, J.V. Effect on bacterial kinetics in the rumen of eliminating rumen protozoa or supplementing with soyabean meal or urea in sheep on a low protein fibrous feed. In *The Roles of Protozoa and Fungi in Ruminant Digestion*; Nolan, J.V., Leng, R.A., Demeyer, D.I., Eds.; Penambul Books: Armidale, Australia, 1989; pp. 199–210.

59. Hill, J.O.; Coates, D.B.; Whitbread, A.M.; Clem, R.L.; Robertson, M.J.; Pengelly, B.C. Seasonal changes in pasture quality and diet selection and their relationship with liveweight gain of steers grazing tropical grass and grass legume pastures in northern Australia. *Anim. Prod. Sci.* **2009**, *49*, 983–993. [CrossRef]

60. Murphy, A.M.; Colucci, P.E. A tropical forage solution to poor quality ruminant diets: A review of Lablab purpureus. *Livest. Res. Rural Dev.* **1999**, *11*, 1999.

61. Osuji, P.O.; Sibanda, S.; Nsahlai, I.V. Supplementation of maize stover for Ethiopian Menz sheep: Effects of cottonseed, noug (Guizotia abyssinica) or sunflower cake with or without maize on the intake, growth, apparent digestibility, nitrogen balance and excretion of purine derivatives. *Anim. Sci.* **1993**, *57*, 429–436. [CrossRef]

62. Gillard, P. Improvement of native pasture with Townsville stylo in the dry tropics of sub-coastal northern Queensland [stylosanthes]. *Aust. J. Exp. Agric.* **1979**, *19*, 325–336. [CrossRef]

63. Rochon, J.J.; Doyle, C.J.; Greef, J.M.; Hopkins, A.; Molle, G.; Sitzia, M.; Scholefield, D.; Smith, C.J. Grazing legumes in Europe: A review of their status, management, benefits, research needs and future prospects. *Grass Forage Sci.* **2004**, *59*, 197–214. [CrossRef]

64. Radrizzani, A.; Nasca, J.A. The effect of Leucaena leucocephala on beef production and its toxicity in the Chaco Region of Argentina. *Trop. Grassl. Forrajes Trop.* **2014**, *2*, 127–129. [CrossRef]

65. Jackson, F.S.; Barry, T.N.; Lascano, C.; Palmer, B. The extractable and bound condensed tannin content of leaves from tropical tree, shrub and forage legumes. *J. Sci. Food Agric.* **1996**, *71*, 103–110. [CrossRef]

66. Piluzza, G.; Sulas, L.; Bullitta, S. Tannins in forage plants and their role in animal husbandry and environmental sustainability: A review. *Grass Forage Sci.* **2014**, *69*, 32–48. [CrossRef]

67. Thi, M.N.; Van Binh, D.; Ørskov, E.R. Effect of foliages containing condensed tannins and on gastrointestinal parasites. *Anim. Feed Sci. Technol.* **2005**, *121*, 77–87.

68. McSweeney, C.S.; Palmer, B.; McNeill, D.M.; Krause, D.O. Microbial interactions with tannins: Nutritional consequences for ruminants. *Anim. Feed Sci. Technol.* **2001**, *91*, 83–93. [CrossRef]

69. Puchala, R.; Min, B.R.; Goetsch, A.L.; Sahlu, T. The effect of a condensed tannin-containing forage on methane emission by goats. *J. Anim. Sci.* **2005**, *83*, 182–186. [CrossRef] [PubMed]

70. Toral, P.G.; Monahan, F.J.; Hervas, G.; Frutos, P.; Moloney, A.P. Review: Modulating ruminal lipid metabolism to improve the fatty acid composition of meat and milk. challenges and opportunities. *Animal* **2018**, *12*, s272–s281. [CrossRef]

71. Jenkins, T.C. Lipid Metabolism in the Rumen. *J. Dairy Sci.* **1993**, *76*, 3851–3863. [CrossRef]

72. Garton, G.A.; Lough, A.K.; Vioque, E. Glyceride Hydrolysis and Glycerol Fermentation by Sheep Rumen Contents. *J. Gen. Microbiol.* **1961**, *25*, 215–225. [CrossRef]

73. Kronberg, S.L.; Scholljegerdes, E.J.; Barceló-Coblijn, G.; Murphy, E.J. Flaxseed treatments to reduce biohydrogenation of α-linolenic acid by rumen microbes in cattle. *Lipids* **2007**, *42*, 1105–1111. [CrossRef]

74. Khiaosa-Ard, R.; Bryner, S.F.; Scheeder, M.R.L.; Wettstein, H.R.; Leiber, F.; Kreuzer, M.; Soliva, C.R. Evidence for the inhibition of the terminal step of ruminal α-linolenic acid biohydrogenation by condensed tannins. *J. Dairy Sci.* **2009**, *92*, 177–188. [CrossRef]

75. Vasta, V.; Makkar, H.P.S.; Mele, M.; Priolo, A. Ruminal biohydrogenation as affected by tannins in vitro. *Br. J. Nutr.* **2009**, *102*, 82–92. [CrossRef]

76. Alves, S.P.; Francisco, A.; Costa, M.; Santos-Silva, J.; Bessa, R.J.B. Biohydrogenation patterns in digestive contents and plasma of lambs fed increasing levels of a tanniferous bush (*Cistus ladanifer* L.) and vegetable oils. *Anim. Feed Sci. Technol.* **2017**, *225*, 157–172. [CrossRef]

77. Campidonico, L.; Toral, P.G.; Priolo, A.; Luciano, G.; Valenti, B.; Hervás, G.; Frutos, P.; Copani, G.; Ginane, C.; Niderkorn, V. Fatty acid composition of ruminal digesta and longissimus muscle from lambs fed silage mixtures including red clover, sainfoin, and timothy. *J. Anim. Sci.* **2016**, *94*, 1550–1560. [CrossRef] [PubMed]

78. Scislowski, V.; Bauchart, D.; Gruffat, D.; Laplaud, P.M.; Durand, D. Effects of dietary n-6 or n-3 polyunsaturated fatty acids protected or not against ruminal hydrogenation on plasma lipids and their susceptibility to peroxidation in fattening steers. *J. Anim. Sci.* **2005**, *83*, 2162–2174. [CrossRef] [PubMed]

79. Niezen, J.H.; Waghorn, T.S.; Charleston, W.A.G.; Waghorn, G.C. Growth and gastrointestinal nematode parasitism in lambs grazing either lucerne ( *Medicago sativa* ) or sulla ( *Hedysarum coronarium* ) which contains condensed tannins. *J. Agric. Sci.* **1995**, *125*, 281–289. [CrossRef]

80. Min, B.; Hart, S. Tannins for suppression of internal parasites. *J. Anim. Sci.* **2003**, *81*, E102–E109.

81. Molan, A.L.; Waghorn, G.C.; Mcnabb, W.C. Condensed tannins and gastro-intestinal parasites in sheep. *57 Proc. N. Z. Grassl. Assoc.* **1999**, *61*, 57–61.

82. Max, R.A.; Buttery, P.J.; Wakelin, D.; Kimambo, A.E.; Kassuku, A.A.; Mtenga, L.A. The potential of controlling gastrointestinal parasitic infections in tropical small ruminants using plants high in tannins or extracts from them. In *The Contribution of Small Ruminants in Alleviating Poverty: Communicating Messages from Research*; Smith, T., Godfrey, S.H., Buttery, P.J., Owen, E., Eds.; Natural Resources International: England, UK, 2004; pp. 115–125.

83. Marley, C.L.; Cook, R.; Keatinge, R.; Barrett, J.; Lampkin, N.H. The effect of birdsfoot trefoil (*Lotus corniculatus*) and chicory (*Cichorium intybus*) on parasite intensities and performance of lambs naturally infected with helminth parasites. *Vet. Parasitol.* **2003**, *112*, 147–155. [CrossRef]

84. Hoskin, S.O.; Wilson, P.R.; Barry, T.N.; Charleston, W.A.G.; Waghorn, G.C. Effect of forage legumes containing condensed tannins on lungworm (*Dictyocaulus* sp.) and gastrointestinal parasitism in young red deer (*Cervus elaphus*). *Res. Vet. Sci.* **2000**, *68*, 223–230. [CrossRef]

85. Mannetje, L. Harry Stobbs Memorial Lecture, 1994: Potential and prospects of legume-based pastures in the tropics. *Trop. Grassl.* **1997**, *31*, 81–94.

86. Meat and Livestock Australia State of the Industry Report: The Australian Red Meat and Livestock Industry. Available online: https://www.mla.com.au/globalassets/mla-corporate/research-and-development/documents/industry-issues/state-of-the-industry-v-1.2-final.pdf (accessed on 29 April 2019).

87. Coates, D.B.; Miller, C.P.; Hendricksen, R.E.; Jones, R.J. Stability and productivity of Stylosanthes pastures in Australia. II. Animal production from Stylosanthes pastures. *Trop. Grassl.* **1997**, *31*, 494–502.

88. Meat and Livestock Australia Legumes for Clay Soils. Available online: https://www.google.com/url?sa=t&rct=j&q=&esrc=s&source=web&cd=1&ved=2ahUKEwjp6eOtsN7iAhUNfysKHVxNC4cQFjAAegQIAxAC&url=https%3A%2F%2Fwww.mla.com.au%2Fdownload%2Ffinalreports%3FitemId%3D874&usg=AOvVaw3MfMlkc2BzojTHeSR7NdVZ (accessed on 29 April 2019).

89. Schlink, A.C.; Burt, R.L. Assessment of the chemical composition of selected tropical legume seeds as animal feed. *Trop. Agric.* **1993**, *70*, 169–173.

90. Cook, B.G.; Pengelly, B.C.; Brown, S.D.; Donnelly, J.L.; Eagles, D.A.; Franco, M.A.; Hanson, J.; Mullen, B.F.; Partridge, I.J.; Peters, M.; et al. Tropical Forages: An Interactive Selection Tool. Available online: http://www.tropicalforages.info/key/forages/Media/Html/entities/index.htm (accessed on 10 June 2019).

91. Clem, R.L. Animal production from legume-based ley pastures in southeastern Queensland. In *Tropical Legumes for Sustainable Farming Systems in Southern Africa and Australia*; Whitbread, A.M., Pengelly, B.C., Eds.; Australian Centre for International Agricultural Research: Canberra, Australia, 2004; pp. 136–144. ISBN 1863204199 (print)r1863204202.

92. Vandermeulen, S.; Singh, S.; Ramírez-Restrepo, C.A.; Kinley, R.D.; Gardiner, C.P.; Holtum, J.A.M.; Hannah, I.; Bindelle, J. In vitro assessment of ruminal fermentation, digestibility and methane production of three species of Desmanthus for application in northern Australian grazing systems. *Crop Pasture Sci.* **2018**, *69*, 797–807. [CrossRef]

93. Gardiner, C.P.; Swan, S.J. Abandoned pasture legumes offer potential economic and environmental benefits in semiarid clay soil rangelands. In Proceedings of the Australian Rangeland Society 15th Biennial Conference Proceedings, Charters Towers, QLD, Australia, 28 September–2 October 2008; p. 93.

94. Gonzalez-V, E.A.; Hussey, M.A.; Ortega-S, J.A. Nutritive value of Desmanthus associated with Kleingrass during the establishment year. *Rangel. Ecol. Manag.* **2005**, *58*, 308–314. [CrossRef]

95. Isbell, R.F. *The Australian Soil Classification*, 2nd ed.; CSIRO: Melbourne, Australia, 2016; ISBN 9781486304639.

96. Gardiner, C.; Bielig, L.; Schlink, A.; Coventry, R.; Waycott, M. Desmanthus—A new pasture legume for the dry tropics. In Proceedings of the 4th International Crop Science Congress, Brisbane, Australia, 26 September–1 October 2004; pp. 1–6.

97. Gardiner, C.; Kempe, C.; Kempe, N.; Campbell, G.; Fleury, H.; Malau-Aduli, A.; Walker, G.; Suybeng, B.; Mwangi, F. Progardes Desmanthus—An update. In Proceedings of the Northern Beef Research Update Conference, Brisbane, Australia, 19–22 August 2019; p. 33.

98. Gardiner, C.; Parker, A. Steer liveweight gains on ProgardesTM Desmanthus/Buffel pastures in Queensland. In Proceedings of the 2nd Australian and New Zealand Societies of Animal Production Joint Conference, England, New Zealand, 2–5 July 2012.

99. Aoetpah, A.; Gardiner, C.; Gummow, B.; Walker, G. Growth and eye muscle area of cross bred Boer goats fed Desmanthus cultivar JCU 1 hay. In Proceedings of the 32nd Biennial Conference of the Australian Society of Animal Production, Wagga Wagga, Australia, 2–4 July 2018; p. 2563

100. Rusiyantono, Y.; Syukur, S.H. The effect of supplementation of different legume leaves on feed intake, digestion and growth of Kacang goats given Mulato grass. *J. Agric. Sci. Technol.* **2017**, *7*, 117–122.

101. Pacheco, P.S.; Pascoal, L.L.; Restle, J.; Vaz, F.N.; Arboitte, M.Z.; Vaz, R.Z.; Santos, J.P.A.; de Oliveira, T.M.L. Risk assessment of finishing beef cattle in feedlot: Slaughter weights and correlation amongst input variables. *Rev. Bras. Zootec.* **2014**, *43*, 92–99. [CrossRef]

102. Meat and Livestock Australia Lotfeeding and Intensive Finishing. Available online: https://www.mla.com. au/research-and-development/feeding-finishing-nutrition/Lotfeeding-intensive-finishing/# (accessed on 29 April 2019).

103. Drouillard, J.S.; Kuhl, G.L. Effects of previous grazing nutrition and management on feedlot performance of cattle. *J. Anim. Sci.* **1999**, *77*, 136–146. [CrossRef]

104. Coffey, K.P.; Lomas, L.W.; Moyer, J.L. Grazing and subsequent feedlot performance by steers that grazed different types of fescue pasture. *J. Prod. Agric.* **2013**, *3*, 415–420. [CrossRef]

105. Burton, R.O.J.; Berends, P.T.; Moyer, J.L.; Coffey, K.P.; Lomas, L.W. Economic analysis of grazing and subsequent feeding of steers from three fescue pasture alternatives. *J. Prod. Agric.* **1994**, *7*, 482–489. [CrossRef]

106. Johnson, H.E.; DiCostanzo, A. A meta-analysis on the effects of backgrounding strategy on feedlot and carcass performance. *J. Anim. Sci.* **2017**, *95*, 49. [CrossRef]

107. Maltin, C.; Balcerzak, D.; Tilley, R.; Delday, M. Determinants of meat quality: Tenderness. *Proc. Nutr. Soc.* **2003**, *62*, 337–347. [CrossRef]

108. Wood, J.D.; Enser, M.; Fisher, A.V.; Nute, G.R.; Richardson, R.I.; Sheard, P.R. Manipulating meat quality and composition. *Proc. Nutr. Soc.* **1999**, *58*, 363–370. [CrossRef] [PubMed]

109. Wood, J.D. Consequences for meat quality of reducing carcass fatness. In *Reducing Fat in Meat Animals*; Wood, J.D., Fisher, A.V., Eds.; Elsevier Applied Science: Barking, UK, 1990; pp. 344–397. ISBN 1851664556.

110. Webb, E.C. Manipulating beef quality through feeding. *S. Afr.J. Food Sci. Nutr.* **2006**, *7*, 5–15.

111. Webb, E.C.; O'Neill, H.A. The animal fat paradox and meat quality. *Meat Sci.* **2008**, *80*, 28–36. [CrossRef] [PubMed]

112. Dujková, R.; Ranganathan, Y.; Dufek, A.; Macák, J.; Bezdíček, J. Polymorphic effects of FABP4 and SCD genes on intramuscular fatty acid profiles in longissimus muscle from two cattle breeds. *Acta Vet. BRNO* **2015**, *84*, 327–336. [CrossRef]

113. Scollan, N.; Hocquette, J.F.; Nuernberg, K.; Dannenberger, D.; Richardson, I.; Moloney, A. Innovations in beef production systems that enhance the nutritional and health value of beef lipids and their relationship with meat quality. *Meat Sci.* **2006**, *74*, 17–33. [CrossRef]

114. French, P.; Stanton, C.; Lawless, F.; O'Riordan, E.G.; Monahan, F.J.; Caffrey, P.J.; Moloney, A.P. Fatty acid composition, including conjugated linoleic acid, of intramuscular fat from steers offered grazed grass, grass silage, or concentrate-based diets. *J. Anim. Sci.* **2000**, *78*, 2849–2855. [CrossRef]

115. Dehghan, M.; Mente, A.; Zhang, X.; Swaminathan, S.; Li, W.; Mohan, V.; Iqbal, R.; Kumar, R.; Wentzel-Viljoen, E.; Rosengren, A.; et al. Associations of fats and carbohydrate intake with cardiovascular disease and mortality in 18 countries from five continents (PURE): A prospective cohort study. *Lancet* **2017**, *390*, 2050–2062. [CrossRef]

116. Astrup, A.; Bertram, H.C.S.; Bonjour, J.P.; De Groot, L.C.P.; De Oliveira Otto, M.C.; Feeney, E.L.; Garg, M.L.; Givens, I.; Kok, F.J.; Krauss, R.M.; et al. WHO draft guidelines on dietary saturated and trans fatty acids: Time for a new approach? *BMJ* **2019**, *366*, 14137. [CrossRef]

117. Guasch-Ferré, M.; Zong, G.; Willett, W.C.; Zock, P.L.; Wanders, A.J.; Hu, F.B.; Sun, Q. Associations of monounsaturated fatty acids from plant and animal sources with total and cause-specific mortality in two us prospective cohort studies. *Circ. Res.* **2019**, *124*, 1266–1275. [CrossRef]

118. Scollan, N.D.; Choi, N.J.; Kurt, E.; Fisher, A.V.; Enser, M.; Wood, J.D. Manipulating the fatty acid composition of muscle and adipose tissue in beef cattle. *Br. J. Nutr.* **2001**, *85*, 115–124. [CrossRef]

119. McAfee, A.J.; McSorley, E.M.; Cuskelly, G.J.; Moss, B.W.; Wallace, J.M.W.; Bonham, M.P.; Fearon, A.M. Red meat consumption: An overview of the risks and benefits. *Meat Sci.* **2010**, *84*, 1–13. [CrossRef]

120. FAO Fats and Fatty Acids in Human Nutrition Report of an Expert Consultation. Available online: http://www.fao.org/3/a-i1953e.pdf (accessed on 15 May 2019).

121. WHO Diet, Nutrition and the Prevention of Chronic Diseases: Report of a Joint WHO/FAO Expert Consultation, 28 January–1 February 2002. Available online: https://www.who.int/dietphysicalactivity/ publications/trs916/en/ (accessed on 15 May 2019).

122. Department of Health, London (United Kingdom). Nutritional aspects of cardiovascular disease. Report of the cardiovascular review group committee on medical aspects of food policy. *Nutr. Asp. Cardiovasc. Dis.* **1994**, *46*, 1–186.

123. Lorenzen, C.L.; Golden, J.W.; Martz, F.A.; Grün, I.U.; Ellersieck, M.R.; Gerrish, J.R.; Moore, K.C. Conjugated linoleic acid content of beef differs by feeding regime and muscle. *Meat Sci.* **2007**, *75*, 159–167. [CrossRef]

124. Muir, P.D.; Deaker, J.M.; Bown, M.D. Effects of forage- and grain-based feeding systems on beef quality: A review. *N. Z. J. Agric. Res.* **1998**, *41*, 623–635. [CrossRef]

125. Hocquette, J.F.; Gondret, F.; Baza, E.; Mdale, F.; Jurie, C.; Pethick, D.W. Intramuscular fat content in meat-producing animals: Development, genetic and nutritional control, and identification of putative markers. *Animal* **2010**, *4*, 303–319. [CrossRef]

126. Jeremiah, L.E. The influence of subcutaneous fat thickness and marbling on beef: Palatability and consumer acceptability. *Food Res. Int.* **1996**, *29*, 513–520. [CrossRef]

127. Steen, R.W.J.; Lavery, N.P.; Kilpatrick, D.J.; Porter, M.G. Effects of pasture and high-concentrate diets on the performance of beef cattle, carcass composition at equal growth rates, and the fatty acid composition of beef. *N. Z. J. Agric. Res.* **2003**, *46*, 69–81. [CrossRef]

128. De Smet, S.; Raes, K.; Demeyer, D. Meat fatty acid composition as affected by fatness and genetic factors: A review. *Anim. Res.* **2004**, *53*, 81–98. [CrossRef]

129. Insausti, K.; Beriain, M.; Alzueta, M.; Carr, T.; Purroy, A. Lipid composition of the intramuscular fat of beef from Spanish cattle breeds stored under modified atmosphere. *Meat Sci.* **2004**, *66*, 639–646. [CrossRef]

130. Špehar, M.; Vincek, D.; Žgur, S. Beef Quality: Factors affecting tenderness and marbling. *Stočarstvo* **2008**, *62*, 463–478.

131. Hwang, I.H.; Thompson, J.M. Effects of pH early postmortem on meat quality in beef longissimus. *Asian Australas. J. Anim. Sci.* **2003**, *16*, 1218–1223. [CrossRef]

132. Smith, G.C.; Dutson, T.R.; Hostetler, R.L.; Carpenter, Z.L. Fatness, rate of chilling and tenderness of lamb. *J. Food Sci.* **1976**, *41*, 748–756. [CrossRef]

133. Sañudo, C.; Nute, G.R.; Campo, M.M.; María, G.; Baker, A.; Sierra, I.; Enser, M.E.; Wood, J.D. Assessment of commercial lamb meat quality by British and Spanish taste panels. *Meat Sci.* **1998**, *48*, 91–100. [CrossRef]

134. Juárez, M.; Aldai, N.; López Campos, Ó.; Dugan, M.; Uttaro, B.; Aalhus, J. Beef texture and juiciness. In *Handbook of Meat and Meat Processing*; Hui, Y.H., Ed.; CRC Press: Boca Raton, FL, USA, 2012; pp. 177–206.

135. Arshad, M.S.; Sohaib, M.; Ahmad, R.S.; Nadeem, M.T.; Imran, A.; Arshad, M.U.; Kwon, J.H.; Amjad, Z. Ruminant meat flavor influenced by different factors with special reference to fatty acids. *Lipids Health Dis.* **2018**, *17*, 223. [CrossRef]

136. Mottram, D.S.; Salter, L.J. Flavor formation in meat-related maillard systems containing phospholipids. In *Thermal Generation of Aromas*; ACS Symposium Series; American Chemical Society: Washington, DC, USA, 1989; pp. 442–451.

137. Mottram, D. Meat flavour. In *Understanding Natural Flavors*; Piggott, J.R., Paterson, A., Eds.; Springer: Boston, MA, USA, 1994; pp. 140–163.

138. Melton, S.L.; Black, J.M.; Davis, G.W.; Backus, W.R. Flavor and Selected Chemical Components of Ground Beef from Steers Backgrounded on Pasture and Fed Corn up to 140 Days. *J. Food Sci.* **1982**, *47*, 699–704. [CrossRef]

139. Lee, J.Y.; Oh, D.Y.; Kim, H.J.; Jang, G.S.; Lee, S.U. Detection of superior genotype of fatty acid synthase in Korean native cattle by an environment-adjusted statistical model. *Asian Australas. J. Anim. Sci.* **2017**, *30*, 765–772. [CrossRef]

140. McMillin, K.W.; Hoffman, L.C. Improving the quality of meat from ratites. In *Improving the Sensory and Nutritional Quality of Fresh Meat*; Kerry, J., Ed.; Elsevier: Amsterdam, The Netherlands, 2009; pp. 418–446. ISBN 9781845693435.

141. Wood, J.D. Fat deposition and the quality of fat tissue in meat animals. In *Fats in Animal Nutrition*; Wiseman, J., Ed.; Elsevier: Amsterdam, The Netherlands, 1984; pp. 407–435. ISBN 9781483100357.

142. Mapiye, C.; Aalhus, J.L.; Turner, T.D.; Rolland, D.C.; Basarab, J.A.; Baron, V.S.; McAllister, T.A.; Block, H.C.; Uttaro, B.; Lopez-Campos, O.; et al. Effects of feeding flaxseed or sunflower-seed in high-forage diets on beef production, quality and fatty acid composition. *Meat Sci.* **2013**, *95*, 98–109. [CrossRef]

143. Whetsell, M.S.; Rayburn, E.B.; Lozier, J.D. Human Health Effects of Fatty Acids in Beef. Available online: https://pdfs.semanticscholar.org/d2b7/462e78b4e9c11c17a99a5aabed57d4e2fb49.pdf (accessed on 14 August 2019).

144. Smith, S.B.; Gill, C.A.; Lunt, D.K.; Brooks, M.A. Regulation of fat and fatty acid composition in beef cattle. *Asian-Australas. J. Anim. Sci.* **2009**, *22*, 1225–1233. [CrossRef]

145. Nürnberg, K.; Ender, B.; Papstein, H.J.; Wegner, J.; Ender, K.; Nürnberg, G. Effects of growth and breed on the fatty acid composition of the muscle lipids in cattle. *Z. Lebensm. Forsch. A* **1999**, *208*, 332–335. [CrossRef]

146. Malau-Aduli, A.E.O.; Siebert, B.D.; Bottema, C.D.K.; Pitchford, W.S. A comparison of the fatty acid composition of triacylglycerols in adipose tissue from Limousin and Jersey cattle. *Aust. J. Agric. Res.* **1997**, *48*, 715–722. [CrossRef]

147. Malau-Aduli, A.E.O.; Siebert, B.D.; Bottema, C.D.K.; Pitchford, W.S. Breed comparison of the fatty acid composition of muscle phospholipids in Jersey and Limousin cattle. *J. Anim. Sci.* **1998**, *76*, 766–773. [CrossRef] [PubMed]

148. Abe, T.; Saburi, J.; Hasebe, H.; Nakagawa, T.; Misumi, S.; Nade, T.; Nakajima, H.; Shoji, N.; Kobayashi, M.; Kobayashi, E. Novel mutations of the FASN gene and their effect on fatty acid composition in Japanese black beef. *Biochem. Genet.* **2009**, *47*, 397–411. [CrossRef] [PubMed]

149. Savage, D.B. Nutritional management of heifers in northern Australia. In *Recent Advances in Animal Nutrition in Australia*; Cronj, P.B., Richards, N., Eds.; Animal Science, University of New England: Armidale, Australia, 2005; Volume 15, pp. 205–214. ISBN 1863899278.

150. Drouillard, J.S. Current situation and future trends for beef production in the United States of America—A review. *Asian Australas. J. Anim. Sci.* **2018**, *31*, 1007–1016. [CrossRef] [PubMed]

151. Priolo, A.; Bella, M.; Lanza, M.; Galofaro, V.; Biondi, L.; Barbagallo, D.; Ben Salem, H.; Pennisi, P. Carcass and meat quality of lambs fed fresh sulla (*Hedysarum coronarium* L.) with or without polyethylene glycol or concentrate. *Small Rumin. Res.* **2005**, *59*, 281–288. [CrossRef]

152. Dierking, R.M.; Kallenbach, R.L.; Grün, I.U. Effect of forage species on fatty acid content and performance of pasture-finished steers. *Meat Sci.* **2010**, *85*, 597–605. [CrossRef]

153. Kook, K.; Choi, B.H.; Sun, S.S.; Garcia, F.; Myung, K.H. Effect of fish oil supplement on growth performance, ruminal metabolism and fatty acid composition of longissimus muscle in Korean cattle. *Asian Australas. J. Anim. Sci.* **2002**, *15*, 66–71. [CrossRef]

154. Dhiman, T.R.; Zaman, S.; Olson, K.C.; Bingham, H.R.; Ure, A.L.; Pariza, M.W. Influence of feeding soybean oil on conjugated linoleic acid content in beef. *J. Agric. Food Chem.* **2005**, *53*, 684–689. [CrossRef]

155. Nuernberg, K.; Dannenberger, D.; Nuernberg, G.; Ender, K.; Voigt, J.; Scollan, N.D.; Wood, J.D.; Nute, G.R.; Richardson, R.I. Effect of a grass-based and a concentrate feeding system on meat quality characteristics and fatty acid composition of longissimus muscle in different cattle breeds. *Livest. Prod. Sci.* **2005**, *94*, 137–147. [CrossRef]

156. Tume, R.K. The effects of environmental factors on fatty acid composition and the assessment of marbling in beef cattle: A review. *Aust. J. Exp. Agric.* **2004**, *44*, 663–668. [CrossRef]

157. Jacques, J.; Chouinard, Y.; Gariépy, C.; Cinq-Mars, D. Meat quality, organoleptic characteristics and fatty acid composition of Dorset lambs fed different forage to concentrate ratio or fresh grass. *Can. J. Anim. Sci.* **2016**, *97*, 290–301. [CrossRef]

158. Poulson, C.S.; Dhiman, T.R.; Ure, A.L.; Cornforth, D.; Olson, K.C. Conjugated linoleic acid content of beef from cattle fed diets containing high grain, CLA, or raised on forages. *Livest. Prod. Sci.* **2004**, *91*, 117–128. [CrossRef]

159. Duckett, S.K.; Neel, J.P.S.; Lewis, R.M.; Fontenot, J.P.; Clapham, W.M. Effects of forage species or concentrate finishing on animal performance, carcass and meat quality. *J. Anim. Sci.* **2013**, *91*, 1454–1467. [CrossRef] [PubMed]

160. Realini, C.E.; Duckett, S.K.; Brito, G.W.; Dalla Rizza, M.; De Mattos, D. Effect of pasture vs. concentrate feeding with or without antioxidants on carcass characteristics, fatty acid composition, and quality of Uruguayan beef. *Meat Sci.* **2004**, *66*, 567–577. [CrossRef]

161. Park, S.J.; Beak, S.H.; Jung, D.J.S.; Kim, S.Y.; Jeong, I.H.; Piao, M.Y.; Kang, H.J.; Fassah, D.M.; Na, S.W.; Yoo, S.P.; et al. Genetic, management and nutritional factors affecting intramuscular fat deposition in beef cattle—A review. *Asian Australas. J. Anim. Sci.* **2018**, *31*, 1043–1061. [CrossRef]

162. Duckett, S.K.; Wagner, D.G.; Yates, L.D.; Dolezal, H.G.; May, S.G. Effects of time on feed on beef nutrient composition. *J. Anim. Sci.* **1993**, *71*, 2079–2088. [CrossRef]

163. Schönfeldt, H.C.; Naudé, R.T.; Boshoff, E. Effect of age and cut on the nutritional content of South African beef. *Meat Sci.* **2010**, *86*, 674–683. [CrossRef]

164. Kelava Ugarković, N.; Ivanković, A.; Konjačić, M. Effect of breed and age on beef carcass quality, fatness and fatty acid composition. *Arch. Anim. Breed.* **2013**, *56*, 958–970. [CrossRef]

165. Bednárová, A.; Mocák, J.; Gössler, W.; Velik, M.; Kaufmann, J.; Staruch, L. Effect of animal age and gender on fatty acid and elemental composition in Austrian beef applicable for authentication purposes. *Chem. Pap.* **2013**, *67*, 274–283. [CrossRef]

166. Chail, A.; Legako, J.F.; Pitcher, L.R.; Griggs, T.C.; Ward, R.E.; Martini, S.; MacAdam, J.W. Legume finishing provides beef with positive human dietary fatty acid ratios and consumer preference comparable with grain-finished beef. *J. Anim. Sci.* **2016**, *94*, 2184–2197. [CrossRef]

167. Ramírez-Retamal, J.; Morales, R.; Martínez, M.E.; Barra, R. de la Effect of the type of pasture on the meat characteristics of Chilote lambs. *Food Nutr. Sci.* **2014**, *05*, 635–644.

168. Moloney, A.P.; Fievez, V.; Martin, B.; Nute, G.R.; Richardson, R.I. Botanically diverse forage-based rations for cattle: Implications for product composition, product quality and consumer health. In Proceedings of the 22nd General Meeting of the European Grassland Federation, Uppsala, Sweden, 9–12 June 2008; pp. 361–374.

169. Lee, M.R.F.; Winters, A.L.; Scollan, N.D.; Dewhurst, R.J.; Theodorou, M.K.; Minchin, F.R. Plant-mediated lipolysis and proteolysis in red clover with different polyphenol oxidase activities. *J. Sci. Food Agric.* **2004**, *84*, 1639–1645. [CrossRef]

170. Ashes, J.R.; Siebert, B.D.; Gulati, S.K.; Cuthbertson, A.Z.; Scott, T.W. Incorporation of n-3 fatty acids of fish oil into tissue and serum lipids of ruminants. *Lipids* **1992**, *27*, 629–631. [CrossRef] [PubMed]

171. Richardson, R.I.; Hallett, K.G.; Ball, R.; Nute, G.R.; Wood, J.D.; Scollan, N.D. Effect of free and ruminally protected fish oils on fatty acid composition, sensory and oxidative characteristics of beef loin muscle. In Proceedings of the 50th International Congress Meat Science and Technology, Heksinki, Finland, 8–13 August 2004; Volume 2, p. 43.

172. Kruk, Z.A.; Bottema, C.D.K.; Davis, J.J.; Siebert, B.D.; Harper, G.S.; Di, J.; Pitchford, W.S. Effects of vitamin A on growth performance and carcass quality in steers. *Livest. Sci.* **2008**, *119*, 12–21. [CrossRef]

173. Oka, A.; Maruo, Y.; Miki, T.; Yamasaki, T.; Saito, T. Influence of vitamin A on the quality of beef from the Tajima strain of Japanese black cattle. *Meat Sci.* **1998**, *48*, 159–167. [CrossRef]

174. Hida, Y.; Kawada, T.; Kayahashi, S.; Ishihara, T.; Fushiki, T. Counteraction of retinoic acid and 1,25-dihydroxyvitamin D3 on up-regulation of adipocyte differentiation with PPARγ ligand, an antidiabetic thiazolidinedione, in 3T3-L1 cells. *Life Sci.* **1998**, *62*, PL205–PL211. [CrossRef]

175. Smith, S.B.; Kawachi, H.; Choi, C.B.; Choi, C.W.; Wu, G.; Sawyer, J.E. Cellular regulation of bovine intramuscular adipose tissue development and composition. *J. Anim. Sci.* **2009**, *87*, E72–E82. [CrossRef]

176. McDowell, L.R. *Vitamins in Animal Nutrition: Comparative Aspects to Human Nutrition*; Academic Press: Gainesville, FL, USA, 1989; ISBN 9780124833722.

177. Kawachi, H. Micronutrients affecting adipogenesis in beef cattle. *Anim. Sci. J.* **2006**, *77*, 463–471. [CrossRef]

178. Ohashi, H. Effect of vitamin C on the quality of Wagyu beef. *Res. Bull. Aichi Agric. Res. Cent.* **2000**, *32*, 207–214.

179. Kawada, T.; Aoki, N.; Kamei, Y.; Maeshige, K.; Nishiu, S.; Sugimoto, E. Comparative investigation of vitamins and their analogues on terminal differentiation, from preadipocytes to adipocytes, of 3T3-L1 cells. *Comp. Biochem. Physiol. Part A Physiol.* **1990**, *96*, 323–326. [CrossRef]

180. Montgomery, J.L.; Blanton, J.R.; Horst, R.L.; Galyean, M.L.; Morrow, K.J.; Wester, D.B.; Miller, M.F. Effects of biological type of beef steers on vitamin D, calcium, and phosphorus status. *J. Anim. Sci.* **2004**, *82*, 2043–2049. [CrossRef]

181. Lee, C.E.; Park, N.K.; Seong, P.N.; Jin, S.H.; Park, B.Y.; Kim, K.I. Effects of deletion of Ca Supplement (limestone) on growth and beef quality in Hanwoo finishing steers. *J. Anim. Sci. Technol.* **2003**, *45*, 455–462. [CrossRef]

182. Vernon, R.G. Lipid metabolism in the adipose tissue of ruminant animals. In *Lipid Metabolism in Ruminant Animals*; Christie, W., Ed.; Elsevier: Ayr, UK, 1981; pp. 279–362. ISBN 9780080237893.

183. Zembayashi, M.; Nishimura, K.; Lunt, D.K.; Smith, S.B. Effect of breed type and sex on the fatty acid composition of subcutaneous and intramuscular lipids of finishing steers and heifers. *J. Anim. Sci.* **1995**, *73*, 3325–3332. [CrossRef] [PubMed]

184. Dannenberger, D.; Nuernberg, K.; Nuernberg, G.; Scollan, N.; Steinhart, H.; Ender, K. Effect of pasture vs. concentrate diet on CLA isomer distribution in different tissue lipids of beef cattle. *Lipids* **2005**, *40*, 589–598. [CrossRef] [PubMed]

185. Laborde, F.L.; Mandell, I.B.; Tosh, J.J.; Buchanan-Smith, J.G.; Wilton, J.W Effect of management strategy on growth performance, carcass characteristics, fatty acid composition, and palatability attributes in crossbred steers. *Can. J. Anim. Sci.* **2011**, *82*, 49–57. [CrossRef]

186. Oka, A.; Iwaki, F.; Dohgo, T.; Ohtagaki, S.; Noda, M.; Shiozaki, T.; Endoh, O.; Ozaki, M. Genetic effects on fatty acid composition of carcass fat of Japanese Black Wagyu steers. *J. Anim. Sci.* **2002**, *80*, 1005–1011. [CrossRef]

187. Pitchford, W.S.; Deland, M.P.B.; Siebert, B.D.; Malau-Aduli, A.F.O.; Bottema, C.D.K. Genetic variation in fatness and fatty acid composition of crossbred cattle. *J. Anim. Sci.* **2002**, *80*, 2825–2832. [CrossRef]

188. Rempel, L.A.; Casas, E.; Shackelford, S.D.; Wheeler, T.L. Relationship of polymorphisms within metabolic genes and carcass traits in crossbred beef cattle. *J. Anim. Sci.* **2012**, *90*, 1311–1316. [CrossRef]

189. Shin, S.C.; Chung, E.R. Association of SNP marker in the leptin gene with carcass and meat quality traits in Korean cattle. *Asian Australas. J. Anim. Sci.* **2006**, *20*, 1–6. [CrossRef]

190. Barendse, W.; Bunch, R.J.; Harrison, B.E.; Thomas, M.B. The growth hormone 1 GH1:c.457C>G mutation is associated with intramuscular and rump fat distribution in a large sample of Australian feedlot cattle. *Anim. Genet.* **2006**, *37*, 211–214. [CrossRef]

191. Bhuiyan, M.S.A.; Yu, S.L.; Jeon, J.T.; Yoon, D.; Cho, Y.M.; Park, E.W.; Kim, N.K.; Kim, K.S.; Lee, J.H. DNA polymorphisms in SREBF1 and FASN genes affect fatty acid composition in Korean cattle (Hanwoo). *Asian Australas. J. Anim. Sci.* **2009**, *22*, 765–773. [CrossRef]

192. Zhang, S.; Knight, T.J.; Reecy, J.M.; Beitz, D.C. DNA polymorphisms in bovine fatty acid synthase are associated with beef fatty acid composition. *Anim. Genet.* **2008**, *39*, 62–70. [CrossRef] [PubMed]

193. Matsuhashi, T.; Maruyama, S.; Uemoto, Y.; Kobayashi, N.; Mannen, H.; Abe, T.; Sakaguchi, S.; Kobayashi, E. Effects of bovine fatty acid synthase, stearoyl-coenzyme A desaturase, sterol regulatory element-binding protein 1, and growth hormone gene polymorphisms on fatty acid composition and carcass traits in Japanese Black cattle. *J. Anim. Sci.* **2011**, *89*, 12–22. [CrossRef] [PubMed]

194. Kaplanová, K.; Dufek, A.; Dračková, E.; Simeonovová, J.; Šubrt, J.; Vrtková, I.; Dvořák, J. The association of CAPN1, CAST, SCD, and FASN polymorphisms with beef quality traits in commercial crossbred cattle in the czech republic. *Czech J. Anim. Sci.* **2013**, *58*, 489–496. [CrossRef]

195. Westerling, D.B.; Hedrick, H.B. Fatty acid composition of bovine lipids as influenced by diet, sex and anatomical location and relationship to sensory characteristics. *J. Anim. Sci.* **1979**, *48*, 1343–1348. [CrossRef]

196. May, S.G.; Sturdivant, C.A.; Lunt, D.K.; Miller, R.K.; Smith, S.B. Comparison of sensory characteristics and fatty acid composition between Wagyu crossbred and Angus steers. *Meat Sci.* **1993**, *35*, 289–298. [CrossRef]

197. Sturdivant, C.A.; Lunt, D.K.; Smith, G.C.; Smith, S.B. Fatty acid composition of subcutaneous and intramuscular adipose tissues and M. longissimus dorsi of Wagyu cattle. *Meat Sci.* **1992**, *32*, 449–458. [CrossRef]

198. Archibeque, S.L.; Lunt, D.K.; Gilbert, C.D.; Tume, R.K.; Smith, S.B. Fatty acid indices of stearoyl-CoA desaturase do not reflect actual stearoyl-CoA desaturase enzyme activities in adipose tissues of beef steers finished with corn-, flaxseed-, or sorghum-based diets. *J. Anim. Sci.* **2005**, *83*, 1153–1166. [CrossRef]

199. Lee, S.H.; Yoon, D.H.; Choi, N.J.; Hwang, S.H.; Cheong, E.Y.; Oh, S.J.; Cheong, I.C.; Lee, C.S. Developmental relationship of unsaturated fatty acid composition and stearoyl-CoA desaturase mRNA level in Hanwoo steers' muscle. *Asian Australas. J. Anim. Sci.* **2005**, *18*, 562–566. [CrossRef]

200. Taniguchi, M.; Utsugi, T.; Oyama, K.; Mannen, H.; Kobayashi, M.; Tanabe, Y.; Ogino, A.; Tsuji, S. Genotype of stearoyl-CoA desaturase is associated with fatty acid composition in Japanese Black cattle. *Mamm. Genome* **2004**, *14*, 142–148. [CrossRef]

201. Roy, R.; Taourit, S.; Zaragoza, P.; Eggen, A.; Rodellar, C. Genomic structure and alternative transcript of bovine fatty acid synthase gene (FASN): Comparative analysis of the FASN gene between monogastric and ruminant species. *Cytogenet. Genome Res.* **2005**, *111*, 65–73. [CrossRef]

202. Morris, C.A.; Cullen, N.G.; Glass, B.C.; Hyndman, D.L.; Manley, T.R.; Hickey, S.M.; McEwan, J.C.; Pitchford, W.S.; Bottema, C.D.K.; Lee, M.A.H. Fatty acid synthase effects on bovine adipose fat and milk fat. *Mamm. Genome* **2007**, *18*, 64–74. [CrossRef] [PubMed]

203. Oh, D.; Lee, Y.; La, B.; Yeo, J.; Chung, E.; Kim, Y.; Lee, C. Fatty acid composition of beef is associated with exonic nucleotide variants of the gene encoding FASN. *Mol. Biol. Rep.* **2012**, *39*, 4083–4090. [CrossRef] [PubMed]

204. Wong, R.H.F.; Chang, I.; Hudak, C.S.S.; Hyun, S.; Kwan, H.Y.; Sul, H.S. A Role of DNA-PK for the Metabolic Gene Regulation in Response to Insulin. *Cell* **2009**, *136*, 1056–1072. [CrossRef] [PubMed]

205. Wong, R.H.F.; Sul, H.S. DNA-PK: Relaying the insulin signal to USF in lipogenesis. *Cell Cycle* **2009**, *8*, 1977–1978. [CrossRef] [PubMed]

206. Kulig, H.; Kowalewska-Łuczak, I.; Kmieć, M.; Wojdak-Maksymiec, K. ANXA9, SLC27A3, FABP3 and FABP4 single nucleotide polymorphisms in relation to milk production traits in Jersey cows. *Czech J. Anim. Sci.* **2010**, *55*, 463–467. [CrossRef]

 *foods*

*Article*

# Dietary Supplementation of Tannin-Extracts to Lambs: Effects on Meat Fatty Acids Composition and Stability and on Microbial Characteristics

**Luisa Biondi** [1], **Cinzia L. Randazzo** [1,*], **Nunziatina Russo** [1], **Alessandra Pino** [1], **Antonio Natalello** [1], **Koenraad Van Hoorde** [2,3] **and Cinzia Caggia** [1]

[1]   Department of Agriculture, Food and Environment, University of Catania, 95123 Catania, Italy;
      lubiondi@unict.it (L.B.); nunziatinarusso83@gmail.com (N.R.); alessandra.pino@unict.it (A.P.);
      antonio.natalello@unict.it (A.N.); ccaggia@unict.it (C.C.)
[2]   Service Foodborne Pathogens, Sciensano, B-1050 Brussels, Belgium; Koenraad.VanHoorde@sciensano.be
[3]   Faculty of Bioscience Engineering, Department of Biotechnology, Laboratory of Brewing Science and
      Technology, Ghent University, B-9000 Ghent, Belgium
*   Correspondence: cranda@unict.it

Received: 9 September 2019; Accepted: 7 October 2019; Published: 10 October 2019

**Abstract:** Two extracts derived from plant material rich in hydrolysable (Tara, T; *Caesalpinia spinosa*) or condensed (Mimosa, M; *Acacia mearnsii*) tannins were added to lamb's diet and their effects on meat quality and on microbial population were evaluated; a diet without tannins represented the Control (C). Meat pH, vitamin E, intramuscular fat content and muscle fatty acid composition were determined. Oxidative stability and microbiological analyses were performed on meat samples after 0, 4 and 7 days of refrigerated storage. Psychrotrophic bacteria were identified through MALDI-TOF MS analysis. Regarding meat fatty acids, Tara treatment decreased the percentage of monounsaturated fatty acids. The counts of all microbial groups were similar among dietary treatments at day 0, while a significant reduction of microbial loads was observed in T-group at day 7. *Pseudomonas fluorescens* group count was significantly affected by T extract supplementation. The MALDI-TOF MS identification revealed the dominance of *Pseudomonas fragi* species in all samples while *Pseudomonas lundensis*, *Brochothrix thermosphacta* and *Candida famata* were revealed only in control ones. In conclusions, the tannin extract supplementation is a promising dietary strategy to preserve lamb meat quality.

**Keywords:** hydrolysable tannins; condensed tannins; meat chemical composition; meat spoilage; *Pseudomonas* spp.; meat shelf-life

---

## 1. Introduction

In the last decade, the demand for natural preservative agents has increased due to the growing concern, among consumers, about the potential toxic effect of the synthetic antioxidants [1]. A number of reviews deal with the addition to feed or meat and meat products of plant extracts as natural antioxidants and antimicrobial agents [1–3]. According to Papuc et al. [3], commercially available polyphenols can extend meat shelf-life not only by improving its oxidative stability, but also by inhibiting bacterial growth. The most studied plant bioactive compounds in livestock feeding are phenolic compounds, such as tannins, and essential oils [4]. Animal responses to dietary tannins have been extensively reviewed mainly focusing on animal nutrition and production [5,6]. In the rumen, tannins can impair dietary fatty acids (FA) biohydrogenation (BH) influencing the FA profile of rumen content [7,8]. Thus, tannins could be exploited to favorably modulate FA composition of ruminant products [4,9]. However, controversial results on the effect of tannins on the meat FA composition have been observed in vivo [10,11]. Such inconsistency could be due to the type of tannin (e.g., hydrolysable or condensed), which can be differently metabolized in the rumen and can differently interact with feeds, bacteria and microbial metabolites [12]. It is well known that tannins are able to exert astringent, antiviral, antibacterial and antioxidant effects. The antimicrobial activity of tannins as well as their toxicity to bacteria, fungi and yeasts have long been recognized and several mechanisms, including inhibition of extracellular microbial enzymes, inhibition of oxidative phosphorylation or metal ions deprivation are involved [13]. It was already established that the use of plants rich in secondary compounds or the supplementation of plant-extract rich in polyphenols could represent a promising strategy for improving the meat oxidative stability, extending product shelf-life [4]. Therefore, the aim of the present study was to investigate the dietary supplementation of a source of hydrolysable (Tara; *Caesalpinia spinosa*) or condensed (Mimosa; *Acacia mearnsii*) tannins on oxidative stability and microbial population of lamb meat during its shelf-life.

## 2. Materials and Methods

### 2.1. Animals, Diets and Samplings Procedures

The applied procedures were in compliance with the European guidelines for the care and use of animals in research (Directive 2010/63/EU). The animals were handled by trained personnel. Fifteen male Sarda × Comisana lambs (body weight 19.6 kg ± 1.6) were selected at the age of 2 months and individually penned indoor in the experimental farm of the University of Catania. Lambs were randomly assigned to 3 dietary treatments ($n$ = 5). The Control group (C) received a conventional concentrate containing (as fed): barley (480 g/kg), wheat bran (230 g/kg), dehydrated alfalfa hay (150 g/kg), soybean meal (100 g/kg), molasses (20 g/kg) and mineral-vitamin premix (20 g/kg). The other two treatments received the same basal diet of control in which 4% (as fed) of Mimosa (M group) or Tara (T group) tannin extracts was added. Tannins from Mimosa (*Acacia mearnsii*) and Tara (*Cesalpinia spinosa*) plants were extracted by maceration in water. The two extracts (commercial name: Mimosa OP® and Tannino T80®, respectively for Mimosa and Tara) were purchased from Silvateam S.p.A. (San Michele Mondovì, Cuneo, Italy). All the diets were supplied in the form of a pellet and the tannin extracts were added to the diet ingredients before pelleting at the temperature of 40 °C. According the procedure described by Natalello et al. [14], total tannin concentration of experimental diets was 1.50, 22.3 and 25.3 g/kg dry matter (tannic acid equivalents). The chemical composition of basal diet is shown in Table 1. After a 9-day-adaptation period, consisting in a gradual introduction of the experimental diet, the lambs received their respective diet ad libitum; fresh water was always available. Individual feed intake and body weight were recorded during the experimental trial. At the end of feeding trial (75 days), all animals were slaughtered on the same day at a commercial abattoir according to the European Union welfare guidelines. After 24 h of storage at 4 °C, carcasses were halved and the entire longissimus thoracis et lumborum (LTL) muscles were removed from the right side, packed under vacuum and stored at −80 °C until tocopherol and fatty acids analysis. The LTL muscle from the left half-carcass was used to measure pH values by a pH meter (HI-110; Hanna Instruments) and then vacuum packaged and aged at +4 °C for 3 days. After that, 2 cm-thickness slices of LTL were prepared (one for each analysis and for each storage time) and stored at 4 °C for 0, 4 and 7 days, pending oxidative stability measurements and microbiological determination.

**Table 1.** Chemical composition of the basal diet.

|  | Basal Diet |
|---|---|
| Dry matter (DM), g/100 g as-fed | 89.65 |
| Crude protein, g/100 g DM | 15.67 |
| Ether extract, g/100 g DM | 2.68 |
| Neutral detergent fibre, g/100 g DM | 30.36 |
| Acid detergent fibre, g/100 g DM | 15.97 |
| Acid detergent lignin, g/100 g DM | 3.62 |
| Ash, g/100 g DM | 7.01 |
| Total tocopherols, µg/g DM | 13.08 |
| α-Tocopherol, % of total tocopherols | 98.75 |
| γ-Tocopherol, % of total tocopherols | 1.16 |
| δ-tocopherol, % of total tocopherols | 0.08 |
| Fatty acids, g/kg DM | |
| 14:0 | 0.06 |
| 16:0 | 5.82 |
| 18:0 | 1.51 |
| 18:1 n-9 | 8.94 |
| 18:2 n-6 | 28.03 |
| 18:3 n-3 | 0.07 |
| 20:0 | 0.16 |

## 2.2. Sampling and Analyses of Feeds

Samples of experimental diets were collected weekly and immediately stored vacuum-packed at −20 °C. At the end of feeding trial, the weekly-collected feed samples were combined together in equal amount and the representative sample was analyzed for dry matter (DM), crude protein, ether extract, ash and fiber fractions (i.e., neutral and acid detergent fiber and acid detergent lignin) as described by Biondi et al. [15]. Furthermore, fatty acids and tocopherols of feedstuffs were extracted and quantified according the procedures reported in detail by Valenti et al. [16].

## 2.3. Vitamin E, Intramuscular Fat (IMF) and Fatty Acid Profile of Meat

The concentration of α-, γ- and δ-tocopherols (i.e., vitamin E) from muscle was determined as described by Luciano et al. [17]. In short, 2 g of muscle were homogenized in an ethanolic KOH solution (60%), containing BHT (0.06%), and incubated at 70 °C for 30 min. Tocopherols were extracted with hexane/ethyl acetate solution. Then, the extracts were dried under nitrogen, resuspended in acetonitrile and injected in a HPLC. The instrument information and chromatograph conditions were reported in [17]. The lipid concentration and fatty acid profile of meat were analysis as described by Natalello et al. [14]. In brief, intramuscular fat (IMF) was extracted from 5 g of muscle using chloroform/methanol (2:1, *v/v*). Then, lipid extract was methylated by a base-catalyzed procedure and the obtained FA methyl esters (FAME) were injected into a gas chromatograph as reported in [14]. Nonadecanoic acid (C19:0) was used as internal standard and FAs were expressed as g/100 g of total methylated fatty acids.

## 2.4. Meat Oxidative Stability Measurements

Meat oxidative stability over aerobic refrigerated storage was assessed on three slices (2 cm thickness) from the left LTL placed in polystyrene trays, covered with PVC film and stored at +4 °C. Each slice was used for measuring lipid oxidation and color stability at days 0 (after 2 h of blooming), 4 and 7. Lipid oxidation was measured as thiobarbituric acid reactive substances (TBARS) values according to the procedure described in Luciano et al. [18]. Meat color parameters were measured using a portable spectrophotometer (Minolta CM-2022, Tokyo, Japan), which recorded the following descriptors: lightness (L*), redness (a*), yellowness (b*), Chroma (C*) and Hue angle (H*), as well as the reflectance (R) spectra from 400 to 700 nm. The accumulation of metmyoglobin (MetMb) on the meat surface during storage was monitored by the ratio $(K/S)_{572} \div (K/S)_{525}$ as described in [18]. This ratio decreases with increasing proportion of MetMb.

*2.5. Microbiological Analysis*

As for oxidative stability measurement, the microbiological analysis was evaluated at 0, 4 and 7 days of storage at 4 °C. At each sampling time, 25 g of sample, aseptically weighed, was transferred into a stomacher bag and homogenized with peptone water 0.1% *w/w* (Oxoid) for 2–3 min. Ten-fold dilutions were made and plated in triplicate on the following agar media and conditions: Plate Count Agar (PCA) incubated at $32 \pm 2$ °C for 48 h and at 4 °C for 7 days, for total mesophilic bacteria and total psychrotrophics, respectively; Pseudomonas Agar, supplemented with Cetrimide, Fucidine and Cephaloridine, incubated at 30 °C for 48 h for *Pseudomonas* spp. detection; Violet Red Bile Glucose agar, aerobically incubated at 37 °C for 24 h, for *Enterobacteriaceae* count; Brilliance Salmonella agar, supplemented with Salmonella selective supplement, incubated at 37 °C for $24 \pm 3$h, was used for *Salmonella* spp. count according to ISO 6579:2002 + A1:2007; Chromogenic *E. coli* incubated at 37 °C for 24 h, for *Escherichia coli* determination; Campylobacter Selective Agar (Preston) was used for *Campylobacter* spp. detection using the selective enrichment broth technique. All media were purchased at Oxoid, Milan, Italy. Results were expressed as $\log_{10}$ CFU/mL.

*2.6. Isolation and Genetic Identification of Psychrotrophic Bacteria*

From each PCA plate of T, M and C samples, previously incubated at 4 °C for 7 days, at 0, 4, and 7 days of refrigerated storage, 20% of the total number of colonies were randomly selected, purified, checked for catalase activity and Gram reaction, and microscopically examined before storing at −80 °C in liquid culture using 20% glycerol. Total genomic DNA of isolates was extracted from overnight cultures according to the method described by Pino et al. [19]. DNA concentration and quality were assessed by measuring the optical density at 260 nm using Fluorometer Qubit (Invitrogen, Carlsbad, CA, USA). All isolates were clustered by PCR-RFLP analysis, using primer pairs, PCR conditions and restriction endonucleases reported by Franzetti and Scarpellini [20]. The isolates of each PCR-RFLP cluster were subsequently subjected to species identification by MALDI-TOF MS analysis according to Russo et al. [21].

*2.7. Statistical Analysis*

The effect of the dietary treatment on muscle pH, intramuscular fat and fatty acids was assessed by means of one-way analysis of variance (ANOVA). Each animal represented an experimental unit. The oxidative stability (color descriptors, metmyoglobin and TBARS) and the microbiological (mesophilic and psychrotrophic bacteria, *E. coli*, *Enterobacteriaceae* and *Pseudomonas* spp.) data in meat were analyzed using a mixed model procedure for repeated measures. The fixed factors in the model were the dietary treatment (C, T and M), the time of storage (Time: days 0, 4, 7) and their interaction (Diet × Time), while the individual animal was included as a random factor. Differences between means were assessed using the Tukey's adjustment for multiple comparisons. Effects and differences were declared significant when $p \leq 0.05$, while trends toward significance where considered when $0.05 < p \leq 0.1$. Statistical analyses were performed with the statistical software Minitab, version 16 (Minitab Inc., State College, PA, USA).

## 3. Results

*3.1. Animal Performance Parameters*

Dietary treatments influenced the dry matter intake ($p = 0.031$), with lower intake for T-fed lambs (1.05 kg/day per lamb) as compared to M group (1.22 kg/day per lamb), while no difference was observed among Control (1.21 kg/day per lamb) and tannin groups ($p > 0.05$). Meanwhile, the other performances parameters were not affected by the experimental diets. Indeed, final body weight (average 35.3 ± 2.70 kg; $p = 0.658$), average daily gain (average 186 ± 29.4 g/day; $p = 0.953$), carcass weight (average 17.3 ± 1.64 kg; $p = 0.261$) and carcass yield (average 49.4 ± 5.3%; $p = 0.582$) were comparable between treatments.

### 3.2. Muscle Chemical Parameters and Oxidative Stability

Intramuscular fat content and muscle pH were not affected by the addition of Tara or Mimosa tannin extract to the basal diet (Table 2). The muscle concentration of γ-Tocopherol was increased by the diet containing Tara extract compared to the other treatments (0.013), while the other detected tocopherols were not affected ($p < 0.05$). Among the main FAs classes (Table 2), only the monounsaturated fatty acids (MUFA) were found at lower concentration ($p < 0.05$) in muscle from Tara group as compared to Control and Mimosa groups. A trend ($p < 0.1$) towards an increased total polyunsaturated fatty acids (PUFA) content in the muscle from Tara lambs is worth of mentioning. The main individual FAs were shown in the Supplementary Table S1. Oleic acid (C18:1 c9) was the most represented fatty acid among MUFA (82.7 ± 1.11% of total MUFA on average in the three groups) and its proportion was lower in T group compared to C lambs ($p = 0.027$). Similar to oleic acid, C16:1 c7, C16:1 c9, C17:0 ante and C17:1 c9 were found at lower proportion in T muscle compared to Control ($p = 0.05$). Moreover, the sum of all identified trans-18:1 was significantly lower in meat from T lambs as compared to the other two groups (2.25% vs. 3.13% and 3.26% FAME respectively for T, C and M groups; $p < 0.05$; data not shown). Among PUFA, linoleic acid (C18:2 c9 c12) showed a trend towards a higher proportion in meat from Tara lambs as compared to Control lambs ($p = 0.08$).

**Table 2.** Effect of the dietary treatment [1] on meat quality parameters.

| Item [2] | C | T | M | SEM [3] | Diet Effect |
|---|---|---|---|---|---|
| IMF³, g/100 g muscle | 2.19 | 1.15 | 1.87 | 0.135 | 0.136 |
| pH | 5.94 | 5.88 | 6.00 | 0.040 | 0.490 |
| α-Tocopherol, ng/g of muscle | 258 | 468 | 340 | 44.40 | 0.149 |
| γ-Tocopherol, ng/g of muscle | 1.82 [b] | 3.40 [a] | 1.83 | 0.278 | 0.013 |
| δ-Tocopherol, ng/g of muscle | 18.6 | 20.2 | 29.0 | 2.530 | 0.200 |
| Σ Tocopherols, ng/g of muscle | 278 | 492 | 371 | 46.00 | 0.167 |
| SFA, g/100g total FAME | 39.5 | 39.9 | 40.4 | 0.593 | 0.841 |
| MUFA, g/100g total FAME | 47.0 [a] | 41.4 [b] | 45.5 [a] | 0.840 | 0.006 |
| PUFA, g/100g total FAME | 10.3 | 15.6 | 10.9 | 1.120 | 0.093 |

[1] C = concentrate-based diet; T and M mean C diet + 4% tannin extract from either Tara or Mimosa. [2] IMF = intramuscular fat; SFA = saturated fatty acids; MUFA = monounsaturated fatty acids; PUFA = polyunsaturated fatty acids; FAME = fatty acid methyl esters. [3] SEM = standard error of mean. [a,b] Within a row, different superscript letters indicate differences ($p \leq 0.05$) between dietary treatments tested using the Tukey's adjustment for multiple comparisons.

Table 3 shows the oxidative stability parameters measured in meat samples from the three dietary treatments during 7 days of aerobic refrigerated storage. The dietary supplementation with Tara or Mimosa tannins did not produce relevant effects ($p > 0.05$) on color descriptors and TBARS. Metmyoglobin percentage was the only meat stability parameter affected by dietary treatment, showing a significantly ($p < 0.048$) lower value in meat from M-fed lambs compared to C-fed lambs. Meat yellowness (b*) was not affected by the diet supplied to lambs or by the time of storage ($p > 0.05$). Meat lightness (L*), redness (a*) and hue angle (H*) values were affected by the time of storage ($p < 0.001$); regarding saturation (C*), a trend ($p = 0.07$) towards a lower value on day 7 as compared to day 0 has been observed. Metmyoglobin percentage and TBARS values were affected by the time of storage, with increasing values from day 0 to day 7.

**Table 3.** Effect of the dietary treatment and time of storage on the oxidative stability parameters of meat.

| | Dietary Treatment (D) [1] | | | Time of Storage (T) [2] | | | SEM [3] | *p* Values [4] | | |
|---|---|---|---|---|---|---|---|---|---|---|
| | C | T | M | 0 | 4 | 7 | | D | T | D × T |
| L* values | 42.5 | 42.9 | 41.7 | 40.8 [b] | 42.8 [ab] | 43.5 [a] | 0.455 | 0.750 | <0.001 | 0.029 |
| a* values | 12.5 | 12.7 | 12.2 | 13.9 [a] | 12.2 [b] | 11.3 [b] | 0.279 | 0.786 | <0.001 | 0.131 |
| b* values | 11.8 | 11.6 | 10.8 | 11.1 | 11.5 | 11.6 | 0.223 | 0.396 | 0.479 | 0.084 |
| C* values | 17.3 | 17.2 | 16.4 | 17.8 | 16.8 | 16.2 | 0.314 | 0.579 | 0.035 [6] | 0.085 |
| H* values | 43.6 | 42.4 | 41.4 | 38.3 [c] | 43.2 [b] | 46.0 [a] | 0.602 | 0.225 | <0.001 | 0.945 |
| MetMb, % of Mb | 48.4 [a] | 46.0 [ab] | 43.8 [b] | 38.3 [c] | 48.3 [b] | 51.6 [a] | 1.02 | 0.048 | <0.001 | 0.499 |
| TBARS [5], mg/kg | 0.76 | 0.83 | 0.71 | 0.18 [c] | 0.64 [b] | 1.48 [a] | 0.106 | 0.857 | <0.001 | 0.884 |

[1] C = concentrate-based diet; T and M mean C diet + 4% tannin extract from either Tara or Mimosa. [2] Time 0, 4, 7 = days of storage at 4 °C under aerobic conditions (raw meat slices) [3] SEM = standard error of mean. [4] *p* values for the effects of the dietary treatment (D), time of storage (T) and of the D × T interaction. [5] Lipid oxidation, measured as TBARS values. [6] No significant differences were found for multiple comparisons using Tukey's method. [a–c] Within row, different superscript letters indicate differences ($p < 0.05$) between dietary treatments or times of storage tested using the Tukey's adjustment for multiple comparisons.

## 3.3. Microbiological Results

The effects of dietary treatment (C, T, M) and of storage condition (days 0, 4 and 7) on microbial counts, expressed as $\log_{10}$ CFU/mL, are reported in Figure 1. As expected, all microbial groups analyzed, increased in meat samples during the refrigerated storage, even a dietary supplementation dependent effect was achieved. In detail, the cell density of all microbial groups was significantly lower in tannin treatments ($p \leq 0.05$) than control one, at 4 days of storage. It is interesting to note that *Enterobacteriaceae* population showed a dramatically decrease in meat from T and M-fed lambs, which was maintained till 7 days, in contrast to control, which exhibited an increasing trend during storage. A similar trend was observed after 7 days of refrigerated storage in meat from T and M-fed lambs, despite a slightly higher cell density than day 4 was registered. Overall, T supplementation was more effective than M treatment, indicating a higher inhibiting effect against meat pathogens, especially versus *Enterobacteriaceae*.

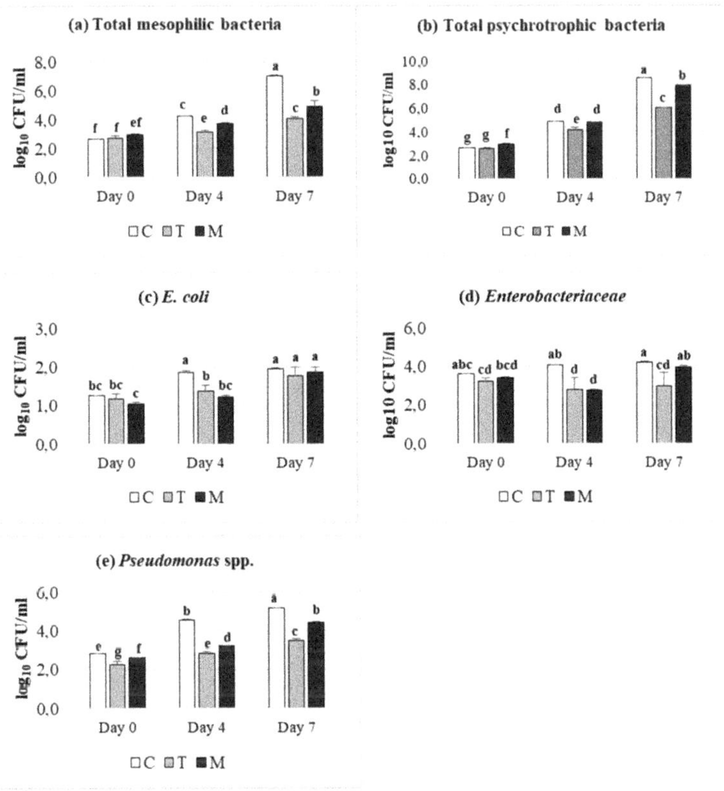



**Figure 1.** Effect of dietary treatment (Control (C), Tara (T) or Mimosa (M)) and time of storage (days 0, 4 and 7) on microbial counts expressed as log10 CFU/mL: (**a**) total mesophilic bacteria; (**b**) total psychrotrophic bacteria; (**c**) *E. coli*; (**d**) *Enterobacteriaceae*; (**e**) *Pseudomonas* spp. **a–g**: Values with different superscripts are significantly different ($p \leq 0.05$).

*3.4. Isolation and Genetic Identification of Psychrotrophic Bacteria*

Seventy-nine randomly selected isolates were subsequently subjected to MALDI-TOF MS analysis for identification at species level. Results are illustrated in Figure 2. Overall, among isolates, 57 (72%) strains were ascribed to *P. fragi*, 7 (9%) to *Brochothrix thermosphacta*, 1 (1%) to *Pseudomonas lundensis*, and 1 (1%) to *Candida famata* (Figure 3). The remaining 13 (17%) strains were ascribed to the members of *Pseudomonas fluorescens* group, representing the most found group in T and M samples. They will be subjected to sequencing of the 16S rDNA in order to confirm their affiliation.

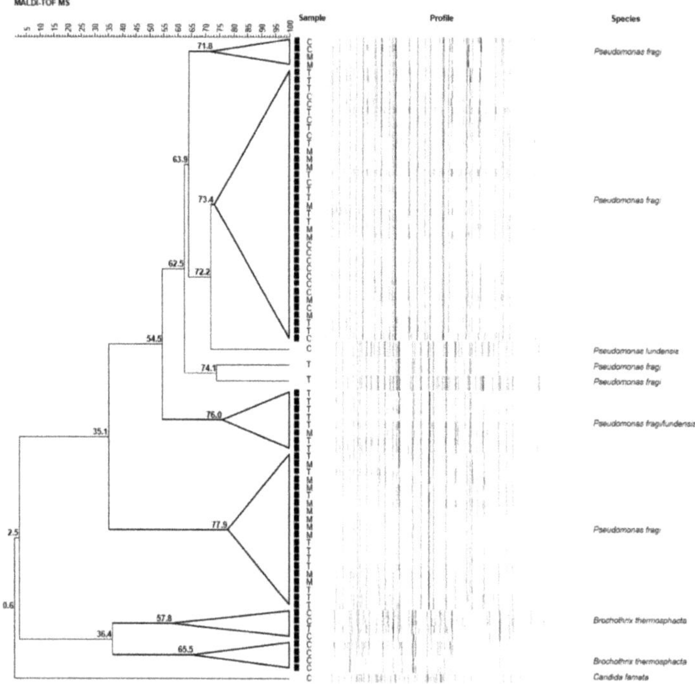

**Figure 2.** UPGMA (unweighted pair group method with arithmetic mean) dendrogram of the MALDI-TOF MS analyses of seventy-nine randomly selected meat samples. Node values indicate the average percentage of similarity based on MALDI-TOF MS profiles. The tree was made with BioNumerics version 5.1.

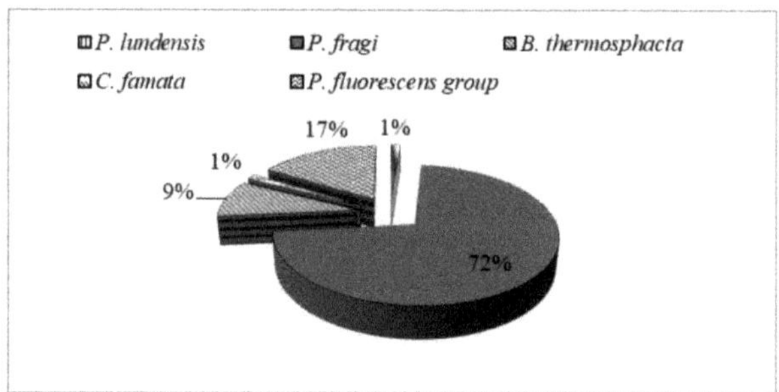


**Figure 3.** Occurrence (percentage, %) of dominant spoilage species in meat samples.

Zooming in to the prevalence of spoilage bacteria in M, T, and C samples, as represented in Figure 4, the *P. fragi* was the most abundant species in all samples, followed by *B. thermosphacta*, *C. famata* and *P. lundensis*. Only in T samples was a significant reduction of spoilage bacteria revealed during storage. On the contrary, a qualitative and quantitative increase of spoilage bacteria was observed in C samples at 7 days of refrigerated storage; indeed, *P. lundensis* and *C. famata* species were detected only in C samples.

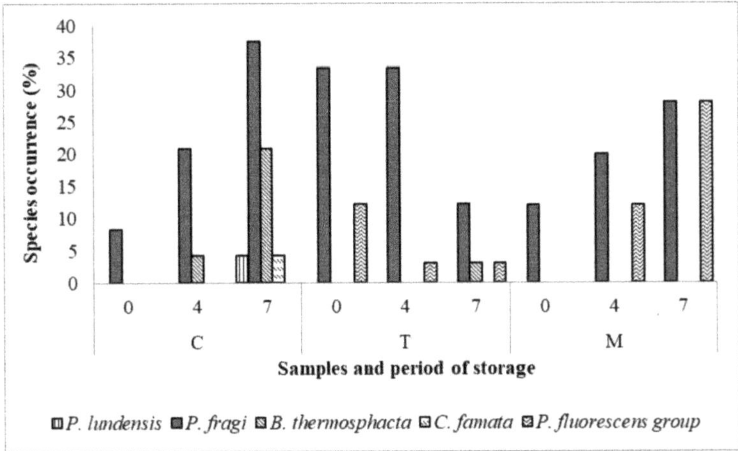



**Figure 4.** Occurrence (percentage, %) of dominant spoilage species in Tara (T), Mimosa (M) and Control (C) meat samples during the refrigerated storage.

## 4. Discussion

### 4.1. Fatty Acid Composition

Several studies have showed that dietary tannins can favorably modulate ruminal biohydrogenation (BH) of dietary polyunsaturated fatty acids. Depending on the steps in which tannins affect the dietary PUFA biohydrogenation process, different results on rumen fatty acids outflow were observed. Literature shows an increase of PUFA [14,22] when the initial steps of the BH are inhibited. On the other side, an increase of the BH intermediates, rumenic, vaccenic and other trans-18:1 acids and a decrease of stearic acid might result from the inhibition of the last step of BH [11,22]. In the barley-based diet supplied to lambs in the present study, linoleic acid represents the main PUFA substrate for BH. The addition of Tara or Mimosa extracts to the diet produced only minor effects on dietary PUFA biohydrogenation. Indeed, no significant ($p > 0.05$) effects were observed on meat fatty acids composition for the main products of linoleic acid BH, vaccenic and rumenic acids. However, other FAs (i.e., C17:0 anteiso, C17:1 c9 and C18:1 t10), also linked to BH process, were depressed by the tannin-hydrolysable supplementation (i.e., Tara). Moreover, a significant effect was observed on the total trans-18:1 FAs, which resulted in a lower content in meat from T lambs as compared to the other two groups. In the light of these findings, it would seem that the hydrolysable tannins have influenced the rumen microbial population. Indeed, variations in C17:0 anteiso and C17:1 c9 indicate changes in microbial growth as they are synthesized by rumen microorganisms and included in their membranes [23], whereas, the effect on C18:1 t10 could be explained by a shift of rumen microorganism community, which can produce mainly C18:1 t10 at the expense of the C18:1 t11 [24]. Differently, condensed tannins from Mimosa did not affect meat fatty acids composition, as assessed by the lack of significance between meat from M and C groups.

The most relevant results on meat fatty acid composition concern the significant effect of T extract addition to lambs' diet on meat total MUFA and oleic acid and the trend observed for linoleic acid. According to Wood et al. [25], an increase in the proportion of oleic acid and a decrease in the proportion of linoleic acid in neutral lipid as fat deposition accelerates is often observed, both in muscle and in adipose tissue. In the present study, the Pearson coefficients between intramuscular fat (IMF) content and oleic and linoleic acids were highly significant (respectively for oleic and linoleic: r = 0.752, $p = 0.001$; r = −0.797, $p < 0.001$). It may be inferred that, in our experimental conditions, the effect of the diet on oleic and linoleic acids percentages could be the result of the slightly but not significantly

($p > 0.05$; Table 2) different IMF level observed in the meat. Indeed, meat from T lambs contained, on average, about fifty percent of intramuscular fat as compared to meat from C lambs.

*4.2. Oxidative Stability of Meat*

As is well known, the meat oxidative stability depends on the balance between muscle oxidizable substrates and antioxidants defenses [26]. Vitamin E is a powerful lipophilic antioxidant, which can effectively delay the deterioration of the meat during storage [27]. In our experiment, a higher concentration of $\gamma$-Tocopherol was found in muscle from Tara treatment compared to Control and Mimosa groups. Considering that the basal diet was the same for all the treatment and the supplemented tannins did not contain tocopherols because they were produced by aqueous maceration, all lambs ingested similar concentration of vitamin E. In turn, a change of tocopherols in muscle was unexpected. However, the increase of $\gamma$-Tocopherol in T treatment did not lead to an improvement in oxidative stability. Indeed, comparable results were observed for color descriptors and lipid oxidation (TBARS values). This lack of results can be partially explained by the very low concentration of $\gamma$-Tocopherol compared to the other vitamin E isomers (i.e., $\alpha$- and $\delta$-tocopherols) and the lower antioxidant action than $\alpha$-tocopherol [27].

Previous studies have reported that tannin supplementation of the ruminant diet has positive effects on the oxidative stability of meat [28,29]. However, in the present study, these desired effects of dietary tannins were not observed. These discrepancies can be due to many factors, such as the type of tannins, the dose used and the interaction with the basal diet, which do not allow for reaching clear conclusions on the effects of tannins on meat oxidative stability [18].

The metmyoglobin formation was reduced by condensed tannins supplementation (M groups) as compared to meat from lambs receiving no tannins or hydrolysable tannins (C and T groups respectively). Similarly, Luciano et al. [28] observed a reduction of metmyoglobin formation during refrigerated storage of meat from lambs fed a diet supplemented with condensed tannin extract (quebracho; *Schinopsis lorentzii*). Nevertheless, unlike the present study, these authors also reported an improvement in color descriptors of meat.

Regardless of the dietary treatments, color descriptors, TBARS values and metmyoglobin percentage changed during time of storage, which was expected and in agreement with literature [16,28,29]. However, it should be underlined that the TBARS values were always below the threshold of 2 mg MDA/kg, at which consumers can perceive off-flavors and rancidity of meat [30].

*4.3. Microbiological Quality of the Lamb Meat*

In recent years, the application of plant extracts to meat, as natural antioxidants and antimicrobial agents, has become a topic of great concern for meat industry [1–3]. On the other hand, few studies have investigated on the relationships between dietary consumption of plant extracts rich in secondary compounds and meat microbial spoilage. An interesting antimicrobial effect has been observed in meat when lambs received rosemary extract [31,32] or thyme [33] or in carcasses from rabbit supplemented with oregano essential oil [34]. Looking at the microbiological quality of the lamb meat, our data demonstrated a positive effect of tannin-rich feed on microbial composition of lamb meat, stored for 7 days at refrigerated condition. A significant decrease of the overall load of microbial groups investigated was observed mostly in T samples, suggesting the antimicrobial properties of hydrolysable tannins or of some product of their degradation, inside the muscular tissue. Indeed, previous studies reviewed by Smeriglio and co-workers [35], have already demonstrated the antimicrobial activity of tannin compounds against a wide range of gram-positive and gram-negative bacterial strains by complexing with proteins through both covalent and non-covalent interactions, and with polysaccharides [36]. Our data demonstrated that the occurrence of *P. fragi* species was dramatically reduced at 7 days of storage. It is well established that *P. fluorescens* group is the main cause of meat spoilage during aerobic storage condition [37]. Together with enterobacteria, these spoilers are well-known to produce several volatile organic compounds causing off-odors upon storage [38]. It is interesting to highlight that both

Tara and Mimosa tannin-extracts supplemented in lamb's diets inhibited the growth of *P. lundensis* and *C. famata* species, which were detected only in the Control groups samples. However, *B. thermosphacta*, well known as a dominant organism in meat spoilage, becoming ubiquitous throughout the meat production chain [39], was detected both the C and the T samples after 7 days of storage. Its ability to grow at refrigeration condition and its tolerance to high-salt and low-pH conditions involve production of organoleptically unpleasant compounds [40] in fresh and cured meats, and fish products [41], playing an important role in shortening the shelf-life of these products [39]. The occurrence of *C. famata* in lamb meat could be originated by spoiled carcasses representing the main responsible of pink pigments formation in many meat products [42]. Even though polyphenols in muscle were not measured in this study, it may be hypothesized that metabolites from Tara or from Mimosa tannins reached the muscular tissue operating an antibacterial activity during meat storage.

## 5. Conclusions

The inclusion of Mimosa and Tara tannins in the lamb's diet showed mild effects on meat fatty acid profile and on oxidative stability of meat. However, the most relevant result was the antimicrobial effect of Tara extract on spoilage bacteria. To the best of our knowledge, this is the first study addressing the positive effects of tannin-extract supplied through the diet on microbial population of lamb meat during the shelf-life. The ability of tannin molecules to exert positive effects on lamb's meat represents a promising strategy to improve the quality and the shelf-life of the final product.

**Supplementary Materials:** The following are available online at http://www.mdpi.com/2304-8158/8/10/469/s1, Table S1: Effect of the dietary treatment[1] on the main fatty acids (g/100 of total FAME) of meat.

**Author Contributions:** Conceptualization, L.B., C.L.R.; Data curation, N.R., A.P., K.V.H., A.N., C.C., C.L.R.; Formal analysis, N.R., K.V.H., A.N.; Methodology, N.R., A.P., A.N., K.V.H.; Supervision, L.B., C.L.R.; Writing—review and editing, L.B., C.L.R., A.N. and C.C.; Funding acquisition, L.B., C.L.R.

**Funding:** This work was partially supported by funding from the Ministero dell'Istruzione, dell'Università e della Ricerca, University of Catania (Italy), under the Research Project 'FIR 14-16' and under the Piano della Ricerca 2016–2018, Linea di intervento 2 "Dotazione ordinaria per attività istituzionali dei dipartimenti" 2016-18.

**Conflicts of Interest:** The authors declare no conflict of interest.

## References

1. Shah, M.A.; Don Bosco, S.J.; Mir, S.H. Plant extracts as natural antioxidants in meat and meat products. *Meat Sci.* **2014**, *98*, 21–33. [CrossRef] [PubMed]

2. Falowo, A.B.; Fayemi, P.O.; Muchenjeet, V. Natural antioxidants against lipid–protein oxidative deterioration in meat and meat products: A review. *Food Res. Int.* **2014**, *64*, 171–181. [CrossRef] [PubMed]

3. Papuc, C.; Goran, G.V.; Predescu, C.N.; Nicorescu, V.; Stefan, G. Plant polyphenols as antioxidant and antibacterial agents for shelf-life extension of meat and meat products: Classification, structures, sources, and action mechanisms. *Compr. Rev. Food Sci. Food Saf.* **2017**, *16*, 1243–1268. [CrossRef]

4. Vasta, V.; Luciano, G. The effects of dietary consumption of plants secondary compounds on small ruminants' products quality. *Small Rumin. Res.* **2011**, *101*, 150–159. [CrossRef]

5. Mueller-Harvey, I. Unravelling the conundrum of tannins in animal nutrition and health. *J. Sci. Food Agric.* **2006**, *86*, 13–37. [CrossRef]

6. Waghorn, G. Beneficial and detrimental effects of dietary condensed tannins for sustainable sheep and goat production progress and challenges. *Anim. Feed Sci. Technol.* **2008**, *147*, 116–139. [CrossRef]

7. Vasta, V.; Makkar, H.P.S.; Mele, M.; Priolo, A. Ruminal biohydrogenation as affected by tannins in vitro. *Br. J. Nutr.* **2009**, *102*, 82–92. [CrossRef] [PubMed]

8. Carreño, D.; Hervás, G.; Toral, P.G.; Belenguer, A.; Frutos, P. Ability of different types and doses of tannin extracts to modulate in vitro ruminal biohydrogenation in sheep. *Anim. Feed Sci. Technol.* **2015**, *202*, 42–51. [CrossRef]

9. Morales, R.; Ungerfeld, E.M. Use of tannins to improve fatty acids profile of meat and milk quality in ruminants: A review. *Chil. J. Agric. Res.* **2015**, *75*, 239–248. [CrossRef]

10. Vasta, V.; Mele, M.; Serra, A.; Scerra, M.; Luciano, G.; Lanza, M.; Priolo, A. Metabolic fate of fatty acids involved in ruminal biohydrogenation in sheep fed concentrate or herbage with or without tannins. *J. Anim. Sci.* **2009**, *87*, 2674–2684. [CrossRef]

11. Willems, H.; Kreuzer, M.; Leiber, F. Alpha-linolenic and linoleic acid in meat and adipose tissue of grazing lambs differ among alpine pasture types with contrasting plant species and phenolic compound composition. *Small Rumin. Res.* **2014**, *116*, 153–164. [CrossRef]

12. Vasta, V.; Daghio, M.; Cappucci, A.; Buccioni, A.; Serra, A.; Viti, C.; Mele, M. Invited review: Plant polyphenols and rumen microbiota responsible for fatty acid biohydrogenation, fiber digestion, and methane emission: Experimental evidence and methodological approaches. *J. Dairy Sci.* **2019**, *102*, 3781–3804. [CrossRef] [PubMed]

13. Scalbert, A. Antimicrobial properties of tannins. *Phytochemistry* **1991**, *30*, 3875–3883. [CrossRef]

14. Natalello, A.; Luciano, G.; Morbidini, L.; Valenti, B.; Pauselli, M.; Frutos, P.; Biondi, L.; Rufino-Moya, P.; Lanza, M.; Priolo, A. Effect of Feeding Pomegranate Byproduct on Fatty Acid Composition of Ruminal Digesta, Liver, and Muscle in Lambs. *J. Agric. Food Chem.* **2019**, *67*, 4472–4482. [CrossRef] [PubMed]

15. Biondi, L.; Luciano, G.; Cutello, D.; Natalello, A.; Mattioli, S.; Priolo, A.; Lanza, M.; Morbidini, L.; Valenti, B. Meat quality from pigs fed tomato processing waste. *Meat Sci.* **2020**, *159*, 107940. [CrossRef] [PubMed]

16. Valenti, B.; Luciano, G.; Pauselli, M.; Mattioli, S.; Biondi, L.; Priolo, A.; Natalello, A.; Morbidini, L.; Lanza, M. Dried tomato pomace supplementation to reduce lamb concentrate intake: Effects on growth performance and meat quality. *Meat Sci.* **2018**, *145*, 63–70. [CrossRef]

17. Luciano, G.; Roscini, V.; Mattioli, S.; Ruggeri, S.; Gravador, R.S.; Natalello, A.; Lanza, M.; De Angelis, A.; Priolo, A. Vitamin E is the major contributor to the antioxidant capacity in lambs fed whole dried citrus pulp. *Animal* **2017**, *11*, 411–417. [CrossRef]

18. Luciano, G.; Natalello, A.; Mattioli, S.; Pauselli, M.; Sebastiani, B.; Niderkorn, V.; Copani, G.; Benhissi, H.; Amanpour, A.; Valenti, B. Feeding lambs with silage mixtures of grass, sainfoin and red clover improves meat oxidative stability under high oxidative challenge. *Meat Sci.* **2019**, *156*, 59–67. [CrossRef]

19. Pino, A.; Liotta, L.; Randazzo, C.L.; Todaro, A.; Mazzaglia, A.; De Nardo, F.; Chiofalo, V.; Caggia, C. Polyphasic approach to study physico-chemical, microbiological and sensorial characteristics of artisanal Nicastrese goat's cheese. *Food Microbiol.* **2018**, *70*, 143–154. [CrossRef]

20. Franzetti, L.; Scarpellini, M. Characterisation of *Pseudomonas* spp. isolated from foods. *Ann. Microbiol.* **2007**, *57*, 39–47. [CrossRef]

21. Russo, N.; Caggia, C.; Pino, A.; Coque, T.M.; Arioli, S.; Randazzo, C.L. *Enterococcus* spp. in Ragusano PDO and Pecorino Siciliano cheese types: A snapshot of their antibiotic resistance distribution. *Food Chem. Toxicol.* **2018**, *120*, 277–286. [CrossRef] [PubMed]

22. Campidonico, L.; Toral, P.G.; Priolo, A.; Luciano, G.; Valenti, B.; Hervás, G.; Frutos, P.; Copani, G.; Ginane, G.; Niderkorn, V. Fatty acid composition of ruminal digesta and *longissimus muscle* from lambs fed silage mixtures including red clover, sainfoin, and timothy. *J. Anim. Sci.* **2016**, *94*, 1550–1560. [CrossRef] [PubMed]

23. Fievez, V.; Colman, E.; Castro-Montoya, J.M.; Stefanov, I.; Vlaeminck, B. Milk odd-and branched-chain fatty acids as biomarkers of rumen function—An update. *Anim. Feed Sci. Technol.* **2012**, *172*, 51–65. [CrossRef]

24. Bessa, R.J.; Alves, S.P.; Santos-Silva, J. Constraints and potentials for the nutritional modulation of the fatty acid composition of ruminant meat. *Eur. J. Lipid Sci. Technol.* **2015**, *117*, 1325–1344. [CrossRef]

25. Wood, J.D.; Enser, M.; Fisher, A.V.; Nute, G.R.; Sheard, P.R.; Richardson, R.I.; Hughes, S.I.; Whittington, F.M. Fat deposition, fatty acid composition and meat quality: A review. *Meat Sci.* **2008**, *78*, 343–358. [CrossRef]

26. Bekhit, A.E.D.A.; Hopkins, D.L.; Fahri, F.T.; Ponnampalam, E.N. Oxidative processes in muscle systems and fresh meat: Sources, markers, and remedies. *Compr. Rev. Food Sci. Food Saf.* **2013**, *12*, 565–597. [CrossRef]

27. Bellés, M.; del Mar Campo, M.; Roncalés, P.; Beltrán, J.A. Supranutritional doses of vitamin E to improve lamb meat quality. *Meat Sci.* **2018**, *149*, 14–23. [CrossRef]

28. Luciano, G.; Monahan, F.J.; Vasta, V.; Biondi, L.; Lanza, M.; Priolo, A. Dietary tannins improve lamb meat colour stability. *Meat Sci.* **2009**, *81*, 120–125. [CrossRef]

29. Luciano, G.; Vasta, V.; Monahan, F.J.; López-Andrés, P.; Biondi, L.; Lanza, M.; Priolo, A. Antioxidant status, colour stability and myoglobin resistance to oxidation of longissimus dorsi muscle from lambs fed a tannin-containing diet. *Food Chem.* **2011**, *124*, 1036–1042. [CrossRef]

30. Howes, N.; Bekhit, A.E.-D.A.; Burritt, D.J.; Campbell, A.W. Opportunities and implications of pasture-based lamb fattening to enhance the long-chain fatty acid composition in meat. *Compr. Rev. Food Sci. Food Saf.* **2015**, *4*, 22–36. [CrossRef]

31. Ortuño, J.; Serrano, R.; Bañón, S. Antioxidant and antimicrobial effects of dietary supplementation with rosemary diterpenes (carnosic acid and carnosol) vs vitamin E on lamb meat packed under protective atmosphere. *Meat Sci.* **2015**, *110*, 62–69. [CrossRef] [PubMed]

32. Ortuño, J.; Serrano, R.; Bañón, S. Incorporating rosemary diterpenes in lamb diet to improve microbial quality of meat packed in different environments. *Anim. Sci. J.* **2017**, *88*, 1436–1445. [CrossRef] [PubMed]

33. Nieto, G.; Díaz, P.; Bañón, S.; Garrido, M.D. Effect on lamb meat quality of including thyme (*Thymus zygis* ssp. gracilis) leaves in ewes' diet. *Meat Sci.* **2010**, *85*, 82–88. [CrossRef] [PubMed]

34. Soultos, N.; Tzikas, Z.E.; Christaki Papageorgiou, K.; Steris, V. The effect of dietary oregano essential oil on microbial growth of rabbit carcasses during refrigerated storage. *Meat Sci.* **2009**, *81*, 474–478. [CrossRef] [PubMed]

35. Smeriglio, A.; Barreca, D.; Bellocco, E.; Trombetta, D. Proanthocyanidins and hydrolysable tannins: Occurrence, dietary intake and pharmacological effects. *Br. J. Pharmacol.* **2017**, *174*, 1244–1262. [CrossRef] [PubMed]

36. Aguilar-Galvez, A.; Noratto, G.; Chambi, F.; Debaste, F.; Campos, D. Potential of tara (*Caesalpinia spinosa*) gallotannins and hydrolysates as natural antibacterial compounds. *Food Chem.* **2014**, *156*, 301–304. [CrossRef] [PubMed]

37. Casaburi, A.; Piombino, P.; Nychas, G.J.; Villani, F.; Ercolini, D. Bacterial populations and the volatilome associated to meat spoilage. *Food Microbiol.* **2015**, *45*, 83–102. [CrossRef]

38. Ercolini, D.; Casaburi, A.; Nasi, A.; Ferrocino, I.; Di Monaco, R.; Ferranti, P.; Mauriello, G.; Villani, F. Different molecular types of *Pseudomonas fragi* have the same overall behavior as meat spoilers. *Int. J. Food Microbiol.* **2010**, *142*, 120–131. [CrossRef]

39. Nychas, G.J.E.; Skandamis, P.N.; Tassou, C.C.; Koutsoumanis, K.P. Meat spoilage during distribution. *Meat Sci.* **2008**, *78*, 77–89. [CrossRef]

40. Stanborough, T.; Fegan, N.; Powell, S.M.; Tamplin, M.; Chandry, P.S. Insight into the genome of *Brochothrix thermosphacta*, a problematic meat spoilage bacterium. *Appl. Environ. Microbiol.* **2017**, *83*, e02786-16. [CrossRef]

41. Mamlouk, K.; Macé, S.; Guilbaud, M.; Jaffrès, E.; Ferchichi, M.; Prévost, H.; Pilet, M.F.; Dousset, X. Quantification of viable *Brochothrix thermosphacta* in cooked shrimp and salmon by real-time PCR. *Food Microbiol.* **2012**, *30*, 173–179. [CrossRef] [PubMed]

42. Nielsen, D.S.; Jacobsen, T.; Jespersen, L.; Koch, G.A.; Arneborg, N. Occurrence and growth of yeasts in processed meat products—Implications for potential spoilage. *Meat Sci.* **2008**, *80*, 919–926. [CrossRef] [PubMed]

 *foods* 

*Article*

# Combined Effect of Dietary Protein, Ractopamine, and Immunocastration on Boar Taint Compounds, and Using Testicle Parameters as an Indicator of Success [†]

Tersia Needham [1,2,*], Rob M. Gous [3], Helet Lambrechts [2], Elsje Pieterse [2] and Louwrens C. Hoffman [2,4]

1   Department of Animal Science and Food Processing, Faculty of Tropical AgriSciences, Czech University of Life Sciences Prague, Kamycka 129, 16500 Prague-Suchdol, Czech Republic
2   Department of Animal Sciences, University of Stellenbosch, Private Bag X1, Matieland, Stellenbosch 7602, South Africa; helet@sun.ac.za (H.L.); elsjep@sun.ac.za (E.P.); louwrens.hoffman@uq.edu.au (L.C.H.)
3   School of Agricultural, Earth and Environmental Sciences, University of KwaZulu-Natal, Scottsville 3209, South Africa; Gous@ukzn.ac.za
4   Center for Nutrition and Food Sciences, Queensland Alliance for Agriculture and Food Innovation (QAAFI), The University of Queensland, Health and Food Sciences Precinct, 39 Kessels Rd, Coopers Plains 4108, Australia
*   Correspondence: needham@ftz.czu.cz; Tel.: +420-224-382-343
†   The manuscript was generated from one chapter within Tersia Needham's MSc thesis.

Received: 29 September 2020; Accepted: 11 November 2020; Published: 14 November 2020

**Abstract:** This study investigates the combined effect of immunocastration, dietary protein level (low, medium or high) and ractopamine hydrochloride supplementation (0 or 10 mg/kg) on the adipose concentrations of androstenone, skatole and indole in pigs, and explores whether body mass, carcass fatness or testicular parameters may be indicators of boar taint in these carcasses. Immunocastration was successful in decreasing testicle functioning, and adipose androstenone and skatole concentrations, in all individuals. Immunocastration decreased testicle weight and length, seminiferous tubule circumference and epithelium thickness. Testicle tissue from immunocastrates was also paler, and less red in color, in comparison to non-castrated controls. Dietary protein level and ractopamine hydrochloride supplementation had no influence on the adipose concentration of androstenone, skatole and indole. Testicle size and color were moderate to strong indicators of androstenone and skatole concentrations in the carcasses, and thus vaccination success. Immunocastration together with the adjustment of dietary protein and ractopamine hydrochloride supplementation, is successful in preventing boar taint while maintaining growth performance.

**Keywords:** androstenone; dietary protein; GnRH; Improvac; ractopamine hydrochloride; skatole; testicles

---

## 1. Introduction

The use of surgical castration in male piglets is currently under ethical scrutiny, motivating the investigation into alternative methods to control boar taint in pork products. Boar taint is described as an unpleasant aroma and flavor in pork, and is the result of an increased production of androstenone (5α-androst-16-en-3-one) [1], skatole (3-methylindole) [2] and to a lesser extent indole [3,4] by the boar as it reaches sexual maturity. The lipophilic pheromone androstenone (5α-androst-16-en-3-one) is produced in the testicles of boars, and has the primary function of stimulating the standing reflex in sows, but it also accumulates in adipose tissue [3]. Thus, castration is used to prevent the production and accumulation of androstenone in male pig carcasses. Although a successful means of preventing boar taint, physical/surgical castration is linked to certain welfare issues, such as acute and chronic pain, wound infection, morbidity and mortalities [5]. Castration also inhibits the anabolic effect of male androgens, resulting in less efficient growth and fatter carcasses. When castration is not considered to be a viable option, entire (intact/non-castrated) pubertal males are slaughtered before attaining sexual maturity, to minimize the occurrence of boar taint in their carcasses. However, this approach results in small carcasses with narrow profit margins for producers, and in fact, does not appear to be commercially successful in decreasing the incidences of boar tainted-carcasses [6]. The production of young, non-castrated male pigs with lean carcasses poses further issues in countries where subcutaneous fat deposition is important for the production of high-quality, traditional dry-cured products [7].

When heated, the odor of androstenone is described as "ruinous", "sweaty" and "sexual" [8], and as "fecal", "boar", "urine" and "perspiration" [9]. Consumer sensitivity to androstenone depends on its concentration [10], as well as the consumer's genotype [11], ethnic group [12], and sex [8,13]. While the sensory detection threshold for androstenone is approximately 0.4 to 0.5 $\mu$g/g [14], consumer sensitivity is highly variable, from consumers being anosmic, to those being highly sensitive to androstenone at low concentrations. Regular exposure to androstenone can also induce the ability to perceive androstenone in those consumers considered to be anosmic, which in turn can further reduce consumer acceptability of pork containing boar taint [15]. Additionally, when both skatole and androstenone are present in pork, the risk of the consumer detecting boar taint and rejecting the product increases further [10]. Factors affecting skatole accumulation in adipose tissue are still largely undescribed, but it is known that skatole is produced by bacterial breakdown of tryptophan in the large intestine [1]. This microbial degradation may be affected by the digestibility of feed, as well as the extent of intestinal cell debris production, and thus skatole is not exclusively produced in male pigs [16]. However, skatole is catabolized by the liver, the enzymatic functionality of which is influenced by steroids produced by the testicles [17]. Skatole in pork produces a fecal odor [2] and typically, consumers' odor detection threshold concentration for skatole is much lower than androstenone, and while the commonly used skatole threshold is 0.2 $\mu$g/g fat, consumers can be sensitive even at as low as 0.026 $\mu$g/g [14]. Compared to androstenone, the majority of consumers are able to perceive skatole in pork, and dislike it [8]. Currently, a number of techniques for identification of carcasses with boar taint in commercial abattoirs exist, with their own cost-implications [18].

Various approaches have been investigated to control boar taint-related compounds [19], with gene-editing showing huge potential, but also a rather lengthy timeline before commercial implementation. While some approaches may address decreasing the incidences of boar taint, they do not necessarily address the welfare issues of entire male production, such as aggressive behavior [20], the meat/product quality issues associated with entire males [21], or the application of these technologies, such as gene editing, is currently limited in various countries. Thus, commercial interest in the use of immunocastration has increased, which entails vaccinating the animal against its own gonadotropin-releasing hormone (GnRH). The production of GnRH-antibodies blocks the functioning of the hypothalamic-pituitary-gonadal axis, which in turn inhibits the production of the gonadotropins luteinizing hormone (LH) and follicle-stimulating hormone (FSH). Both of these gonadotropins stimulate steroidogenesis in the testicles, and thus immunocastration results in the inhibition of testicular androstenone production. Together with the decrease in testicular steroid production, the growth, nutrient requirements and fat deposition of immunocastrated pigs are influenced [22]. However, feeding a higher protein diet, together with ractopamine hydrochloride ($\beta$-adrenoreceptor agonist), improves feed efficiency of immunocastrated pigs, offsetting fat deposition and slowing growth rates [22]. Seeing as diet has an influence on skatole production, and the fact that most consumers are capable of perceiving this compound, the potential effect of dietary changes and feed additives on boar taint levels cannot be ignored. Thus, the combined effects of such *ante-mortem* strategies for preventing boar taint while maintaining high levels of growth productivity should be investigated, prior to their application in commercial settings. In addition to this, correlations between changes in body mass, carcass fatness, and testicular functioning in relation to adipose androstenone and skatole concentrations should be investigated. By examining these correlations, simple slaughter-line methodology for identifying carcasses with a higher predisposition for boar taint may be developed.

Thus, the aims of this study were to determine the combined effect of immunocastration, dietary protein level and ractopamine hydrochloride supplementation on adipose concentrations of androstenone, skatole and indole, as well as to establish whether body mass, carcass fatness or testicular parameters may be possible indicators of boar taint in these carcasses.

## 2. Materials and Methods

### 2.1. Animals, Housing and Feeding

Ethical approval for the study was obtained from the Research Ethics Committee: Animal Care and Use of Stellenbosch University (SU-ACUM13-00022). A growth performance trial was conducted with 120 entire PIC® male pigs (PIC© Large White × Landrace × White Duroc maternal line bred with a PIC© 410 terminal sire), at the Elsenburg boar testing facilities (Stellenbosch, South Africa). Each individual pen consisted of a concrete sleeping area with pine wood shavings, and a separate dunging area (free from shavings) with a nipple drinker. The pens were cleaned daily, and food and water were provided *ad libitum*. Each pig was allocated randomly to one of 12 treatment combinations (10 pigs per treatment), according to a 2 × 2 × 3 factorial design. The main effects included castration status (immunocastrated (IC) versus non-castrated/entire (E) males), ractopamine hydrochloride (RAC; Paylean, Elanco™ Animal Health, Pretoria, South Africa) supplementation (0 or 10 mg/kg), and the dietary protein inclusion level (low, medium and high; Tables 1 and 2). The medium protein diet was established by mixing the low and high protein diets at a ratio of 50:50.

**Table 1.** The ingredient and nutrient composition (as-is basis) of the various dietary protein finisher diets fed to immunocastrated and entire male pigs from 20 to 24 weeks of age, with or without ractopamine hydrochloride supplementation (*n* = 120).

| Ingredient Composition, g/kg | Dietary Protein Level | | |
|---|---|---|---|
| | Low | Medium | High |
| Maize | 327 | 266 | 206 |
| Wheat bran | 295 | 210 | 124 |
| Barley meal | 150 | 150 | 150 |
| Soya oil cake (470 g CP [1]/kg) | 127 | 213 | 300 |
| Sunflower oil cake (360 g CP [1]/kg) | 50 | 100 | 150 |
| Canola oil | 25.0 | 27.5 | 30 |
| Limestone | 14.2 | 13.4 | 12.7 |
| Salt | 4.3 | 4.36 | 4.42 |
| L-lysine HCL | 2 | 1.9 | 1.8 |
| Vitamin and mineral premix [2] | 2 | 2 | 2 |
| Monocalcium phosphate | 1.9 | 1.0 | 0 |
| Mycotoxin binder | 1 | 1 | 1 |
| L-threonine | 0.53 | 0.36 | 0.18 |
| Phytase enzyme | 0.5 | 0.5 | 0.5 |
| DL-methionine | 0.27 | 0.32 | 0.37 |
| Choline chloride liquid | 0.13 | 0.13 | 0.13 |
| Xylanase and β-glucanase enzyme combination | 0.1 | 0.1 | 0.1 |
| Maize gluten meal (600 g CP [1]/kg) | 0 | 8.34 | 16.7 |

[1] CP: crude protein; [2] Vitamin and Mineral premix: Vitamin A: 5489.5 IU/kg, Vitamin D: 1005.3 IU/kg, Vitamin E: 27.6 IU/kg, Vitamin K: 2.8 mg/kg, Niacin: 22.0 mg/kg, Riboflavin 4.9 mg/kg, d-Pantothenate: 16.5 mg/kg, Vitamin B12: 22.0 mcg/kg, Zinc: 100 mg/kg, Iron: 66 mg/kg, Manganese: 25 mg/kg, Copper: 10 mg/kg, Iodine: 0.33 mg/kg and Selenium: 0.25 mg/kg.

**Table 2.** The nutrient composition (as-is basis) of the various dietary protein finisher diets fed to immunocastrated and entire male pigs from 20 to 24 weeks of age, with or without ractopamine hydrochloride supplementation. All amino acids shown are provided as their digestible values.

| Calculated Nutrient Composition | Dietary Protein Level | | |
|---|---|---|---|
| | Low | Medium | High |
| NE [1] pig, MJ/kg | 9.2 | 9.2 | 9.2 |
| DE [2] pig, MJ/kg | 13.29 | 13.56 | 13.83 |
| Crude protein, g/kg | 161 | 208 | 256 |
| Crude starch, g/kg | 359 | 315 | 271 |
| Crude fiber, g/kg | 60.8 | 64.4 | 68.1 |
| Crude fat, g/kg | 49.1 | 48.3 | 47.4 |
| Ash, g/kg | 59.3 | 65.2 | 71.0 |
| Lysine, g/kg | 7.50 | 9.79 | 12.1 |
| Methionine, g/kg | 2.47 | 3.28 | 4.09 |
| TSAA [3], g/kg | 4.74 | 6.11 | 7.48 |
| Tryptophan, g/kg | 1.59 | 2.12 | 2.66 |
| Threonine, g/kg | 4.88 | 6.36 | 7.85 |
| Arginine, g/kg | 9.24 | 12.7 | 16.1 |
| Isoleucine, g/kg | 5.08 | 7.09 | 9.09 |
| Leucine, g/kg | 10.3 | 13.8 | 17.2 |
| Valine, g/kg | 6.11 | 8.12 | 10.1 |
| Histidine, g/kg | 3.48 | 4.55 | 5.61 |
| Calcium, g/kg | 7.51 | 7.5 | 7.49 |
| Total phosphorus, g/kg | 6.86 | 7.14 | 7.42 |
| Available phosphorus, g/kg | 2.5 | 2.52 | 2.53 |
| Sodium, g/kg | 2.0 | 2.0 | 2.0 |
| Potassium, g/kg | 9.9 | 11.4 | 12.9 |

[1] NE: net energy; [2] ME: metabolizable energy; [3] TSSA: total sulphur-containing amino acids.

The sixty pigs allocated to the IC treatment group, received 2 mL Improvac® (Zoetis™ Animal Health, Sandton, South Africa) at 16 weeks of age, and again at 20 weeks. Up until 20 weeks of age, all pigs received a commercial grower diet, after which they were fed one of three experimental balanced protein diets (7.50, 9.79 and 12.07 g digestible lysine/kg; Table 1) with RAC supplementation at 0 or 10 mg/kg, for the last 28 days of growth.

*2.2. Slaughtering and Testicle Measurements*

All pigs were slaughtered at 24 weeks of age, i.e., four weeks after the administration of the second Improvac® injection. The pigs were transported to a commercial abattoir and slaughtered according to standard practices, which involved electrical stunning followed by exsanguination. According to the live weight at slaughter, 96 pigs were sampled by selecting eight pigs from the midweight-range in each treatment. The backfat depths of the 96 pigs were determined between the second and third last rib (counted from the cranial end) and 45 mm from the spine (P2 position), using a Hennessy Grading Probe (Hennessy Grading Systems, Auckland, New Zealand). A summary of the initial live weight at the start of the trial (16 weeks old), the slaughter weight at 24 weeks of age and backfat depth at slaughter for the 96 selected pigs only can be found in Table 3. Further details regarding the growth performance, carcass traits and meat quality of the pigs may be found in Needham et al. [22–24].

Their testicles were collected on the slaughter-line prior to evisceration, placed in marked plastic bags, and transported on ice to the laboratory for further processing. Upon arrival at the laboratory, the epididymis and connective tissue were removed, and each individual testicle was weighed on a RADWAG PS750/C/2 scale (Wagi Elektroniczne, Radwag, Radom, Poland; accurate to 0.001 g). Testicle measurements (length and width) were taken using a calibrated engineering caliper (150 mm Electronic Digital Vernier Caliper CE ROHS), and testicle volume was determined by water displacement [25].

**Table 3.** Summary of the initial live weight, the slaughter weight and backfat depth at slaughter for 96 pig carcasses, which were selected from a larger growth study (*n* =120) investigating the effects of varying dietary protein levels, with or without ractopamine hydrochloride supplementation.

| Treatment | | Initial Live Weight at 16 Weeks Old (kg) | Slaughter Weight at 24 Weeks Old (kg) | Hennessey Grading Probe Backfat Depth (mm) |
|---|---|---|---|---|
| **Castration Status** | Entire (*n* = 48) | 57.7 ± 0.70 | 130.2 ± 1.18 | 16.9 ± 0.42 |
| | Immunocastrated (*n* = 48) | 57.5 ± 0.65 | 127.5 ± 1.21 | 17.4 ± 0.43 |
| | *p* | 0.727 | 0.134 | 0.444 |
| **Ractopamine Hydrochloride Supplementation** | 0 mg/kg (*n* = 48) | 57.3 ± 0.74 | 127.7 ± 1.21 | 17.8 ± 0.43 |
| | 10 mg/kg (*n* = 48) | 57.9 ± 0.59 | 129.9 ± 1.18 | 16.6 ± 0.42 |
| | *p* | 0.481 | 0.182 | 0.05 |
| **Dietary Protein** | Low (*n* = 32) | 57.2 ± 0.89 | 128.7 ± 1.47 | 17.6 ± 0.53 |
| | Medium (*n* = 32) | 57.9 ± 0.88 | 128.7 ± 1.44 | 17.5 ± 0.52 |
| | High (*n* = 32) | 57.7 ± 0.71 | 129.1 ± 1.47 | 16.5 ± 0.53 |
| | *p* | 0.821 | 0.979 | 0.305 |

For histological evaluation, 32 pairs of testicles were sub-sampled by selecting eight of the mid-weight animals from the following treatment combinations, fed only the medium protein diet: E fed RAC, E fed no RAC, IC fed RAC, and IC fed no RAC. Each testicle was cut alongside the widest axis, and the surface color was measured, without bloom time [25], according to the International Commission on Illumination (CIE) Lab color system. A Color-guide 45/0 colorimeter (Catalogue number 6801, BYK-Gardner GmbH, Geretsried, Germany) was used, with an aperture diameter size of 11mm and an illuminant/observer angle of 65/10°. Calibration of the colorimeter was done prior to measurement, using the black calibration standard, white calibration standard, green checking reference, and high gloss standard. The hue angle and chroma value were calculated as follows:

$$Hue-angle\ (°)\ =\ tan^{-1}\left(\frac{b^*}{a^*}\right);\ Chroma\ (C*) = \left(a^{*\,2} + b^{*\,2}\right)^{-0.5}$$

Thereafter, testicle tissue samples (approximately 1 cm × 1 cm × 1 cm) were taken, preserved in 10% neutral buffered formalin, and stored for later histological processing. Slides were prepared from the preserved tissue, and stained using haematoxylin and eosin. Histology slides were evaluated at 40× magnification, using an Olympus IX70 microscope (Olympus Corporation, Tokyo, Japan). One hundred seminiferous tubules were measured per sample, and their circumference and epithelium thickness were determined using the Olympus Image Analysis Software package (Olympus Corporation, Tokyo, Japan).

### 2.3. Chemical Analyses of Androstenone (5α-Androst-16-en-3-One), Skatole (3-Methylindole) and Indole

At approximately 24 h *post-mortem*, subcutaneous backfat samples were taken from the same 96 selected pigs by removing a 2 cm thick strip of fat from the loin, at the position of the third-last rib. Samples were frozen at −20 °C, until simultaneous analysis for androstenone (5α-androst-16-en-3-one), skatole (3-methylindole) and indole following an adapted methodology [26]. At the time of analysis, the adipose tissue samples were thawed, and 5 g of each sample was cut into thin flakes before placing them into stomacher bags. An internal standard was prepared by adding 2-methylindole to methanol at a concentration of 200 µg/kg, and 5 mL of this was added to each stomacher bag. The samples were then homogenized for two minutes, using a stomacher BagMixer® 400 W (Interscience, Saint-Nom-la-Bretèche, France). The supernatant was transferred from the stomacher bag into a sterile Cellstar® tube (Greiner Bio-One, Kremsmünster, Austria), and cooled by submersing the tube in liquid nitrogen for approximately one minute. Subsequently, the samples were centrifuged for six minutes at 5000 rpm. Following removal from the centrifuge, the samples were frozen within their tubes using liquid nitrogen, to clear the upper phase, and then filtered using a 0.22 µm syringe filter. After this filtration, 300 µL of the extract was placed into vials for analysis, and diluted with 200 µL of 1% acetic acid.

The samples were analyzed for 5α-androst-16-en-3-one using tandem mass spectrometry (Waters Xevo TQ triple quadruple mass spectrometer, Waters Corporation, Milford, CT, USA), and for skatole and indole using ultra-performance liquid chromatography with fluorescence (Waters ACQUITY UPLC FLR Detector, Waters Corporation, Milford, CT, USA). A Kinetex C18 column was used (2.6 um, 150 × 2.1 mm, Phenomenex Inc., Torrance, CA, USA) with two solvent gradients: 7.5% formic acid and 49:49:2 methanol:acetonitrile:isopropanol. Sample injection volume was 10 µL, and the column temperature was set to 40 °C. Standard curves were established for each of the compounds analyzed, such that the calibration range and limit of quantification for 5α-androst-16-en-3-one was 0.01 to 13 µg/g and 0.02 µg/g, respectively, whereas the calibration range and limit of detection for both skatole and indole was 0.008 to 0.08 µg/g and 0.004 µg/g, respectively.

*2.4. Statistical Analyses*

Data were analyzed using STATISTICA (Version 13.5.0, StatSoft Inc., Tulsa, OK, USA), together with integrated R software (R Foundation, Vienna, Austria). Normality of the data was evaluated, and those which were not normally distributed (testes volume, indole and skatole concentrations) underwent Box-Cox transformation. One-way analysis of variances (ANOVAs) were established for each parameter using the R "lm" function, with the treatments (castration status, dietary protein level and ractopamine supplementation) as the fixed effects, and the animal as the random effect. In the case of testicles color and histology data, only castration status was included as a fixed effect. Treatment means were compared using Fishers LSD post-hoc testing. Correlations between all parameters were established in Statistica, using Spearman's Correlation Coefficients, and visually displayed using a correlation heatmap with cluster dendrograms, generated with R software (function: "heatmap.2"). A significance level of 5% was used throughout. Further descriptive statistics were performed for the average concentration range and percentage of samples above the analytical detection limit for androstenone, skatole and indole.

## 3. Results

Immunocastration decreased the testicles' weight ($p < 0.01$) and length ($p < 0.01$), but did not influence testicles' volume and width (Table 4). Dietary protein level and RAC had no influence on testicle size. Furthermore, immunocastration decreased seminiferous tubule circumference ($p < 0.001$) and epithelium thickness ($p = 0.002$), resulting in increased lumen size and deformation of the seminiferous tubules (Figure 1). The results for the CIE Lab color values indicated that the testicles from immunocastrates had higher $L^*$ values ($p < 0.001$), higher $b^*$ values ($p < 0.001$), and lower $a^*$ values ($p < 0.001$; Table 4). Thus, they were paler, more yellow and less red when compared to the surface color of testicles from entire male pigs. Supplementation with RAC had no influence on testicles color and histology.

Immunocastration decreased the subcutaneous backfat concentrations of androstenone ($p < 0.01$) and skatole ($p < 0.01$), but not indole (Table 5). For both the immunocastrated and entire male carcasses, no adipose tissue had androstenone concentrations over the sensory threshold (0.426 µg/g fat) [14]. However, 48% of entire males had mean adipose skatole concentrations above the sensory threshold (0.026 µg/g fat) [14], while the average immunocastrated adipose tissue skatole concentrations were below that of the sensory threshold, with only two animals exceeding the sensory threshold with concentrations of 0.034 and 0.028 µg/g fat. There was no effect of RAC supplementation or dietary balanced protein level on the levels of skatole.

**Table 4.** The testicle size, CIE Lab colour value and tissue morphology traits (Least Square (LS) Mean ± Standard Error of Mean (SEM)) of immunocastrated and entire male pigs.

| Parameters | Castration Status | | $p$ |
| --- | --- | --- | --- |
| | Entire | Immunocastrated | |
| **Testicle size (*n* = 96)** | **(*n* = 48)** | **(*n* = 48)** | |
| Paired weight (g) | 536 ± 14.0 | 282 ± 14.4 | <0.001 |
| Paired volume (mL) | 1173 ± 23.1 | 1186 ± 23.6 | 0.914 |
| Individual length (mm) | 113 ± 1.48 | 93.6 ± 1.52 | <0.001 |
| Individual width (mm) | 61.7 ± 1.36 | 62.7 ± 1.39 | 0.599 |
| **Testicle surface colour (*n* = 32)** | **(*n* = 16)** | **(*n* = 16)** | |
| $L^*$ | 52.6 ± 0.66 | 59.1 ± 0.72 | <0.001 |
| $a^*$ | 9.98 ± 0.26 | 6.95 ± 0.42 | <0.001 |
| $b^*$ | 11.0 ± 0.14 | 12.4 ± 0.21 | <0.001 |
| **Seminiferous tubule (*n* = 32)** | **(*n* = 16)** | **(*n* = 16)** | |
| Circumference (μm) | 1080 ± 23.4 | 921 ± 21.5 | <0.001 |
| Epithelium thickness (μm) | 76.1 ± 2.27 | 65.4 ± 2.65 | 0.002 |

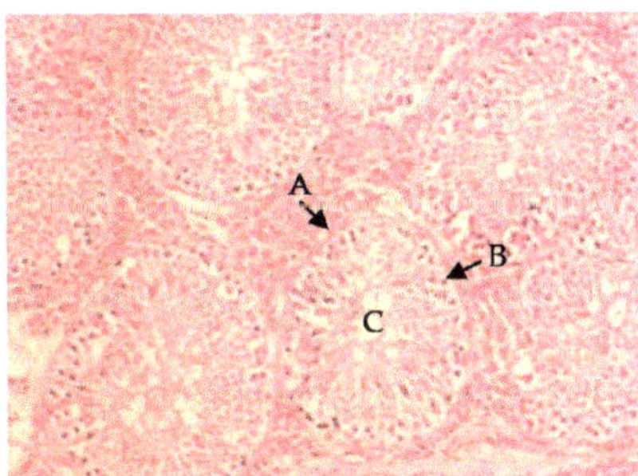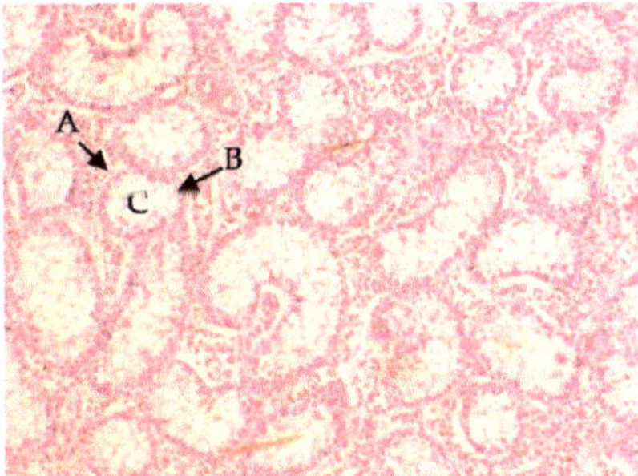

**Figure 1.** Micrographs of testicular tissue for entire (**left**) and immunocastrated (**right**) pigs slaughtered at 24 weeks of age, 4 weeks after the second immunocastration vaccination with Improvac (40× magnification). Immunocastration resulted in atrophy of the seminiferous tubules (A), loss of spermatocytes (B), and increased lumen size (C).

**Table 5.** The concentrations (μg/g) of androstenone, skatole and indole in the adipose tissue of immunocastrated and entire male pigs, fed varying protein diets, with or without ractopamine hydrochloride for the last 28 days of finishing. Results are reported as LS Mean ± SEM.

| | | Androstenone | Skatole | Indole |
| --- | --- | --- | --- | --- |
| **Castration Status** | **Entire (*n* = 48)** | 0.098 ± 0.008 | 0.038 ± 0.004 | 0.009 ± 0.0015 |
| | **Immunocastrated (*n* = 48)** | 0.029 ± 0.009 | 0.012 ± 0.004 | 0.004 ± 0.0015 |
| | $p$ | <0.001 | <0.001 | 0.116 |
| **Ractopamine Hydrochloride Supplementation** | **0 mg/kg (*n* = 48)** | 0.061 ± 0.009 | 0.024 ± 0.004 | 0.006 ± 0.0015 |
| | **10 mg/kg (*n* = 48)** | 0.067 ± 0.008 | 0.026 ± 0.004 | 0.007 ± 0.0015 |
| | $p$ | 0.590 | 0.567 | 0.334 |
| **Dietary Protein** | **Low (*n* = 32)** | 0.071 ± 0.0104 | 0.023 ± 0.0054 | 0.008 ± 0.0018 |
| | **Medium (*n* = 32)** | 0.061 ± 0.0102 | 0.025 ± 0.0053 | 0.006 ± 0.0018 |
| | **High (*n* = 32)** | 0.06 ± 0.0204 | 0.027 ± 0.0053 | 0.005 ± 0.0018 |
| | $p$ | 0.671 | 0.754 | 0.571 |

Correlations between all parameters are represented in Figure 2, with the variables showing three primary groupings. The first grouping (average testicle length, combined testicle weight, androstenone concentration, skatole concentration and $a^*$ color values), showed moderate to strong positive correlations with each other, as well as chroma values. The second grouping (indole concentration, P2 backfat thickness, average testicle width, testicle volume and live mass) showed no, or weak, correlations with all other variables. The third grouping (*b** color values, hue angle and *L** color values) showed moderate to strong negative correlations with the first group of variables (average testicle length, combined testicle weight, androstenone concentration, skatole concentration and *a** color values). The linkage distance of the clustering analysis indicates how closely the correlations of these variables match one another. For example, average testicle length and combined testicles weight showed the most similar correlations to one another. Regarding potential indicators of androstenone, skatole and indole concentrations, testicle parameters (weight, length and color values) were moderately to strongly correlated; however, live mass and carcass fatness show weak correlations with boar taint compound levels (Figure 2).

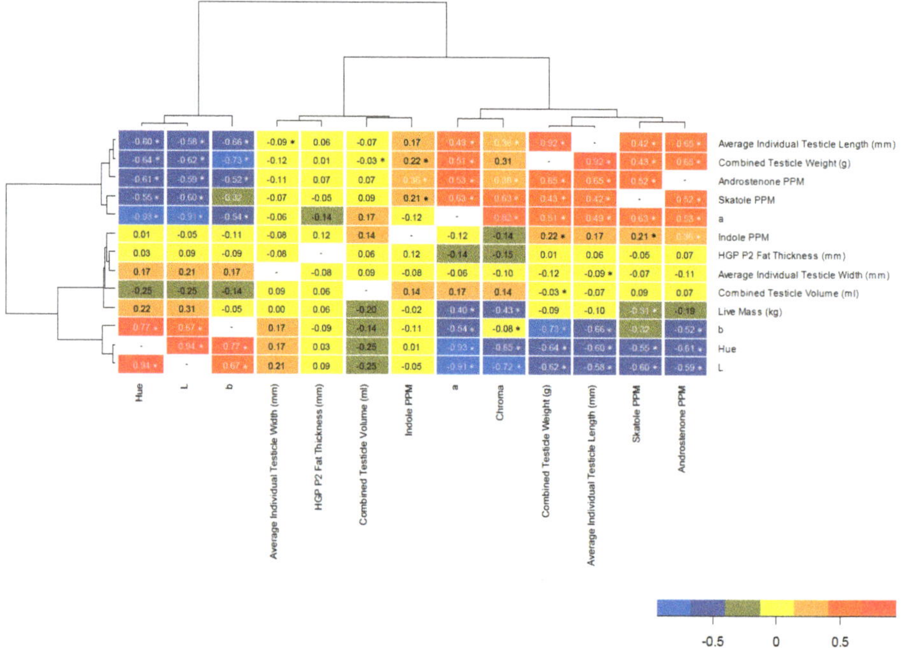

**Figure 2.** Pearson's correlation coefficient heatmap for body mass, Hennessy Grading Probe backfat thickness (HPG at P2 location), testicle size and color (*L**, *a**, *b**, hue and chroma) parameters and backfat concentrations of androstenone, skatole and indole. Linkage distance within the clustering analysis indicates how closely the correlations of these variables match one another. The legend indicates the strength and direction of the correlations between the parameters, and Spearman's correlation coefficients are shown between parameters. Significant differences are indicated by asterisks, at $p < 0.05$.

## 4. Discussion

The immunocastration vaccination schedule used was successful in decreasing testicular functioning, as indicated by the decrease in weight, disruption of the seminiferous tubule morphology, color changes and decreased androstenone production. Previous results have shown seminiferous tubule atrophy and spermatocyte loss in immunocastrated pigs [27], as well as sheep [28–30], and cattle [31,32]. Immunocastration also caused the testicle tissue surface color to become paler, less red and more yellow, as similarly reported [25]. These changes in testicle tissue color are likely indicative of their decreased functioning after immunocastration and, together with testicle weights, may provide an indication of vaccination success [25] and subsequently risk of boar taint in the carcass. Immunocastration decreased adipose androstenone concentrations, to values comparable with previous studies using similar vaccination schedules [25,33]. Immunocastration also decreased adipose skatole levels, which is expected, as androstenone inhibits the skatole-induced expression of the CYP2E1 enzyme involved in skatole metabolism [34]. However, diet and supplementation with ractopamine hydrochloride had no influence on skatole levels in the present study, which was somewhat unexpected, as changes in dietary protein and fiber contents may influence the gut pH and microflora [35] and intestinal cell debris production [16], all of which are thought to contribute to the amount of skatole production in the gut. Furthermore, the use of feed additives may also potentially have an influence on the enterohepatic recirculation of androstenone as well as on gut microflora, influencing skatole production [19]. However, skatole synthesis and metabolism is a complex process of which the influence of various factors on this remain poorly understood [36], and it is likely that the dietary factors included in the present study were not detrimental to the gut pH, microflora and metabolism of boar taint-related compounds.

The adipose androstenone concentrations in the carcasses of entire males within the present study were also low, despite their slaughter weight (~125 kg), compared to the concentrations reported in the control males of other immunocastration studies [25,33]. Whilst variation in these reported values are expected, considering the different analytical methodologies used, it is likely that genotype influenced the androstenone levels reported in the present study compared to other studies. For example, Weiler et al. [33] used progeny from a Large White x Landrace maternal line and Pietrain terminal sire line in their study, and reported an average fat androstenone level of 1.75 µg/g in entire males. Whereas Lealiifano et al. [25] used Large White × Landrace boars, and reported an average fat androstenone level of 0.91 µg/g. A recent study also describes differences in adipose androstenone levels of slaughter pigs from three different sire lines at slaughter (after immunocastration), indicating that differences in growth rate, maturity at slaughter weight and lean growth potential of each genotype influences the adipose androstenone levels [37]. It is also accepted that heavier pigs have a higher risk of elevated androstenone levels, as described by Pieterse [38], who found that when slaughter weight was increased from between 102–113 kg to 133 kg over five different genotypes of boars, adipose androstenone levels increased concomitantly. Despite differences in genotype and weights between studies, it is also possible that in the absence of female pigs in the housing system in the present study, the boars were not stimulated to produce testicular steroids and thus androstenone. Nonetheless, despite the low levels of androstenone in entire males in the present study, immunocastration successfully decreased the adipose skatole levels compared to entire males, the latter having nearly 50% of their adipose tissue skatole levels above the sensory threshold. This is particularly important, as the majority of consumers are able to perceive skatole in pork, and dislike it [8].

Thus, using genotype, environment, age and weight alone is not a good indication of boar taint in carcasses. Even though technology is continuously improving for the identification of boar taint on slaughter lines, there remain abattoirs which cannot afford such equipment or the trained staff required to operate them, and thus still rely on basic indicators such as weight, fatness and age of the animal. According to the correlations investigated in the present manuscript, testicle weight and tissue color are better indicators of androstenone and skatole concentrations in the adipose tissue than live weight and subcutaneous backfat depth. As these compounds are affected by the degree of testicular activity, it was thus expected that indicators of change in testicular activity would be more reliable estimators of immunocastration success and of boar taint in carcasses. While these factors should be investigated on a large scale and integrated into models which may set sorting limits for successfully immunocastrated pigs, or high-risk boar taint carcasses according to the popular genotypes before they may be used in abattoirs which cannot afford higher technologies, these physiological factors may also be integrated into other precision livestock management tools to improve the power of their indirect prediction. It may also be possible to develop color cards that can be used to distinguish testicles that might be indicative of boar taint, as is typically used in beef abattoirs for meat color and fat grading of Wagyu. These correlations should also be expanded over higher concentrations of androstenone and skatole

Such information could be used as a pre-screening tool, without cutting or damaging the carcass itself, for those carcasses requiring further investigation before determination of their sale potential.

Nonetheless, the results of the present study indicate that immunocastration was successful in suppressing testicular functioning and preventing the accumulation of androstenone and skatole in the adipose tissue of male slaughter pigs (Large White × Landrace × White Duroc maternal line and PIC© 410 terminal sire line), regardless of changes in dietary protein and ractopamine hydrochloride supplementation. Thus, in comparison to alternatives, such as entire male production and surgical castration, immunocastration, together with the provision of adequate dietary protein and use of ractopamine hydrochloride, provides a welfare-friendly technique with potentially low-cost implications, when considering the cost (and variable) effectivity of anesthesia, as well as the losses associated with the processing (or rejection) of carcasses with boar taint.

## 5. Conclusions

Immunocastration was successful in decreasing testicular functioning, resulting in 100% of the treated animals having androstenone concentrations below the defined sensory threshold, and decreased skatole concentrations in comparison to the entire males. The dietary protein levels used in this study and ractopamine hydrochloride supplementation had no influence on the accumulation of skatole in the adipose tissue, and thus may be commercially considered to support optimum growth of immunocastrated pigs. Commercially used indicators for potential boar taint in carcasses, including body mass and carcass fatness, were not reliable indicators thereof, but testicle weight and color were better correlated with boar taint compounds. Thus, by further examining the correlations of testicular activity indictors with androstenone and skatole concentrations using mass data and incorporating this information with other physiological factors, potential pre-sorting limits may be established for identifying successful immunocastration or carcasses with a higher predisposition for boar taint.

**Author Contributions:** Conceptualization, investigation, data collection, funding acquisition, project execution, data interpretation, original draft preparation and editing: T.N.; Methodology, funding acquisition, supervision and manuscript editing: L.C.H.; Conceptualization, methodology, and manuscript editing: R.M.G.; Methodology, supervision and manuscript editing: E.P.; Data collection and manuscript editing: H.L. All authors have read and agreed to the published version of the manuscript.

**Funding:** This research was funded by the South African Pork Producers' Organisation and the South African Research Chairs in Meat Science Initiative (UID: 84633). Any opinion, finding, conclusion or recommendation that is expressed in this material is that of the author(s) and the National Research Foundation does not accept any liability.

**Acknowledgments:** The authors would like to thank M. Kidd for assistance with the statistical analyses, M. Stander for assistance with the laboratory analyses, and Elanco™ and Zoetis™ for their provision of their products, and the Agricultural Research Council (South Africa) for the use of their facilities. The support of the Internal Grant Agency (IGA) FTZ CZU, Prague (no. 20195011) is hereby acknowledged. The authors would also like to thank the students, and Mr John Morris, who were involved in the data collection of this study.

**Conflicts of Interest:** The authors declare no conflict of interest.

## References

1. Patterson, R.L.S. 5a-Androst-16-ene-3-one: Compound responsible for taint in boar fat. *J. Sci. Food Agric.* **1968**, *19*, 31–37. [CrossRef]

2. Vold, E. Fleischproduktionseigenschaften bei Ebern und Kastraten. IV. Organoleptische und Gaschromatographische Untersuchungen Wasserdampfflüchtiger Stoffe des Rückenspeckes von Ebern. *Meld. Nor. Landbrukshogsk.* **1970**, *49*, 1–25.

3. Bonneau, M. Compounds responsible for boar taint, with special emphasis on androstenone: A review. *Livest. Prod. Sci.* **1982**, *9*, 687–705. [CrossRef]

4. Garcia-Regueiro, J.A.; Diaz, I. Evaluation of the contribution of skatole, indole, androstenone and androstenols to boar-taint in back fat of pigs by HPLC and capillary gas chromatography (CGC). *Meat Sci.* **1989**, *25*, 307–316. [CrossRef]

5. Rault, J.L.; Lay, D.C., Jr.; Marchant-Fordel, J.N. Castration induced pain in pigs and other livestock. *Appl. Anim. Behav. Sci.* **2011**, *135*, 214–225. [CrossRef]

6. Needham, T.; van Zyl, K.; Hoffman, L.C. Correlations between PORCUS classification and androstenone in boars, and effects of cooking methods thereon. *S. Afr. J. Anim. Sci.* **2020**, *50*, 120–128. [CrossRef]

7. Fuchs, T.; Nathues, H.; Koehrmann, A.; Andrews, S.; Brock, F.; Sudhaus, N.; Klein, G.; Beilage, E.G. A comparison of the carcase characteristics of pigs immunized with a 'gonadotrophin-releasing factor (GnRF)' vaccine against boar taint with physically castrated pigs. *Meat Sci.* **2009**, *83*, 702–705. [CrossRef]

8. Font-i-Furnols, M.; Gispert, M.; Diestre, A.; Oliver, M.A. Acceptability of boar meat by consumers depending on their age, gender, culinary habits, sensitivity and appreciation of androstenone smell. *Meat Sci.* **2003**, *64*, 433–440. [CrossRef]

9. Bonneau, M.; Denmat, M.; Vaudelet, C.; Veloso-Nunes, J.R.; Mortensen, A.B.; Mortensen, H.P. Contribution of fat androstenone and skatole to boar taint I. Sensory attributes of fat and pork meat. *Livest. Prod. Sci.* **1992**, *32*, 63–80. [CrossRef]

10. Font-i-Furnols, M. Consumer studies on sensory acceptability of boar taint: A review. *Meat Sci.* **2012**, *92*, 319–329. [CrossRef]

11. Keller, A.; Zhuang, H.; Chi, Q.; Vosshall, L.B.; Matsunami, H. Genetic variation in a human odorant receptor alters odour perception. *Nature* **2007**, *449*, 468–472. [CrossRef] [PubMed]

12. De Kock, H.L.; van Heerden, S.M.; Heinze, P.H.; Dijksterhuis, B.B.; Minnaar, A. Reaction to boar odour by different South African consumer groups. *Meat Sci.* **2001**, *59*, 353–362. [CrossRef]

13. Gilbert, A.N.; Wysocki, C.J. Results of the smell survey. *Natl. Geogr.* **1987**, *172*, 514–525.

14. Annor-Frempong, I.E.; Nute, G.R.; Whittington, F.W.; Wood, J.D. The problem of taint in pork: 1. Detection thresholds and odour profiles of androstenone and skatole in a model system. *Meat Sci.* **1997**, *46*, 45–55. [CrossRef]

15. Wysocki, C.J.; Dorries, K.M.; Beauchamp, G.K. Ability to perceive androstenone can be acquired by ostensibly anosmic people. *Proc. Natl. Acad. Sci. USA* **1989**, *86*, 7976–7978. [CrossRef]

16. Bonneau, M.; Weiler, U. Pros and Cons of Alternatives to Piglet Castration: Welfare, Boar Taint, and Other Meat Quality Traits. *Animals* **2019**, *9*, 884. [CrossRef]

17. Zamaratskaia, G.; Gilmore, W.J.; Lundström, K.; Squires, E.J. Effect of testicular steroids on catalytic activities of cytochrome P450 enzymes in porcine liver microsomes. *Food Chem. Toxicol.* **2007**, *45*, 676–681. [CrossRef]

18. Font-i-Furnols, M.; Martín-Bernal, R.; Aluwé, M.; Bonneau, M.; Haugen, J.-E.; Mörlein, D.; Mörlein, J.; Panella-Riera, N.; Škrlep, M. Feasibility of on/at Line Methods to Determine Boar Taint and Boar Taint Compounds: An Overview. *Animals* **2020**, *10*, 1886. [CrossRef]

19. Squires, E.J.; Bone, C.; Cameron, J. Pork Production with Entire Males: Directions for Control of Boar Taint. *Animals* **2020**, *10*, 1665. [CrossRef]

20. Needham, T.; Lambrechts, H.; Hoffman, L. Castration of male livestock and the potential of immunocastration to improve animal welfare and production traits: Invited Review. *S. Afr. J. Anim. Sci.* **2017**, *47*, 731–742. [CrossRef]

21. Škrlep, M.; Tomašević, I.; Mörlein, D.; Novaković, S.; Egea, M.; Garrido, M.D.; Linares, M.B.; Peñaranda, I.; Aluwé, M.; Font-i-Furnols, M. The Use of Pork from Entire Male and Immunocastrated Pigs for Meat Products—An Overview with Recommendations. *Animals* **2020**, *10*, 1754. [CrossRef] [PubMed]

22. Needham, T.; Hoffman, L.C.; Gous, R. Growth responses of entire and immunocastrated male pigs to dietary protein with and without ractopamine hydrochloride. *Animal* **2017**, *11*, 1482–1487. [CrossRef] [PubMed]

23. Needham, T.; Hoffman, L.C. Carcass traits and cutting yields of entire and immunocastrated pigs fed increasing protein levels with and without ractopamine hydrochloride supplementation. *J. Anim. Sci.* **2015**, *93*, 4545–4556. [CrossRef] [PubMed]

24. Needham, T.; Hoffman, L.C. Physical meat quality and chemical composition of the *Longissimus thoracis* of entire and immunocastrated pigs fed varying dietary protein levels with and without ractopamine hydrochloride. *Meat Sci.* **2015**, *110*, 101–108. [CrossRef] [PubMed]

25. Lealiifano, A.; Pluske, J.R.; Nicholls, R.; Dunshea, F.; Campbell, R.; Hennessy, D.; Miller, D.; Hansen, C.F.; Mullan, B. Reducing the length of time between slaughter and the secondary gonadotropin-releasing factor immunization improves growth performance and clears boar taint compounds in male finishing pigs. *J. Anim. Sci.* **2011**, *89*, 2782–2792. [CrossRef] [PubMed]

26. Verheyden, K.; Noppe, H.; Aluwé, M.; Millet, S.; Vanden Bussche, J.; De Brabander, H. Development and validation of a method for simultaneous analysis of the boar taint compounds indole, skatole and androstenone in pig fat using liquid chromatography-multiple mass spectrometry. *J. Chromatogr. A* **2007**, *1174*, 132–137. [CrossRef] [PubMed]

27. Kubale, V.; Batorek, N.; Škrlep, M.; Prunier, A.; Bonneau, M.; Fazarinc, G.; Čandek-Potokar, M. Steroid hormones, boar taint compounds, and reproductive organs in pigs according to the delay between immunocastration and slaughter. *Theriogenology* **2013**, *79*, 69–80. [CrossRef]

28. Ülker, H.; Kanter, M.; Gökdal, Ö.; Aygün, T.; Karakus, F.; Sakarya, M.; De Avila, D.; Reeves, J. Testicular development, ultrasonographic and histological appearance of the testis in ram lambs immunized against recombinant LHRH fusion proteins. *Anim. Reprod. Sci.* **2005**, *86*, 205–219. [CrossRef]

29. Ülker, H.; Küçük, M.; Yilmaz, A.; Yörük, M.; Arslan, L.; deAvila, D.; Reeves, J. LHRH fusion protein immunization alters testicular development, ultrasonographic and histological appearance of ram testis. *Reprod. Domest. Anim.* **2009**, *44*, 593–599. [CrossRef]

30. Needham, T.; Lambrechts, H.; Hoffman, L. Extending the interval between second vaccination and slaughter: II. Changes in the reproductive capacity of immunocastrated ram lambs. *Animal* **2019**, *13*, 1962–1971. [CrossRef]

31. Huxsoll, C.C.; Price, E.O.; Adams, T.E. Testis function, carcass traits, and aggressive behavior of beef bulls actively immunized against gonadotropin-releasing hormone. *J. Anim. Sci.* **1998**, *76*, 1760–1766. [CrossRef] [PubMed]

32. Janett, F.; Gerig, T.; Tschuor, A.C.; Amatayakul-Chantler, S.; Walker, J.; Howard, R.; Bollwein, H.; Thun, R. Vaccination against gonadotropin-releasing factor (GnRF) with Bopriva significantly decreases testicular development, serum testosterone levels and physical activity in pubertal bulls. *Theriogenology* **2012**, *78*, 182–188. [CrossRef] [PubMed]

33. Weiler, U.; Götz, M.; Schmidt, A.; Otto, M.; Müller, S. Influence of sex and immunocastration on feed intake behavior, skatole and indole concentrations in adipose tissue of pigs. *Animal* **2013**, *7*, 300–308. [CrossRef] [PubMed]

34. Doran, E.; Whittington, F.W.; Wood, J.D.; McGivan, J.D. Cytochrome P450IIE1 (CYP2E1) is induced by skatole and this induction is blocked by androstenone in isolated pig hepatocytes. *Chem. Biol. Interact.* **2002**, *140*, 81–92. [CrossRef]

35. Jensen, M.T.; Cox, R.P.; Jensen, B.B. 3-Methylindole (skatole) and indole production by mixed populations of pig fecal bacteria. *Appl. Environ. Microb.* **1995**, *61*, 3180–3184. [CrossRef]

36. Wesoly, R.; Weiler, U. Nutritional Influences on Skatole Formation and Skatole Metabolism in the Pig. *Animals* **2012**, *2*, 221–242. [CrossRef]

37. Heyrman, E.; Kowalski, E.; Millet, S.; Tuyttens, F.A.M.; Ampe, B.; Janssens, S.; Buys, N.; Wauters, J.; Vanhaecke, L.; Aluwé, M. Monitoring of behavior, sex hormones and boar taint compounds during the vaccination program for immunocastration in three sire lines. *Res. Vet. Sci.* **2019**, *124*, 293–302. [CrossRef]

38. Pieterse, E. Effects of Increased Slaughter Weight of Pigs on Pork Production. Ph.D. Thesis, University of Stellenbosch, Western Cape, South Africa, April 2006.

**Publisher's Note:** MDPI stays neutral with regard to jurisdictional claims in published maps and institutional affiliations.

 *foods*

*Article*

# Proximate Composition, Amino Acid Profile, and Oxidative Stability of Slow-Growing Indigenous Chickens Compared with Commercial Broiler Chickens

**Antonella Dalle Zotte** [1,*], **Elizabeth Gleeson** [1], **Daniel Franco** [2], **Marco Cullere** [1] and **José Manuel Lorenzo** [2]

[1]  Department of Animal Medicine, Production and Health, University of Padova, Agripolis 16, Viale dell'Università, 35020 Legnaro, PD, Italy; elizabethyvonne.gleeson@studenti.unipd.it (E.G.); marco.cullere@unipd.it (M.C.)

[2]  Centro Tecnológico de la Carne de Galicia, Rúa Galicia No 4, Parque Tecnológico de Galicia, San Cibrán das Viñas, 32900 Ourense, Spain; danielfranco@ceteca.net (D.F.); jmlorenzo@ceteca.net (J.M.L.)

*  Correspondence: antonella.dallezotte@unipd.it

Received: 30 March 2020; Accepted: 24 April 2020; Published: 1 May 2020

**Abstract:** The increased demand for chicken meat products has led to chickens with increased growth rates and heavier slaughter weights. This has had unintentional negative effects on the genetics of these animals, such as spontaneous, idiopathic muscle abnormalities. There has also been a shift in customer preference towards products from alternative farming systems such as organic and free-range. Indigenous purebred chickens, such as the Polverara, show potential in these systems as they are adapted to more extensive systems. The aim of the present study was to characterize the meat quality traits of the Polverara, by comparing the proximate composition and amino acid profile with that of a commercial Hybrid. In addition, the lipid and protein oxidation was analyzed after eight days of storage. A total of 120 leg meat samples, 60 Polverara and 60 Hybrid were analyzed. Polverara exhibited higher protein content, lower lipid content, and a better amino acid profile. These results indicate that the Polverara has better nutritional meat quality. However, Polverara also showed higher levels of lipid and protein oxidation. Therefore, further research is needed, especially in regards to the fatty acid profile and mineral content of the meat, which is known to affect oxidative stability.

**Keywords:** hybrid; meat quality; lipid oxidation; Polverara breed; poultry; protein oxidation

---

## 1. Introduction

In recent decades there has been an increase in chicken meat consumption. There are numerous reasons for this increase in consumer interest. It is perceived as a healthy source of animal protein, it has lower costs associated with it compared to other meat species, it is suitable for further processing, and there are no religious or cultural constraints associated with its consumption [1]. All these characteristics have contributed to the continuous increase in its consumption. Due to this increased demand, the poultry industry has had to adapt its production strategies. Therefore, the poultry industry has incorporated a set of selection criteria for broiler chickens to meet the demand of the ever-increasing world population [2]. Furthermore, to decrease variability in characteristics such as growth rates and slaughter weights, the chickens used in the poultry industry are obtained from a small number of selected genetic lines. As a consequence, the hybrids used in commercial poultry farming have faster growth rates and increased body weights at slaughter than indigenous purebreds that have not undergone the same genetic selection [3–5]. This strategy has mainly focused on production traits, and as a consequence, the intensive selection has had unintentional negative effects on some muscle growth and meat quality attributes of commercial hybrids. Effects such as spontaneous, idiopathic muscle abnormalities and increased susceptibility to stress-induced myopathy are found with increased frequency in fast growing broiler chickens which are typically used in the poultry industry [6,7]. To try and counteract these negative effects, the scientific community has begun to focus on identifying the causes of these abnormalities, and possible breeding strategies that can mitigate these negative effects [8,9]. The research done on indigenous purebred chicken breeds has focused mostly on their role in the conservation of poultry genetics and biodiversity. As indigenous purebred chicken breeds are not commercially used, they do not receive much attention in regards to conservation. The preservation of indigenous purebred populations is mostly limited to hobby or vanity poultry farmers, which has resulted in a progressive loss of biodiversity [10,11]. One possible conservation strategy has been to identify the niche poultry product markets. There is a growing awareness of the health and nutrition regarding animal products amongst consumers. Consequently, there has been an increase in consumer interest in products produced in alternative faming systems such are free range and organic systems [12]. Consumers are willing to pay higher prices for poultry products that are perceived as natural or environmentally friendly, and that are produced on farms with high animal welfare standards and animal nutrition [13]. Indigenous purebred chickens show promise in these types of poultry production systems due to their slower growth rate compared to commercial hybrids and their natural preference for more extensive farming systems. This can be attributed to the fact that they have not been selected for intensive farming and their nature is to be more physically active [7]. One of these indigenous purebred chicken breeds is the Polverara. The Polverara chicken is a medium sized, slow-growing indigenous chicken breed (average slaughter age: 180 days) that has its origin in the Veneto region of Italy [7,14]. Only a small number of studies have been conducted on this breed and its potential for use in the poultry industry and conservation. Recent research such as the studies done by [7,14,15], have been focused on characterizing the breed's meat quality traits, with the aim of promoting its use in conservation and poultry production. Amino acids are the key constituents of protein, and play important roles as regulators in metabolic pathways which are important for maintenance, growth, reproduction and immunity [16,17]. Protein quality and thus the amino acid profile of the protein contributes to the quality of the meat. Due to the importance of amino acids in optimal animal production and human nutrition, it is beneficial to the scientific advancement to analyze the amino acid profiles of the meat species that are being studied. Another factor that contributes to meat quality is the oxidation of the proteins and lipids. Oxidation causes quality deterioration during meat processing and storage [18]. The alterations due to oxidation can influence the physical and chemical properties of meat including water-holding capacity and meat tenderness. It can also decrease the bioavailability of amino acid residues and the digestibility of the protein. This in turn negatively affects the nutritional value of the meat [19]. The aim of the present study was to assist in characterizing the meat quality traits of the Polverara breed, by comparing the proximate composition and amino acid profile with that of a commercially used Hybrid. In addition, the lipid and protein oxidation was assessed.

## 2. Materials and Methods

### 2.1. Sampling Procedure and Experimental Groups

The experiment was conducted at the Department of Animal Medicine, Production and Health (MAPS), University of Padova (Italy). A total of $n = 60$ chickens were sampled: $n = 30$ broiler chickens (Hybrid) obtained from a commercial poultry farm, and $n = 30$ slow-growing, medium sized indigenous chickens (Polverara) which were obtained from the Agricultural Professional High School "Duca degli Abruzzi" (Padova). All chickens were males at their respective slaughter ages, 40 days for the Hybrid group and 180 days for the Polverara group. The samples for both groups were collected on the same day. The farming specifications were reported in the study by [14] and the nutritional composition of the finisher diets fed to the two chicken genotypes is given in Table 1. Hybrids were fed a conventional broiler diet whereas Polverara birds were fed an organic diet for growing chickens. All chickens were processed by an authorized commercial abattoir which consisted of electrical stunning (120 V, 200 Hz) and exsanguination. Thereafter, carcasses underwent soft-scalding (2 min at 53 °C) and evisceration. Subsequently, the carcasses were air-chilled (precooling at 5 °C for 6 min, followed by chilling at 0 °C for 90 min). The legs were then dissected from the carcasses, sorted and collected by a working team. Once sampled, the legs were packaged in food-grade plastic bags and transported in refrigerated conditions (4 ± 1 °C) to the MAPS Department where they were frozen (−18 °C) and transported frozen within 48 h to the Centro Tecnolóxico da Carne (Ourense, Spain) for meat quality evaluations, where the leg samples were thawed during 24 h at 4 ± 1 °C and skinned before analysis. The proximate composition, amino acid profile and day 0 oxidative stability analyses were performed on the right leg samples whereas day 8 oxidative stability analysis was performed on the counterpart samples (left legs).

**Table 1.** Nutrient composition (g/kg as fed) and energy content (MJ/kg as fed) of finisher diets fed to Hybrid and Polverara chickens.

| | Diets | |
| --- | --- | --- |
| **Nutrient Composition** | **Hybrid** | **Polverara** |
| Dry matter (DM) | 895 | 895 |
| Crude protein (CP) | 197 | 168 |
| Ether extract (EE) | 71.5 | 40.7 |
| Crude fiber (CF) | 36.8 | 39.4 |
| Nitrogen-free extract (NFE) [1] | 546 | 579 |
| Ash | 44.5 | 68.2 |
| Gross energy [2] | 17.6 | 16.3 |
| Calcium | 6.30 | 14.4 |
| Phosphor | 5.65 | 7.33 |
| L-Lysine | 11.9 | 8.63 |
| DL-Methionine | 4.10 | 4.17 |

[1] 100 − (water + crude protein + crude fat + crude fiber + ash). [2] (NFE × 4.11) + (CP × 5.64) + (EE × 9.44) + (CF × 4.78) × 10.

## *2.2. Proximate Composition*

The proximate composition of the right leg samples was evaluated according to International Organization for Standards (ISO), where protein [20], moisture [21] and ash [22] content were determined, while total fat was determined according to the Approved Procedure Am 5–04, established by the American Oil Chemists' Society [23].

## *2.3. Meat Amino Acid Profile*

The amino acid profile of the right leg samples was assessed according to the method described by [24]. In short, a sample (100 mg) in a glass ampoule and 6 N hydrochloric acid solution (5 mL) was mixed, sealed and stored at 110 °C for 24 h. After protein hydrolysis was completed, the hydrolysate was diluted with distilled water (200 mL) and filtered through a 0.45 µm filter (Filter Lab, Barcelona, Spain). Tryptophan content was not determined as it transforms into ammonium under acidic conditions. The derivatization of standards and samples was carried out according to Gálvez et al. [25]. The identification of amino acids was done through high performance liquid chromatography (Alliance 2695 model, Waters, Milford, MA, USA), using a scanning fluorescence detector (model 2475, Waters) according to Munekata et al. [26]. The quantification was done using the external standard technique with amino acid standard (Amino Acid Standard H, Thermo, Rockford, IL, USA). The results are expressed as g per 100 g protein

### 2.4. Storage Conditions

For oxidative stability analyses, chicken legs (left and right) were individually placed in 300 mm thick polyethylene-ethylene vinyl alcohol-polyethylene (PET-EVOH-PE) trays and were packaged directly by sealing with multilayer PE-EVOH-PE film (74 mm thick, permeability < 2 mL/m$^2$ bar/day (Viduca, Alicante, Spain) upon the tray (OVERWRAP) using a heat sealer (LARI3/Pn T-VG-R-SKIN, Ca.Ve.Co., Palazzolo, Italy). The trays were stored at 2 ± 1 °C under light to simulate supermarket conditions, being placed on metal shelving and receiving lux values in the range of 15–20, depending on the tray position (HT 306, Digital luxometer, Italy). The light source was conventional, so any wavelength or range, in this case UV, was not filtered. The samples in the chamber were rotated every 24 h to minimize light intensity differences and possible temperature variations on the surface of the meat. Sixty samples (thirty from each experimental group) were removed from the chamber at 0 (left legs) and 8 (right legs) days of storage for lipid and protein oxidation analysis.

### 2.5. Meat Lipid Oxidation

The lipid and protein oxidation was assessed after 0 and 8 days of storage at 4 °C. Lipid oxidation was assessed using the Thiobarbituric Acid Reactive Substances (TBARS) with the method proposed by [27]. In short, a chicken leg meat sample (2 g) was dispersed in 5% trichloroacetic acid (10 mL) and homogenized with an Ultra-Turrax (IkaT25 basic, Staufen, Germany) for 2 min. The homogenate was kept at −10 °C for 19 min and then centrifuged (2360 g for 10 min). The supernatant was then filtered through a Whatman No. 1 filter paper. The filtrate (5 mL) was mixed with a 0.02 M TBA solution (5 mL) and placed in a water bath (96 °C for 40 min). Thereafter the absorbance was measured at 532 nm. The TBARS value was calculated from a standard curve of malonaldehyde with 1,1,3,3-tetraethoxypropane (TEP) and expressed as mg malonaldehyde per kg of sample.

### 2.6. Meat Protein Oxidation

Protein oxidation was measured with the method outlined by [28] with modifications [29]. Measurements were taken for carbonyl and protein quantification to calculate protein oxidation. A sample (2.5 g) was homogenized with 0.6 M NaCl solution (20 mL) and treated with 10% trichloroacetic acid (1 mL) to obtain a homogenate (100 μL). Thereafter, it was centrifuged for 5 min at 5000× $g$. The supernatant was derivatized for carbonyl quantification with 2 M HCl (1 mL) with 0.2% 2,4-dinitrophenyl hydrazine (DNPH). For protein quantification 2 M HCl (1 mL) was added. A pellet was obtained and washed with 1:1 ethanol/ethyl acetate (1 mL) three times. It was then dissolved in 20 mM sodium phosphate buffer (1.5 mL) with 6 M guanidine hydrochloride. The carbonyls and protein concentrations were measured with a spectrophotometer at 370 nm and 280 nm, respectively. The protein concentrations were calculated according to a standard curve and bovine serum albumin was used to calculate a protein concentration standard. The results are expressed as nmol carbonyl per mg protein.

### 2.7. Statistical Analysis

All data were analyzed using SAS 9.1.3 statistical software package for Windows (SAS, 2008). Proximate composition and amino acid profile of chicken leg meat were analyzed by a one-way ANOVA testing the effect of the genotype (Hybrid, Polverara). Lipid and protein oxidation of chicken leg meat were analyzed by a two-way ANOVA testing the effects of the genotype and the day of storage (day 0, day 8) as fixed effects, and their interaction. Least square means were obtained using a Bonferroni test, and the significance was calculated at a 5% confidence level.

## 3. Results

### 3.1. Proximate Composition

Results regarding the proximate composition analysis showed that the Polverara leg meat had higher water ($p$ = 0.0202), protein ($p$ < 0.0001) and ash contents ($p$ < 0.0001), and lower lipid content ($p$ < 0.0001) when compared to the Hybrid leg meat (Table 2).

**Table 2.** Effect of chicken genotype (Hybrid vs. Polverara) on the proximate composition of leg meat.

| | Genotype | | *p*-Value | RSD [1] |
|---|---|---|---|---|
| | Hybrid | Polverara | | |
| N. | 30 | 30 | | |
| Water (%) | 72.6 | 73.7 | 0.0202 | 1.67 |
| Protein (%) | 18.5 | 21.5 | <0.0001 | 0.93 |
| Lipids (%) | 7.28 | 2.25 | <0.0001 | 1.69 |
| Ash (%) | 1.19 | 1.31 | <0.0001 | 0.06 |

[1] Residual standard deviation.

### 3.2. Amino Acid Profile

In regards to the amino acid profile analysis (Table 3), the Polverara leg meat exhibited the highest content for all of the amino acids and significantly higher values were exhibited for some of the essential and non-essential amino acids, which included isoleucine, leucine, phenylalanine, threonine and valine (essential) and glycine, proline and tyrosine (non-essential).

**Table 3.** Effect of chicken genotype (Hybrid vs. Polverara) on the amino acid profile of leg meat.

| | Genotype | | *p*-Value | RSD [1] |
|---|---|---|---|---|
| | Hybrid | Polverara | | |
| N. | 30 | 30 | | |
| *Essential amino acids* | | | | |
| *(g/100 g meat)* | | | | |
| Arginine | 1.45 | 1.55 | 0.0614 | 0.21 |
| Histidine | 0.61 | 0.63 | 0.2686 | 0.07 |
| Isoleucine | 0.94 | 1.00 | 0.0273 | 0.11 |
| Leucine | 1.52 | 1.64 | 0.0134 | 0.19 |
| Lysine | 1.79 | 1.90 | 0.0910 | 0.24 |
| Methionine | 0.35 | 0.39 | 0.0719 | 0.09 |
| Phenylalanine | 0.76 | 0.84 | 0.0023 | 0.09 |
| Threonine | 0.83 | 0.89 | 0.0122 | 0.09 |
| Valine | 0.92 | 0.98 | 0.0313 | 0.10 |
| *Non-essential amino acids* | | | | |
| *(g/100 g meat)* | | | | |
| Alanine | 1.09 | 1.15 | 0.0533 | 0.12 |
| Aspartic acid | 1.75 | 1.85 | 0.1122 | 0.23 |
| Cysteine | 0.21 | 0.21 | 0.7863 | 0.04 |
| Glutamic acid | 2.91 | 3.08 | 0.0991 | 0.37 |
| Glycine | 0.87 | 0.95 | 0.0311 | 0.15 |
| Proline | 0.75 | 0.84 | 0.0007 | 0.10 |
| Serine | 0.94 | 0.95 | 0.8677 | 0.28 |
| Tyrosine | 0.65 | 0.71 | 0.0049 | 0.08 |

[1] Residual standard deviation.

### 3.3. Lipid and Protein Oxidation

Table 4 represents the results of the oxidative status of chicken leg meat evaluated over an 8-day period of refrigerated storage. For lipid oxidation, meat from the Polverara breed exhibited higher levels of oxidation at both day 0 ($p < 0.0001$) and day 8 ($p < 0.0001$) of refrigerated storage compared to meat from the Hybrid chicken. Storage time was also shown to have an effect on TBARS values as its level measured at Day 8 of storage was significantly higher ($p < 0.0001$) than that recorded at day 0. Lipid oxidation exhibited a significant interaction between Genotype and storage time ($p = 0.0330$), as Polverara showed a significant increase in TBARS value from Day 0 to Day 8 and Hybrid did not. Genotype effect was also observed for protein oxidation, where Polverara exhibited higher values at both day 0 ($p < 0.0001$) and day 8 ($p < 0.0001$) of refrigerated storage. In contrast from what was observed for lipid oxidation, protein oxidation was not affected by storage time, observing similar values at day 0 and day 8 of storage.

**Table 4.** Effect of genotype (Hybrid vs. Polverara) and day of storage (0 vs. 8) and their interaction on the TBARS values and protein oxidation of leg meat over 8 days period.

| Storage Time (T) | Day 0 | | Day 8 | | *p*-Values | | | SE [1] |
|---|---|---|---|---|---|---|---|---|
| Genotype (G) | Hybrid | Polverara | Hybrid | Polverara | (G) | (T) | (G) × (T) | |
| N. | 30 | 30 | 30 | 30 | | | | |
| TBARS values (mg MDA/kg meat) | 0.08 [D] | 0.21 [B,C] | 0.14 [C,D] | 0.40 [A] | <0.0001 | <0.0001 | 0.0330 | 0.03 |
| Protein oxidation (nmol/mg protein) | 1.91 [B] | 3.05 [A] | 2.04 [B] | 3.18 [A] | <0.0001 | 0.3812 | 0.9950 | 0.14 |

[1] Standard error, [A,B,C,D] means in the same row with different superscripts significantly differ ($p < 0.0001$).

## 4. Discussion

Lower lipid and higher protein content was found for Polverara compared to Hybrid. This is possibly attributable to the higher level of locomotory activity, which is characteristic of this breed [7]. Locomotory activity is known to favor myogenesis over lipogenesis [12]. The significantly higher ash content found for Polverara leg meat resides in differences in mineral composition. In particular, it was recently observed that the meat from Polverara legs are unexpectedly rich in heme iron, approximately four times higher than that of hybrid chickens [30]. The Polverara leg meat proximate composition reported in the present study differs to the results reported by [15]. This is attributed to the fact that the latter study included male and female chickens, characterized by marked sexual dimorphism at slaughter age, thus resulting in differences in the average values. Moreover, the Polverara breed has not been subjected to genetic selection for productive performance and meat quality traits and thus more variability is expected when compared to commercially used hybrids. There are variations between the proximate compositions reported for different indigenous chicken breeds [31–34]. These variations are however expected as the indigenous breeds originated in different geographic locations and have different genetic potential, diets and feeding behaviors. These factors can possibly affect protein and lipid deposition in the meat and can increase the variation between results from studies on different indigenous chicken breeds. This also makes it difficult to compare the results from the present study to studies done on other indigenous breeds.

To the authors knowledge, the study done by [7] is the only previous study which has analyzed the amino acid profile of Polverara breast meat and the present study is the first to analyze the amino acid profile of Polverara leg meat. Both studies confirm the higher content for all the amino acids compared to the Hybrid chicken meat. Factors such as age, affect protein digestibility and deposition, and diet is known to have an effect on the amino acid profile of meat [35]. There are variations among the amino acid analysis results from previous studies involving different indigenous chicken breeds, with some of the studies reporting no differences in amino acid profiles when comparing indigenous chicken breeds to hybrids [32,36]. However, when considering the quantitative contribution of the single amino acid intake per 100 g of meat, the higher content of amino acids in Polverara chicken meat depends on its leanness, irrespective of the age of the animal.

Humans have nine essential amino acids that need to be obtained through their diets [37]. The results indicate that Polverara was the better protein source, as it contained higher amounts of all the amino acids essential in human nutrition, and overall has superior nutritional meat quality compared to the Hybrid. In the case of a human with a body weight of 60 kg, 100 g of Polverara leg meat (vs. 100 g of Hybrid leg meat) contains more of the daily requirements for essential amino acids, with 106% (vs. 102%) for histidine, 84% (vs. 78%) for isoleucine, 70% (vs. 65%) for leucine, 105% (vs. 100%) for lysine, 67% (vs. 62%) for methionine + cysteine, 104% (vs. 94%) for phenylalanine + tyrosine, 99% (vs. 92%) for threonine, and 63% (vs. 59%) for valine.

Oxidation causes quality deterioration in meat and can lead to decreased shelf life of meat products [38]. It also adversely affects the flavor of the meat due to oxidative rancidity which causes "off-flavor" [18]. It is understood that the lipid content and the fatty acid profile of the lipids in meat can affect the oxidative susceptibility of meat [39]. Higher meat lipid content increases oxidative susceptibility. The results from the present study showed that the Polverara leg meat had a lower lipid content compared to the Hybrid. This would suggest that the Polverara leg meat would possibly be less susceptible to oxidation due to the lower lipid content compared to the Hybrid [40]. The findings in the present study depict a different scenario with the Polverara meat exhibiting more oxidation than the Hybrid meat. The observed interaction between Genotype and storage time for TBARS values indicates that the rate of oxidation in Polverara meat is higher than in Hybrid meat. The fatty acid profile, specifically the proportion of polyunsaturated (PUFA) and monounsaturated fatty acids (MUFA), affects the extent of lipid oxidation [18]. Polyunsaturated fatty acids are more susceptible to lipid oxidation than monounsaturated fatty acids, and therefore the extent of lipid oxidation increases with increased proportion of PUFA. Research has shown that this proportion differs between commercial broilers and indigenous chicken breeds [41]. Moreover, these results could also depend on the higher heme iron content found in Polverara meat in previous studies [30]. Heme iron is known to be a catalyst for lipid oxidation [42]. There was significantly more lipid oxidation detected at Day 8 compared to Day 0 for Polverara. This is expected as lipid oxidation increases over time [43].

Protein oxidation may cause discoloration of fresh meat and influences quality during storage and processing [19]. The extent and rate of protein oxidation is, among other things, influenced by the amino acid profile, lipid content and quality of the meat. Among the amino acids, cysteine and methionine have the highest oxidative susceptibility [19]. Other amino acids such as tyrosine, phenylalanine, tryptophan, histidine, proline, arginine and lysine are also seen as particularly susceptible to oxidation. The results indicate that the Polverara meat had significantly higher levels of protein oxidation at both Day 0 and Day 8. These results may be explained, in part, by the amino acid profile, in which Polverara exhibited higher concentrations of all the previously mentioned amino acids. The significantly higher level of lipid oxidation found in Polverara may also have contributed to the significantly higher level of protein oxidation.

## 5. Conclusions

Polverara chicken outperformed the commercially used Hybrid in both the proximate composition, where it exhibited higher protein content and lower lipid content, and the amino acid profile, where it exhibited higher content for all of the single amino acids that were analyzed. It did, however, exhibit more lipid and protein oxidation, which could negatively affect the oxidative stability and processing of the meat products. This highlights the need for further research into the meat quality characteristics of the breed, especially in regards to the fatty acid profile and mineral content of the Polverara leg meat. Based on the results obtained until now, the Polverara chicken shows potential as a possible breed to be used in alternative farming systems and in conservation efforts.

**Author Contributions:** Conceptualization, A.D.Z.; methodology, A.D.Z. and J.M.L.; formal analysis, D.F.; investigation, A.D.Z..; resources, J.M.L.; data curation, M.C.; writing—original draft preparation, E.G.; writing—review and editing, A.D.Z.; supervision, A.D.Z.; project administration, A.D.Z.; funding acquisition, A.D.Z. and J.M.L. All authors have read and agreed to the published version of the manuscript.

**Funding:** This research was funded by University of Padova funds, grant number BIRD164094, by GAIN (Axencia Galega de Innovación), grant number IN607A2019/01 and by the Healthy Meat network, funded by CYTED (ref. 119RT0568). The research was supported by the Erasmus+/KA107 mobility funds. The APC was funded by University of Padova funds, grant number DALL_FINA_P14_02.

**Acknowledgments:** The authors are grateful to Gabriele Baldan for his technical support.

**Conflicts of Interest:** The authors declare no conflicts of interest.

## References


1. Petracci, M.; Bianchi, M.; Mudalal, S.; Cavani, C. Functional ingredients for poultry meat products. *Trends Food Sci. Technol.* **2013**, *33*, 27–39. [CrossRef]
2. Petracci, M.; Mudalal, S.; Soglia, F.; Cavani, C. Meat quality in fast-growing broiler chickens. *World Poult. Sci. J.* **2015**, *71*, 363–374. [CrossRef]
3. Rizzi, C.; Contiero, B.; Cassandro, M. Growth patterns of Italian local chicken populations. *Poult. Sci.* **2013**, *92*, 2226–2235. [CrossRef]
4. Franco, D.; Rois, D.; Vázquez, J.A. Comparison of growth performance, carcass components, and meat quality between Mos rooster (Galician indigenous breed) and Sasso T-44 line slaughtered at 10 months. *Poult. Sci.* **2012**, *91*, 1227–1239. [CrossRef]
5. Franco, D.; Rois, D.; Vázquez, J.A.; Purriños, L.; González, R.; Lorenzo, J.M. Breed effect between Mos rooster (Galician indigenous breed) and Sassa T-44 line and finishing feed effect of commercial fodder or corn. *Poult. Sci.* **2012**, *91*, 487–498. [CrossRef] [PubMed]
6. Julian, R.J. Rapid growth problems: Ascites and skeletal deformities in broilers. *Poult. Sci.* **1998**, *77*, 1773–1780. [CrossRef]
7. Dalle Zotte, A.; Ricci, R.; Cullere, M.; Serva, L.; Tenti, S.; Marchesini, G. Research note: Effect of chicken genotype and white striping-wooden breast condition on breast meat proximate composition and amino acid profile. *Poult. Sci.* **2019**, *99*, 1797–1803. [CrossRef]
8. Hocking, P.M. Unexpected consequences of genetic selection in broilers and turkeys: Problems and solutions. *Br. Poult. Sci.* **2014**, *55*, 1–12. [CrossRef]
9. Dawkins, M.S.; Layton, R. Breeding for better welfare: Genetic goals for broiler chickens and their parents. *Anim. Welfare* **2012**, *21*, 147–155. [CrossRef]
10. Cassandro, M.; De Marchi, M.; Penasa, M.; Rizzi, C. Carcass characteristics and meat quality traits of Padovana chicken breed, a commercial line, and their cross. *Ital. J. Anim. Sci.* **2015**, *14*, 304–309. [CrossRef]
11. Hoffmann, I. The global plan of action for animal genetic resources and the conservation of poultry genetic resources. *World Poult. Sci. J.* **2009**, *65*, 286–297. [CrossRef]
12. Fanatico, A.C.; Pillai, P.B.; Emmert, J.L.; Owens, C.M. Meat quality of slow- and fast-growing chicken genotypes fed low-nutrient or standard diets and raised indoors or with outdoor access. *Poult. Sci.* **2007**, *86*, 2245–2255. [CrossRef] [PubMed]
13. Lusk, J.L. Consumer preferences for and beliefs about slow growth chicken. *Poult. Sci.* **2018**, *97*, 4159–4166. [CrossRef]
14. Tasoniero, G.; Cullere, M.; Baldan, G.; Dalle Zotte, A. Productive performances and carcass quality of male and female Italian Padovana and Polverara slow-growing chicken breeds. *Ital. J. Anim. Sci.* **2017**, *17*, 530–539. [CrossRef]
15. Dalle Zotte, A.; Tasoniero, G.; Baldan, G.; Cullere, M. Meat quality of male and female Italian Padovana and Polverara slow-growing chicken breeds. *Ital. J. Anim. Sci.* **2019**, *18*, 398–404. [CrossRef]
16. Wu, G. Amino acids: Metabolism, functions, and nutrition. *Amino Acids* **2009**, *37*, 1–17. [CrossRef]
17. Brosnan, J.T. Amino Acids, Then and now—A reflection on Sir Hans Krebs' contribution to nitrogen metabolism. *Life* **2001**, *52*, 265–270. [CrossRef]

18. Domínguez, R.; Pateiro, M.; Gagaoua, M.; Barba, F.J.; Zhang, W.; Lorenzo, J.M. A comprehensive review on lipid oxidation in meat and meat products. *Antioxidants* **2019**, *8*, 429. [CrossRef]

19. Zhang, W.; Xiao, S.; Ahn, D.U. Protein oxidation: Basic principles and implications for meat quality. *Crit. Rev. Food Sci. Nutr.* **2013**, *53*, 1191–1201. [CrossRef]

20. ISO. ISO 937: Meat and Meat Products—Determination of Nitrogen Content. Available online: https://www.sis.se/api/document/preview/600939/ (accessed on 28 March 2020).

21. ISO. ISO 1442: Meat and Meat Products—Determination of Moisture Content. Available online: https://www.sis.se/api/document/preview/601282/ (accessed on 28 March 2020).

22. ISO. ISO 936 Meat and Meat Products—Determination of Ash Content. Available online: https://www.sis.se/api/document/preview/615693/ (accessed on 28 March 2020).

23. American Oil Chemists' Society. *Official Procedure Am5-04. Rapid Determination of Oil/Fat Utilizing High Temperature Solvent Extraction*; AOCS: Urbana, IL, USA, 2005.

24. Domínguez, R.; Borrajo, P.; Lorenzo, J.M. The effect of cooking methods on nutritional value of foal meat. *J. Food Compos. Anal.* **2015**, *43*, 61–67. [CrossRef]

25. Gálvez, F.; Maggiolino, A.; Domínguez, R.; Pateiro, M.; Gil, S.; De Palo, P.; Carballo, J.; Franco, D.; Lorenzo, J.M. Nutritional and meat quality characteristics of seven primal cuts from 9-month-old female veal calves: A premilinary study. *J. Sci. Food Agric.* **2019**, *99*, 2947–2956. [CrossRef]

26. Munekata, P.E.S.; Pateiro, M.; Domínguez, R.; Zhou, J.; Barba, F.J.; Lorenzo, J.M. Nutritional characterization of sea bass processing by-products. *Biomolecules* **2020**, *10*, 232. [CrossRef]

27. Vyncke, W. Evaluation of the direct thiobarbituric acid extraction method for determining oxidative rancidity in mackerel (*Scomber scombrus L.*). *Eur. J. Lipid Sci. Technol.* **1975**, *77*, 239–240. [CrossRef]

28. Oliver, C.N.; Ahn, B.W.; Moerman, E.J.; Goldstein, S.; Stadtman, E.R. Age-related changes in oxidized proteins. *J. Biol. Chem.* **1987**, *262*, 5488–5491. [PubMed]

29. Vourela, S.; Salminen, H.; Mäkelä, M.; Kivikari, R.; Karonene, M.; Heinonen, M. Effect of plant phenolics on protein and lipid oxidation in cooked pork meat patties. *J. Agric. Food Chem* **2005**, *53*, 8492–8497. [CrossRef] [PubMed]

30. Pellattiero, E.; Tasoniero, G.; Cullere, M.; Baldan, G.; Marangon, A.; Gleeson, E.; Contiero, B.; Dalle Zotte, A. Are meat quality traits and sensory attributes in favour of slow-growing chickens? *Foods.* Submitted.

31. Wattanachant, S. Factors affecting the quality characteristics of Thai indigenous chicken meat. *Suranaree J. Sci. Technol.* **2008**, *15*, 317–322.

32. Choe, J.H.; Nam, K.; Jung, S.; Kim, B.; Yun, H.; Jo, C. Differences in the quality characteristics between commercial Korean native chickens and broilers. *Korean J. Food Sci. Ani. Resour.* **2010**, *30*, 13–19. [CrossRef]

33. Van Marle-Köster, E.; Web, E.C. Carcass characteristics of South African native chicken lines. *S. Afr. J. Anim. Sci.* **2000**, *30*, 53–56. [CrossRef]

34. Franco, D.; Rois, D.; Vázquez, J.A.; Lorenzo, J.M. Carcass morphology and meat quality from roosters slaughtered at eight months affected by genotype and finishing feeding. *Span. J. Agric. Res.* **2013**, *11*, 382–393. [CrossRef]

35. Wu, G.; Bazer, F.W.; Dai, Z.; Li, D.; Wang, J.; Wu, Z. Amino acid nutrition in animals: Protein synthesis and beyond. *Annu. Rev. Anim. Biosci.* **2014**, *2*, 387–417. [CrossRef] [PubMed]

36. Wattanachant, S.; Benjakul, S.; Ledward, D.A. Composition, color, and texture of Thai indigenous and broiler chicken muscles. *Poult. Sci.* **2004**, *83*, 123–128. [CrossRef] [PubMed]

37. Food and Agriculture Organization of the United Nations; World Health Organization; United Nations University. *Protein and Amino Acid Requirements in Human Nutrition: Report of a Joint FAO/WHO/UNU Expert Consultation*, 1st ed.; WHO Press: Geneva, Switzerland, 2007, pp. 135–151. Available online: https://apps.who.int/iris/handle/10665/43411 (accessed on 20 March 2020).

38. Lorenzo, J.M.; Gómez, M. Shelf life of foal meat under MAP, overwrap and vacuum packaging conditions. *Meat Sci.* **2012**, *92*, 610–618. [CrossRef] [PubMed]

39. Maraschiello, C.; Sárraga, C.; García Regueiro, J.A. Glutathione peroxidase activity, TBARS, and α-tocopherol in meat from chickens fed different diets. *J. Agric. Food Chem.* **1999**, *47*, 867–872. [CrossRef]

40. Mottram, D.S. Flavour formation in meat and meat products: A review. *Food Chem.* **1998**, *62*, 415–424. [CrossRef]

41. Tang, H.; Gong, Y.Z.; Wu, C.X.; Jiang, J.; Wang, Y.; Li, K. Variation of meat quality traits among five genotypes of chicken. *Poult. Sci.* **2009**, *88*, 2212–2218. [CrossRef]
42. Lorenzo, J.M.; Pateiro, M. Influence of fat content on physico-chemical oxidative stability of foal liver pâté. *Meat Sci.* **2013**, *95*, 330–335. [CrossRef]
43. Dal Bosco, A.; Mattioli, S.; Cullere, M.; Szendrő, Z.; Gerencsér, Z.; Matics, Z.; Castellini, C.; Szin, M.; Dalle Zotte, A. Effect of diet and packaging system on the oxidative status and polyunsaturated fatty acid content of rabbit meat during retail display. *Meat Sci.* **2018**, *143*, 46–51. [CrossRef]

 *foods*

*Article*

# The Effect of Algae or Insect Supplementation as Alternative Protein Sources on the Volatile Profile of Chicken Meat

Vasiliki Gkarane [1,*] , Marco Ciulu [1] , Brianne A. Altmann [1] , Armin O. Schmitt [2,3] and Daniel Mörlein [1]

[1]  Department of Animal Sciences, University of Göttingen, Kellnerweg 6, 37077 Göttingen, Germany; marco.ciulu@uni-goettingen.de (M.C.); brianne.altmann@agr.uni-goettingen.de (B.A.A.); daniel.moerlein@uni-goettingen.de (D.M.)
[2]  Breeding Informatics Group, Department of Animal Sciences, Georg-August University, Margarethe von Wrangell-Weg 7, 37075 Göttingen, Germany; armin.schmitt@uni-goettingen.de
[3]  Center for Integrated Breeding Research (CiBreed), University of Göttingen, Albrecht-Thaer-Weg 3, 37075 Göttingen, Germany
*  Correspondence: vasiliki.gkarane@uni-goettingen.de or v.gkarane@qub.ac.uk

Received: 17 July 2020; Accepted: 1 September 2020; Published: 4 September 2020

**Abstract:** The aim of this study was to investigate the differences in the volatile profile of meat from chickens fed with alternative protein diets (such as algae or insect) through two different trials. In Trial 1, broiler chicken at one day of age were randomly allocated to three experimental groups: a basal control diet (C) and two groups in which the soybean meal was replaced at 75% (in the starter phase) and 50% (in the grower phase) with partially defatted *Hermetia illucens* (HI) larvae or *Arthrospira platensis* (SP). In Trial 2, broiler chickens were housed and reared similar to Trial 1, with the exception that the experimental diets replaced soybean meal with either 100% partially defatted HI or 100% SP. In both trials, chickens were slaughtered at day 35. Per group, 10 chickens were submitted to volatile analysis by using solid-phase microextraction (HS-SPME) and gas chromatography–mass spectrometry (GC-MS) analysis. Results in both trials showed that levels of several lipid-derived compounds were found to be lower in chickens fed an HI diet, which could be linked to a possibly lower level of polyunsaturated fatty acid content in HI-fed chicken. In addition, the dietary treatments could be discriminated based on the volatile profile, i.e., the substitution of soy with HI or SP distinctively affected the levels of flavor compounds.

**Keywords:** volatile compounds; sustainable feeds; spirulina; *Arthrospira platensis*; insect; *Hermetia illucens*

---

## 1. Introduction

Intake of high quality protein, in the range of 1.2–1.6 g per kg body weight per day, has been associated with health benefits, with poultry identified as one of the main sources [1]. Chicken meat consumption is popular for a number of reasons, not least of which are its nutritional value (i.e., the lower content of saturated fat and the higher proportion of polyunsaturated fatty acids (PUFA) compared to other types of meat), its affordable price, and its lack of religious and cultural constraints associated with its consumption [2].

Chicken meat composition and quality are highly associated with animal diet [3], which once supplemented with functional ingredients (e.g., antioxidants, n-3 PUFA) may enhance the nutritional value of meat [4]. To date, corn, soybean, and fishmeal are the main conventional components of poultry diets [5]. However, despite the high protein content and the well balanced amino acid profile that these ingredients provide [6,7], they are often linked with a negative environmental footprint (i.e., land degradation, water deprivation, greenhouse gas emissions, or marine overexploitation) [8]. In light of the forthcoming increase of the global population (expected to reach nine billion by 2050) [9], the rising consumer demand for high quality protein, along with the on-going environmental deterioration and climate change, there is an urgent need to switch to more sustainable feeds in animal production systems [10,11].

The use of microalga and insects have gained increasing attention in the last 30 years, holding the potential to act as alternative economical and more sustainable protein source for the livestock sector [12,13]. One of the most widely studied insect species is the black soldier fly (*Hermetia illucens*; HI), the larvae of which contains high levels of protein (approximately 45% dry matter (DM)) [14,15] and has an optimal amino acid composition [16]. In addition, studies in poultry meat show that growth performance, carcass traits, feeding efficiency [17,18], and several meat quality traits (pH, meat color, drip loss, and proximate or sensory analysis parameters) [18–20] are not negatively affected when soybean meal is partially or totally replaced by HI larvae.

*Arthrospira platensis* (commonly known as spirulina (SP)) is an edible microalga species with high nutritional value characterized by a 55–70% protein content [21]. Moreover, it has γ-linolenic acid, n-3 long chain, essential amino acids, vitamins, minerals, antioxidants, and carotenoids [22]. Khan et al. [23] summarizes the information available on the rich biological activity of spirulina including its antimicrobial, antioxidant, antiviral, anti-inflammatory, and immune-modulating role. Inclusion of microalgae in the poultry diet can affect the performance, oxidative stability, and meat quality parameters of broiler chickens [24–26].

Animal feeding strategies can modify the fatty acid profile of meat, as reflected in its volatile profile [27]. These interventions can influence the nutritional value, the eating quality, and consumer acceptance of meat [28]. Volatile compounds, formed mainly through after meat is cooked [29], have been widely used in meat provenance investigation studies as a way to trace, identify, or discriminate animal feeding systems [27,30]. Interestingly, differences in the volatile profile of chicken meat deriving from conventional vs. the aforementioned alternative protein sources have not yet been reported. Thus, the aim of this study was to investigate the volatile profile of chicken breast meat produced with a traditional soybean meal based diet (control, C diet) or diets with a partial substitution (75–50% in Trial 1) and complete substitution (100% in Trial 2) of soybean meal with either defatted HI or SP.

## 2. Materials and Methods

### 2.1. Birds and Diet

The studies were carried out at the Department of Animal Sciences, University of Göttingen, and approved by the Ethics Committee of the Lower Saxony Office for Consumer Protection and Food Safety (LAVES; #33.9-42502-04-15/2027), Germany. The trials were part of a larger animal nutrition study, which investigated the effects of both alternative protein feeds on performance and apparent digestibility [31]. The trials were analyzed separately, as they were not exact replicates (i.e., substitution levels, spirulina source, sampling time, and storage temperature differed among the two trials). Ross-308 male broiler chickens were slaughtered at 35 days of age at the University of Göttingen poultry slaughterhouse, which is regulated by article 4 of the European Union's (EG) NR. 853/2004 37. Immediately following slaughter, the carcasses were weighed and butchered, where the breasts (m. pectoralis major) were skinned and cooled to 4 °C (5 h), and then later frozen until further analysis. Average carcass weighed 1.77 ± 0.30 kg and 1.75 ± 0.24 kg for Trial 1 and Trial 2, respectively.

### 2.2. Trial 1 (75% to 50% Replacement)

The experiment was divided into a starter feeding phase (1–21 days) and a grower feeding phase (22–34 days). One-day-old chicks were randomly allocated to floor pens with 5.8 birds m$^{-2}$ (i.e., 7 birds per pen). Average body weights per pen were similar at the start of the study. Feed and water were available ad libitum. The control diet (C) was based on wheat, corn, and soybean meal as the main ingredients. The experimental diets replaced soybean meal with either 75% partially defatted HI or SP in the starter phase and either 50% HI or SP in the grower phase. HI meal was sourced from a commercial producer (Hermetia Baruth GmbH, Baruth/Mark, Germany). The larvae were fattened on a rye and wheat bran substrate, dried at 65–70 °C, and partially-defatted with a screw press, then ground into a meal until an ultimate crude protein content 60.8% of DM and crude lipid content 14.1% of DM. Spirulina were sourced commercially from Myanmar, harvested, and sun-dried prior to packaging (crude protein content 58.8% of DM and crude lipid content 4.3% of DM). Amino acids were supplemented according to breeder guidelines. A total of 30 chicken breast samples were analyzed (10 C, 10 HI, and 10 SP). Samples originated from the chicken breast (m. pectoralis major) and were vacuum-packed in polyamide/polyethylene (PA/PE) bags and stored at −72 °C for 31 months until analysis. Full details of the animal and production characteristics and composition of feeds were described by Neumann et al. [31] (experiment 2).

### 2.3. Trial 2 (100% Replacement)

This experiment was also divided into a starter (1–21 days) and a grower feeding phase (22–34 days), and chicks were housed and reared as in Trial 1. One major modification is that the experimental diets replaced soybean meal with either 100% partially defatted HI or 100% SP in both feeding phases. HI meal was produced by a commercial producer (Hermetia Baruth GmbH, Baruth/Mark, Germany) with an ultimate crude protein content 60.8% of DM and crude lipid content 14.1% of DM. Spirulina were sourced commercially from China; the product was harvested, rinsed, and spray-dried prior to packaging (crude protein content 68.9% of DM and crude lipid content 6.3% of DM). A total of 30 chicken breast samples (10 C, 10 HI and 10 SP) were analyzed. Samples (m. pectoralis major) were vacuum-packed in PA/PE bags and stored at −20 °C for 26 months until analysis. Full details of the animal and production characteristics and composition of feeds were described by Neumann et al. [31] (experiment 3).

### 2.4. Sample Preparation and Volatile Compound Analysis

Volatile compounds were analyzed using headspace solid phase microextraction (HS-SPME). Before use, chicken breasts were thawed at room temperature by immersion of the frozen vacuum-packed samples in water, for 10 min. White tendons were removed and samples were minced by means of a mini chopper (DS Produkte GmbH, Gallin, Germany). Next, 4.5 ± 0.05 g of meat were weighed and transferred into 20 mL glass headspace vials together with 5 µL of a methanolic solution of 5 ppm 1,2-dichlorobenzene (Sigma-Aldrich, Munich, Germany), as internal standard. Vials were sealed with polytetrafluoroethylene (PTFE)-faced silicone septum (Macherey-Nagel, Düren, Germany). Meat samples were heated at 70 °C for 10 min in a TriPlus RSH$^{TM}$ autosampler (Thermo Fisher Scientific, Waltham, MA, USA) for equilibration, before exposing the 30/50 µm Divinylbenzene/Carboxen/Polydimethylsiloxane (DVB/CAR/PDMS) fiber (Supelco, Bellefonte, PA, USA) into the headspace, where it was held for 20 min under constant stirring. Before its first use, the solid phase microextraction (SPME) fiber was thermally pre-conditioned at 270 °C for 1 h in accordance with the manufacturer recommendation.

### 2.5. Gas Chromatography-Mass Spectrometry (GC-MS) Analyses

The GC–MS analyses of the samples were performed using a TRACE$^{TM}$ 1310 gas-chromatograph coupled with an ISQ-LT single quadrupole mass spectrometer (Thermo Fisher Scientific, Waltham, MA, USA). The SPME fiber was thermally desorbed in a programmed temperature vaporizing injector at 250 °C in a splitless mode for 2 min while a split ratio of 1:20 was adopted for the remaining time of the chromatographic run. Inlet temperature was set at 250 °C while a desorption time of 7 min was adopted. Separation of compounds was achieved by a TG-5SILMS column (30 m × 0.25 mm × 0.25 µm) provided by Thermo Scientific. Helium was used as carrier gas operating at 1 mL min$^{-1}$. Oven temperature was kept at 40 °C during the first 5 min, increased to 250 °C at 4 °C min$^{-1}$, and was maintained at 250 °C during the last 2 min. After desorption, a fiber bake out was carried out in a bake out unit for 20 min at 260 °C to avoid carry-over phenomena among subsequent samples. The total chromatographic run time was 59.5 min. The MS detector operated in ion scan mode (45–230 amu), ion source, and transfer line temperature were maintained at 250 °C and 270 °C, respectively. The electron energy was 70 eV. Identification of the compounds was performed comparing (a) experimental mass spectra to those contained in the National Institute of Standards and Technology database (NIST/EPA/NIH Mass Spectral Library, Version 2.2, 2014) and (b) linear retention indexes (LRIs) based on a homologous series of n-alkanes (C7-C30, Sigma-Aldrich, Munich, Germany) with those reported in other studies. Data were expressed in ng per g meat using the formula provided by Wang et al. [32].

*2.6. Data Analysis*

GC–MS data were aligned per trial based on retention time in the R Core Team [33] using the package 'CGalignR' [34]. For a better match of slightly shifted peaks, a full join function provided by the package 'fuzzyjoin' [35] was applied, to allow for peak alignment until a maximum difference of 0.03 was achieved in both directions.

Volatile data did not meet the assumption of normal distribution. For this reason, the non-parametric Kruskal–Wallis test was applied in order to test the effect of dietary treatment on the volatile profile of chicken meat in each trial. In the case of a significant result of the Kruskal–Wallis test, the Wilcoxon rank-sum-test was applied to compare the medians of two groups. Differences were considered to be significant when $p < 0.05$. The results are presented as medians. Median absolute deviation was reported as measure of dispersion, defined as the median [absolute($x_i$ − median($x$))], where $x_i$ is an individual observation and median($x$) is the median of values x. Linear discriminant analysis (LDA) was applied to investigate the potential separation of the dietary treatments within each trial [36] and to detect variables (compounds) that were more discriminant among feeding treatments. In order to detect the discriminatory efficiency of LDA, a confusion matrix followed by Cohen's Kappa coefficient value was computed [37]. An individual animal was considered as the experimental unit.

## 3. Results

In total, 61 compounds in Trial 1 and 65 compounds in Trial 2, appearing in at least 60% of the samples (i.e., in 18 or more out of the 30 samples in each trial) were tentatively identified (Supplementary Table S1). The compounds that showed significant differences in the medians in Kruskal–Wallis test and the compounds constituting the minimal discriminatory set are marked in Table 1 (for Trial 1) and Table 2 (for Trial 2). Four compounds in each trial were not identified.

**Table 1.** The composition of the volatile profile of chicken breast meat fed with three different diets in Trial 1 ((Control (C) vs. Spirulina (SP) vs. *Hermetia illucens* (HI)) (results expressed as ng/g).

| | C [1] | SP [1] | HI [1] | Statistical Analysis | *p*-Level |
|---|---|---|---|---|---|
| Alcohols | | | | | |
| 2-methoxy-ethanol | 9.36 (13.17) | 9.50 (5.57) | 8.45 (3.60) | | |
| 1-Penten-3-ol | 4.48 (9.87) | 6.15 (5.33) | 7.86 (4.12) | | |
| 1-Pentanol | 2.13 [a] (3.17) | 2.54 [a] (1.95) | 0. 11 [b] (0.11) | KW; LDA | 0.026 |
| 1-Hexanol | 1.26 [ab] (1.57) | 1.40 [a] (0.59) | 0.71 [b] (0.23) | KW; LDA | 0.030 |
| 1-Heptanol | 0.87 [a] (0.92) | 0.73 [a] (0.31) | 0.32 [b] (0.08) | KW; LDA | 0.027 |
| 1-Octen-3-ol | 21.3 [ab] (21.6) | 31.4 [a] (10.1) | 13.2 [b] (1.77) | KW; LDA | 0.034 |
| 2-Ethyl-2-hexenol | 4.02 (4.13) | 2.17 (0.86) | 1.57 (0.85) | | |
| 4-Ethyl-cyclohexanol | 0.00 (0.53) | 0.00 (0.00) | 0.00 (0.00) | | |
| 2-Ethyl-1-hexanol | 4.70 (6.76) | 5.26 (3.16) | 4.41 (3.15) | | |
| 2-Ethyl-1-decanol | 1.11 (1.11) | 0.77 (0.40) | 0.51 (0.21) | | |
| 2-Octen-1-ol | 0.75 (0.77) | 0.88 (0.47) | 0.34 (0.24) | | |
| 1-Octanol | 4.02 (4.25) | 3.19 (0.88) | 2.91 (0.77) | | |
| Benzenemethanol, α, α-dimethyl | 0.61 (0.78) | 0.46 (0.34) | 0.56 (0.51) | | |
| (Z)-2-Nonen-1-ol | 1.14 (1.01) | 0.79 (0.38) | 0.54 (0.20) | | |
| 1-Nonanol | 0.21 (0.25) | 0.25 (0.07) | 0.26 (0.16) | | |

**Table 1.** *Cont.*

| | C [1] | SP [1] | HI [1] | Statistical Analysis | p-Level |
|---|---|---|---|---|---|
| **Aldehydes** | | | | | |
| Pentanal | 4.16 (7.53) | 7.08 (4.95) | 3.52 (3.52) | | |
| Hexanal | 148.1 [ab] (143.4) | 164.3 [a] (42.8) | 78.62 [b] (41.7) | KW; LDA | 0.043 |
| Heptanal | 3.48 (4.33) | 4.89 (1.46) | 2.94 (1.13) | | |
| Methional | nd | nd | nd | | |
| 2-Heptenal | 0.52 [ab] (1.12) | 0.75 [a] (0.45) | 0.13 [b] (0.13) | KW; LDA | 0.022 |
| Benzaldehyde | 0.49 (0.66) | 0.00 (0.00) | 1.0 (0.80) | | |
| Octanal | 7.91 (8.01) | 7.74 (3.00) | 5.98 (1.07) | | |
| 2-Octenal | 0.84 (1.09) | 1.01 (0.45) | 0.70 (0.37) | | |
| Nonanal | 23.8 (18.2) | 23.7 (10.1) | 17.5 (5.73) | | |
| (*E*)-2-Nonenal | 0.30 (0.32) | 0.38 (0.14) | 0.24 (0.05) | | |
| (*Z*)-4-Decenal | 0.00 [a] (0.00) | 0.00 [a] (0.00) | 0.00 [b] (0.00) | KW; LDA | 0.026 |
| Decanal | 0.96 (1.37) | 1.37 (0.74) | 0.89 (0.49) | | |
| 2-Decenal | 0.20 (0.22) | 0.29 (0.13) | 0.15 (0.05) | | |
| (*E,E*)-2,4-Decadienal | 0.00 [b] (0.18) | 0.17 [a] (0.11) | 0.06 [ab] (0.06) | KW; LDA | 0.028 |
| (*E*)-2-Undecenal | 0.10(0.12) | 0.12 (0.05) | 0.06 (0.02) | | |
| Dodecanal | 0.24 (0.24) | 0.30 (0.12) | 0.20 (0.06) | | |
| Tridecanal | 0.04 [ab] (0.05) | 0.07 [a] (0.04) | 0.01 [b] (0.01) | KW; LDA | 0.029 |
| Tetradecanal | 0.06 [ab] (0.06) | 0.08 [a] (0.03) | 0.03 [b] (0.03) | KW; LDA | 0.031 |
| **Ketones** | | | | | |
| 2-Heptanone | 0.79 (1.03) | 0.81 (0.34) | 0.83 (0.23) | | |
| Butyrolactone | nd | nd | nd | | |
| 2-Methyl-6-heptanone | 0.13 (0.55) | 0.03 (0.03) | 0.00 (0.00) | | |
| 2-Nonanone | 0.08 [ab] (0.12) | 0.00 [b] (0.00) | 0.38 [a] (0.32) | KW; LDA | 0.005 |
| **Hydrocarbons** | | | | | |
| Toluene | nd | nd | nd | | |
| 1,2,4-Trimethyl-cyclopentane | nd | nd | nd | | |
| Propyl-cyclohexane | 0.00 (2.16) | 0.00 (0.00) | 0.00 (0.00) | | |
| 4-Methyl-nonane | nd | nd | nd | | |
| 2,2,6-Trimethyl-octane | nd | nd | nd | | |
| 2,2,4,6-Pentamethyl-heptane | 40.4 (33.5) | 39.7 (16.0) | 27.2 (8.66) | | |
| Decane | 0.41 (0.42) | 0.29 (0.24) | 0.18 (0.14) | | |
| 2,2,4,4-Tetramethyl-octane | 3.16 (3.42) | 2.40 (1.05) | 2.23 (1.06) | | |
| 2,6,7-Trimethyl-decane | 0.67 (0.63) | 0.51 (0.35) | 0.34 (0.34) | | |
| 2-Methyl-decane | 0.00 (0.09) | 0.00 (0.00) | 0.00 (0.00) | | |
| 5-Undecene | nd | nd | nd | | |
| Undecane | 1.10 (1.18) | 1.09 (0.21) | 1.02 (0.42) | | |
| 2,8-Dimethyl-4-methylene-nonane | nd | nd | nd | | |
| Pentyl-cyclohexane | nd | nd | nd | | |
| 3-Methylene-undecane | 0.00 (0.07) | 0.00 (0.00) | 0.00 (0.00) | | |
| Dodecane | 1.16 (1.23) | 1.22 (0.45) | 0.95 (0.39) | | |
| 2,6,11-Trimethyl-dodecane | 0.17 (0.19) | 0.30 (0.10) | 0.23 (0.09) | | |
| Tridecane | 0.55 (0.62) | 0.79 (0.35) | 0.58 (0.20) | | |
| 2,3,5,8-Tetramethyl-decane | 0.19 (0.19) | 0.23 (0.08) | 0.20 (0.06) | | |
| Tetradecane | 0.52 (0.50) | 0.61 (0.22) | 0.48 (0.14) | | |
| Pentadecane | 0.36 [b] (0.66) | 0.57 [ab] (0.13) | 1.18 [a] (0.75) | KW; LDA | 0.013 |
| 2,6,10-Trimethyl-tetradecane | 0.07 (0.08) | 0.10 (0.03) | 0.08 (0.05) | | |
| Hexadecane | 0.21 (0.21) | 0.28 (0.06) | 0.20 (0.06) | | |
| Heptadecane | 0.03 [b] (0.03) | 0.28 [a] (0.10) | 0.00 [b] (0.00) | KW; LDA | 0.001 |
| **Thiols** | | | | | |
| 2-Ethyl-1-hexanethiol | nd | nd | nd | | |
| 2-Methyl-2-heptanethiol | nd | nd | nd | | |
| **Esters** | | | | | |
| Pentanoic acid,2,2,4-trimethyl-3-hydroxy-, isobutyl ester | 0.06 (0.09) | 0.08 (0.08) | 0.15 (0.09) | | |
| Carbamodithioic acid, diethyl-, methyl ester | 0.24 (0.27) | 0.35 (0.14) | 0.14 (0.14) | | |
| Dimethyl phtalate | 4.42 (4.74) | 5.73 (1.43) | 4.70 (0.92) | | |
| Pentanoic acid, 2,2,4-trimethyl-3-carboxyisopropy, isobutyl ester | 0.00 (0.01) | 0.00 (0.00) | 0.00 (0.00) | | |
| **Lactone** | | | | | |
| γ-nonalactone | 0.12 [b] (0.12) | 0.20 [a] (0.07) | 0.09 [b] (0.03) | KW; LDA | 0.046 |
| **Acid** | | | | | |
| Dodecanoic acid | 0.00 (0.01) | 0.00 (0.02) | 0.00 (0.00) | | |
| **Nitrile** | | | | | |
| 4-Cyano-cyclohexene | 0.00 (0.23) | 0.00 (0.00) | 0.16 (0.16) | | |

**Table 1.** *Cont.*

|  | C [1] | SP [1] | HI [1] | Statistical Analysis | *p*-Level |
|---|---|---|---|---|---|
| Azide |  |  |  |  |  |
| 2-Azido-2,4,4,6,6-pentamethyl-heptane | 0.15 (0.26) | 0.05 (0.05) | 0.15 (0.11) |  |  |
| Unknown |  |  |  |  |  |
| Unknown (RT:17.96 min) | 0.00 [b] (0.00) | 0.00 [b] (0.00) | 0.00 [a] (0.00) | KW; LDA | 0.040 |
| Unknown (RT:23.76 min) | 0.01 (0.03) | 0.00 (0.00) | 0.09 (0.07) | LDA |  |
| Unknown (RT:28.62 min) | 0.00 [a] (0.00) | 0.00 [b] (0.00) | 0.00 [a] (0.00) | KW; LDA | 0.012 |
| Unknown (RT:37.40 min) | 0.00 (0.02) | 0.00 (0.00) | 0.00 (0.00) | LDA |  |

[1], Values are expressed as: Median (Median absolute deviation); [a, b], medians assigned different superscripts indicate significant differences ($p < 0.05$) between the dietary treatments; KW: Compounds that were found to be significantly different ($p < 0.05$) due to different dietary treatment, following the Kruskal–Wallis Test; LDA: Compounds that belong to the minimal set of 18 compounds that lead to a complete separation of the three dietary groups in linear discriminant analysis, i.e., to a confusion matrix with kappa = 1; nd: not detected.

**Table 2.** The composition of the volatile profile of chicken breast meat fed with three different diets in Trial 2 ((Control (C) vs. Spirulina (SP) vs. *Hermetia illucens* (HI)) (results expressed as ng/g).

|  | C [1] | SP [1] | HI [1] | Statistical Analysis | *p*-Level |
|---|---|---|---|---|---|
| Alcohols |  |  |  |  |  |
| 2-Methoxy-ethanol | 3.43 (2.11) | 1.62 (1.58) | 1.71 (1.10) |  |  |
| 1-Penten-3-ol | 3.98 (2.84) | 2.93 (1.30) | 2.58 (2.49) |  |  |
| 1-Pentanol | 8.88 [a] (5.76) | 9.89 [a] (3.72) | 2.02 [b] (1.08) | KW | 0.016 |
| 1-Hexanol | 4.20 [a] (2.59) | 4.69 [a] (2.72) | 1.53 [b] (0.38) | KW; LDA | 0.003 |
| 1-Heptanol | 1.44 (0.83) | 2.49 (0.86) | 1.52 (0.49) |  |  |
| 1-Octen-3-ol | 30.8 [a] (13.7) | 36.0 [a] (9.44) | 11.6 [b] (0.93) | KW; LDA | 0.001 |
| 2-Ethyl-2-hexenol | nd | nd | nd |  |  |
| 4-Ethyl-cyclohexanol | 4.39 [ab] (4.39) | 6.30 [a] (1.65) | 2.53 [b] (0.60) | KW | 0.009 |
| 2-Ethyl-1-hexanol | 12.2 (3.02) | 19.4 (4.95) | 20.3 (2.75) |  |  |
| 2-Ethyl-1-decanol | 3.38 (1.66) | 3.90 (1.99) | 4.78 (1.32) |  |  |
| 2-Octen-1-ol | 0.32 (0.32) | 1.34 (1.34) | 1.53 (0.73) |  |  |
| 1-Octanol | 4.05 [ab] (2.02) | 4.83 [a] (1.15) | 2.61 [b] (0.71) | KW | 0.045 |
| Benzenemethanol,α,α-dimethyl | 0.21 (0.21) | 0.81 (0.37) | 0.68 (0.20) |  |  |
| (Z)-2-Nonen-1-ol | 4.80 [ab] (1.53) | 6.52 [a] (1.95) | 4.17 [b] (1.06) | KW | 0.041 |
| 1-Nonanol | 0.05 (0.05) | 0.24 (0.24) | 0.24 (0.16) |  |  |
| Aldehydes |  |  |  |  |  |
| Pentanal | 8.04 [a] (3.71) | 7.40 [ab] (2.21) | 2.85 [b] (1.23) | KW | 0.026 |
| Hexanal | 283.7 [a] (135.8) | 230.5 [a] (123.7) | 44.7 [b] (20.8) | KW; LDA | 0.001 |
| Heptanal | 5.64 [a] (2.44) | 9.90 [a] (3.12) | 2.30 [b] (0.51) | KW; LDA | 0.002 |
| Methional | 0.20 (0.20) | 0.08 (0.08) | 0.27 (0.15) |  |  |
| 2-Heptenal | 0.71 [a] (0.42) | 0.54 [a] (0.25) | 0.08 [b] (0.07) | KW; LDA | 0.001 |
| Benzaldehyde | 3.07 (1.31) | 2.71 (0.57) | 3.14 (0.84) |  |  |
| Octanal | 16.0 [a] (6.11) | 15.8 [a] (6.35) | 6.18 [b] (1.36) | KW; LDA | 0.004 |
| 2-Octenal | 1.23 [b] (0.68) | 3.02 [ab] (1.22) | 3.50 [a] (1.13) | KW | 0.045 |
| Nonanal | 24.8 [a] (8.06) | 31.6 [a] (8.50) | 11.8 [b] (2.52) | KW; LDA | 0.008 |
| (E)-2-Nonenal | 0.53 (0.26) | 0.54 (0.13) | 0.33 (0.05) |  |  |
| (Z)-4-Decenal | 0.00 (0.00) | 0.28 (0.10) | 0.23 (0.09) |  |  |
| Decanal | 0.87 [a] (0.25) | 0.96 [a] (0.37) | 0.09 [b] (0.09) | KW | 0.008 |
| (E)-2-Decenal | 0.06 [ab] (0.06) | 0.19 [a] (0.07) | 0.00 [b] (0.00) | KW; LDA | 0.003 |
| (E,E)-2,4-Decadienal | 0.06 (0.06) | 0.09 (0.04) | 0.03 (0.03) |  |  |
| (E)-2-Undecenal | 0.00 [ab] (0.00) | 0.07 [b] (0.07) | 0.00 [a] (0.00) | KW | 0.022 |
| Dodecanal | 0.05 (0.05) | 0.04 (0.04) | 0.00 (0.00) |  |  |
| Tridecanal | nd | nd | nd |  |  |
| Tetradecanal | nd | nd | nd |  |  |
| Ketones |  |  |  |  |  |
| 2-Heptanone | 0.35 (0.31) | 0.28 (0.18) | 0.21 (0.08) |  |  |
| Butyrolactone | 0.42 (0.40) | 0.97 (0.72) | 0.24 (0.16) |  |  |
| 2-Methyl-6-heptanone | 0.00 (0.00) | 0.00 (0.00) | 0.00 (0.00) |  |  |
| 2-Nonanone | 0.00 (0.00) | 0.00 (0.00) | 0.00 (0.00) |  |  |
| Hydrocarbons |  |  |  |  |  |
| Toluene | 0.17 [a] (0.17) | 0.03 [ab] (0.03) | 0.00 [b] (0.00) | KW | 0.048 |
| 1,2,4-Trimethyl-cyclopentane | 0.68 (0.68) | 0.32 (0.23) | 0.16 (0.16) |  |  |
| Propyl-cyclohexane | 0.35 (0.28) | 0.34 (0.07) | 0.27 (0.08) |  |  |
| 4-Methyl-nonane | 0.00 [b] (0.00) | 0.00 [b] (0.00) | 0.22 [a] (0.16) | KW; LDA | 0.002 |

**Table 2.** *Cont.*

| | C [1] | SP [1] | HI [1] | Statistical Analysis | *p*-Level |
|---|---|---|---|---|---|
| 2,2,6-trimethyl-octane | 0.00 [a] (0.00) | 0.00 [ab] (0.00) | 0.48 [b] (0.48) | KW; LDA | 0.004 |
| 2,2,4,6-Pentamethyl-heptane | 53.5 [a] (35.0) | 25.8 [a] (4.86) | 19.3 [b] (6.90) | KW | 0.011 |
| Decane | 4.24 [a] (2.96) | 2.43 [a] (1.59) | 1.50 [b] (0.44) | KW | 0.034 |
| 2,2,4,4-Tetramethyl-octane | 4.34 (1.86) | 1.24 (1.24) | 0.73 (0.73) | | |
| 2,6,7-Trimethyl-decane | 1.43 [b] (1.43) | 8.24 [a] (2.20) | 10.8 [a] (3.84) | KW; LDA | 0.000 |
| 2-Methyl-decane | 0.00 [b] (0.00) | 2.42 [a] (0.95) | 3.18 [a] (1.27) | KW; LDA | 0.000 |
| 5-Undecene | 0.29 (0.29) | 2.37 (1.33) | 2.50 (1.09) | | |
| Undecane | 1.57 [a] (1.46) | 0.12 [b] (0.08) | 0.09 [b] (0.04) | KW | 0.024 |
| 2,8-dimethyl-4-methylene-nonane | 0.00 [b] (0.00) | 0.07 [ab] (0.07) | 0.20 [a] (0.19) | KW | 0.029 |
| Pentyl-cyclohexane | 0.28 (0.24) | 0.39 (0.12) | 0.64 (0.21) | | |
| 3-Methylene-undecane | 0.25 (0.23) | 0.26 (0.14) | 0.39 (0.14) | | |
| Dodecane | 0.93 (0.79) | 0.15 (0.14) | 0.09 (0.09) | | |
| 2,6,11-trimethyl-dodecane | 0.05 (0.05) | 0.03 (0.03) | 0.00 (0.00) | | |
| Tridecane | 0.50 (0.29) | 0.28 (0.07) | 0.25 (0.11) | | |
| 2,3,5,8-tetramethyl-decane | 0.02 (0.02) | 0.00 (0.00) | 0.01 (0.01) | | |
| Tetradecane | 0.30 (0.09) | 0.32 (0.07) | 0.22 (0.05) | | |
| Pentadecane | 0.21 (0.08) | 0.24 (0.10) | 0.24 (0.09) | | |
| 2,6,10-trimethyl-tetradecane | 0.00 (0.00) | 0.00 (0.00) | 0.00 (0.00) | | |
| Hexadecane | 0.12 (0.07) | 0.14 (0.08) | 0.08 (0.03) | | |
| Heptadecane | 0.00 (0.00) | 0.05 (0.05) | 0.00 (0.00) | | |
| Thiols | | | | | |
| 2-Ethyl-1-hexanethiol | 0.58 [b] (0.58) | 5.86 [a] (1.33) | 8.94 [a] (3.24) | KW | 0.009 |
| 2-Methyl-2-heptanethiol | 4.48 [b] (1.00) | 7.29 [ab] (1.64) | 9.68 [a] (2.53) | KW | 0.009 |
| Esters | | | | | |
| Pentanoic acid,2,2,4-trimethyl-3-hydroxy-, isobutyl ester | 0.00 (0.00) | 0.00 (0.00) | 0.00 (0.00) | | |
| Carbamodithioic acid, diethyl-, methyl ester | nd | nd | nd | | |
| Dimethyl phtalate | 1.85 (0.53) | 1.93 (0.95) | 1.35 (0.41) | | |
| Pentanoic acid, 2,2,4-trimethyl-3-carboxyisopropyl, isobutyl ester | 0.04 (0.04) | 0.10 (0.06) | 0.07 (0.04) | | |
| Lactone | | | | | |
| γ-Nonalactone | nd | nd | nd | | |
| Acid | | | | | |
| Dodecanoic acid | 0.00 (0.00) | 0.00 (0.00) | 0.00 (0.00) | KW | 0.040 |
| Nitrile | | | | | |
| 4-Cyano-cyclohexene | 4.35 [a] (2.31) | 0.98 [ab] (0.98) | 0.31 [b] (0.31) | KW | 0.041 |
| Azide | | | | | |
| 2-Azido-2,4,4,6,6-pentamethyl-heptane | nd | nd | nd | | |
| Unknown | | | | | |
| Unknown (RT = 13.40 min) | 0.35 [a] (0.22) | 0.57 [a] (0.33) | 0.00 [b] (0.00) | KW; LDA | 0.006 |
| Unknown (RT = 15.26 min) | 1.78 [b] (1.06) | 8.07 [a] (2.91) | 10.2 [a] (4.58) | KW; LDA | 0.000 |
| Unknown (RT = 16.51 min) | 0.30 [a] (0.30) | 0.00 [b] (0.00) | 0.00 [ab] (0.00) | KW | 0.020 |
| Unknown (RT = 18.15 min) | 0.26 [ab] (0.26) | 0.30 [a] (0.16) | 0.00 [b] (0.00) | KW | 0.045 |

[1], Values presented as: Median (Median absolute deviation); [a,b], medians assigned different superscripts indicate significant differences ($p < 0.05$) between the dietary treatments; KW: Compounds that were found to be significantly different ($p < 0.05$) due to different dietary treatment, following the Kruskal–Wallis Test; LDA: Compounds that belong to the minimal set of 14 compounds that lead to a complete separation of the three dietary groups in linear discriminant analysis, i.e., to a confusion matrix with kappa = 1; nd: not detected.

### 3.1. Volatile Compounds in Chicken Meat of Trial 1

In total, 210 volatile compounds were detected in chicken breast meat produced with the three different diets in Trial 1. The results of the Kruskal–Wallis test, presented in Table 1, indicate that 16 compounds showed significant difference among the treatments, 6 of which were detected in trace amounts (i.e., <0.30 ng/g). The main groups of compounds affected by diet were alcohols (4), aldehydes (6), ketones (1), hydrocarbons (2), lactones (1), while 2 compounds were not identified (Retention time (RT): 17.96 min and RT = 28.62 min). Two alcohols, 1-pentanol and 1-heptanol, were detected in lower levels in chicken breast meat produced with the HI diet compared to meat from the other two diets ($p < 0.05$), which did not differ from each other ($p > 0.05$). Levels of 1-hexanol, 1-octen-3-ol, hexanal and 2-heptenal, were lower under HI diet than under SP diet, neither of which differed from the C diet. In contrast, levels of 2-nonanone and pentadecane were higher in the HI diet compared to the other treatments, although levels were similar ($p > 0.05$) to the C diet and the SP diet, respectively.

Multivariate analysis was applied to investigate differences among the volatile profiles. LDA revealed that groups were clearly separated (Figure 1) with 18 compounds according to the confusion matrix and the kappa coefficient deriving from it. Four of these compounds were not identified (RTs: 17.96 min; 23.86 min; 28.62 min and 37.40 min; Supplementary Table S1) and were detected in traces. The first component (explaining 74% of the variation) separated mainly the HI group (located on the left lower quadrant) from the SP group (right lower quadrant). The compounds that contributed mostly in the separation (i.e., factor loadings higher than 1.0) were 1-pentanol, hexanal, 1-hexanol, 2-heptenal, 1-heptanol, 1-octen-3-ol, 2-nonanone, 2,4-decadienal, γ-nonalactone, pentadecane, heptadecane. The second component (explaining 26%) of the variation separated the C group (located on the upper side of the plot) from HI and SP group (bottom side of the plot). The compounds that contributed mostly to the separation (i.e., factor loadings higher than 1.0) were hexanal, 1-hexanol, 2-heptenal, 1-heptanol, 1-octen-3-ol, 2,4-decadienal and tridecanal (Supplementary Table S2).

### 3.2. Volatile Compounds in Chicken Meat of Trial 2

One hundred sixty-six volatile compounds were detected in the chicken breast meat of Trial 2. The results of the Kruskal–Wallis test (presented in Table 2) indicated that 33 compounds showed significant differences among treatments. The main groups of compounds that showed differences were alcohols (6), aldehydes (10), hydrocarbons (9), thiols (2), nitriles (1), and acids (1) while 4 compounds were not identified (unknown). In accordance with Trial 1, there were 7 compounds detected in trace amounts (i.e., <0.30 ng/g). Levels of 1-pentanol, 1-hexanol, 1-octen-3-ol, hexanal, heptanal, 2-heptenal, octanal, nonanal, decanal, 2,2,4,6-pentamethyl-heptane, decane, and one unknown (RT: 13.40 min) were lower ($p < 0.05$) in chickens from the HI-dietary treatment compared to SP-fed and C-fed chicken. Levels of 4-ethyl-cyclohexanol, 1-octanol, 2-nonen-1-ol, pentanal, toluene, 2,2,6-trimethyl-octane, undecane, 4-cyano-cyclohexene, and one unknown (RT: 18.15) were lower ($p < 0.05$) in chickens fed the HI-diet than the ones from C and/or the SP diet, which, in most cases, did not differ ($p > 0.05$). On the contrary, levels of 2-octenal, 2-undecenal, 4-methyl-nonane, 2,6,7-trimethyl-decane, 2-methyl-decane, 2,8-dimethyl-4-methylene-nonane, 2-ethyl-1-hexanethiol, 2-methyl-2-heptanethiol, 2-methyl-decane, and two unknown (RT:15.26 min and RT:16.51 min) were higher ($p < 0.05$) in HI-fed chickens in comparison to C-fed and/or the SP-fed chickens.

Linear discriminant analysis of the volatile data separated the dietary groups of Trial 2 with 14 compounds according to the kappa-coefficient (Figure 2). The first component (explaining 61.1% of the variation) separated the HI group (left side of the plot) from the C group (right side of the plot). The compounds that contributed mostly in the separation (i.e., factor loadings higher than 1.0) were hexanal, 1-hexanol, heptanal, 2-heptenal, 4-methyl-nonane, 2,2,6-trimethyl-octane, 1-octen-3-ol, 2,6,7-trimethyl-decane, and 2-decenal. The second component explained 39% of the variation separated the SP group from the HI and C groups (Figure 2). The compounds that contributed to the separation were hexanal, 1-hexanol, heptanal, 2-heptenal, 4-methyl nonane, 2,2,6-trimethyl octane, 1-octen-3-ol, 2-ethyl-2-hexenol, 2,6,7-trimethyl-decane, 2-methyl-decane, and nonanal (Supplementary Table S3).

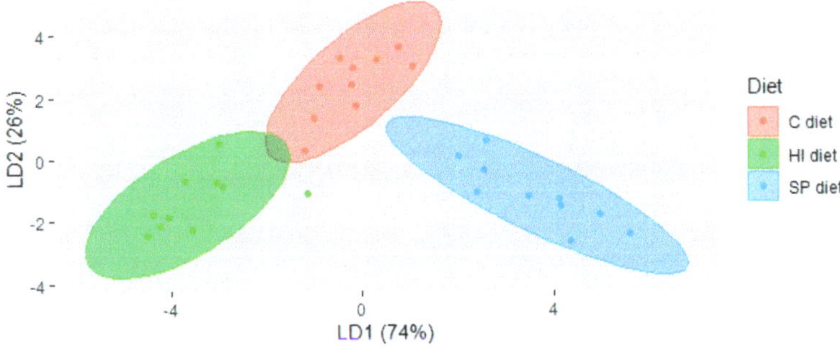

**Figure 1.** Linear Discriminant Analysis plot of the volatile compounds of chicken meat under three dietary treatments in Trial 1. C diet: Soybean meal- based diet; SP diet: Soybean meal-based diet partially supplemented with *Arthrospira platensis*; HI diet: Soybean meal-based diet partially supplemented with *Hermetia illucens* larvae.

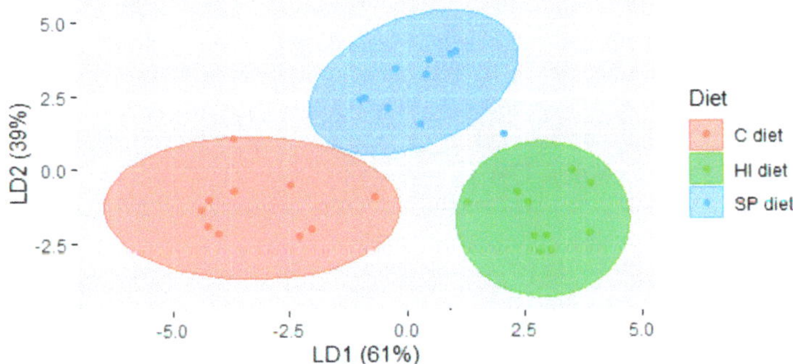

**Figure 2.** Linear discriminant analysis plot of the volatile compounds of chicken meat under three dietary treatments in Trial 2. C diet: Soybean meal-based diet; SP diet: Soybean meal-based diet supplemented with *Arthrospira platensis*; HI diet: Soybean meal-based diet supplemented with *Hermetia illucens* larvae.

## 4. Discussion

Several studies have investigated the effect of dietary treatment on chicken meat quality [20,38–40]. Results indicated that the volatile profile was affected by dietary treatments, with the majority of the compounds having been reported in literature on chicken meat aroma (Supplementary Table S1). Chicken meat is prone to lipid oxidation due to its high content of PUFAs; thus, volatile compounds deriving from lipid oxidation (i.e., aldehydes, alcohols, and ketones) [41] were expected.

Although a substantial number of compounds differed significantly among treatments of Trial 1, only half of them were detected in considerable quantities (i.e., higher than 0.30 ng/g). Results of Trial 2 indicated that the complete substitution of soybean meal with microalga or insects in the starter and grower phase strongly influenced the formation of lipid-derived compounds. Although acknowledging the fact that the two trials were independent and non-directly comparable, there was still a similar oxidation pattern observed among dietary treatments, i.e., the levels of compounds in meat produced with the HI-diet were usually lower than in the other dietary treatments, with most of the compounds deriving from linoleic acid (1-pentanol, 1-hexanol, 1-octen-3-ol, 2-nonen-1-ol, hexanal, heptanal, 2-heptenal) [42] or oleic acid (1-heptanol, 1-octanol, octanal, nonanal, decanal) [43].

The explanation of the results could be associated with the fatty acid (FA) profile of the diets since, in monogastric animals like chickens, the FA profile of the diet reflects the intramuscular FA composition (mainly the triglycerides, as the phospholipid composition is less affected [44]).

The higher levels of the compounds in SP-fed and C-fed chicken compared to HI-fed chicken could be attributed to the possibly high level of PUFA in these diets [45–48], which could have promoted lipid oxidation. Cortinas Hernández et al. [38] attributed the linear increase of lipid oxidation in cooked chicken meat to the increasing PUFA content in raw meat. In this regard, Bonos, et al. [49] reported that dietary supplementation with 5 or 10 g spirulina per kg feed influenced the fatty acid composition of thigh muscle (by enhancing the PUFA content), but not of the breast muscle which had similar PUFA, saturated fatty acid (SFA), and mono unsaturated fatty acid (MUFA) content with the control diet. In contrast, El-Bahr et al. [24] reported that inclusion of 1 g spirulina per kg feed in a corn-soybean basal diet increased the major long chain n-3 PUFA (eicosapentaenoic, docosahexaenoic, and arachidonic acid) and total PUFA content in breast muscle, while the SFA, MUFA, and PUFA/SFA ratio remained unchanged.

In the case of an insect-based diet, the FA composition of chicken meat depends on the fatty acid profile of the insect lipids, which in turn may be affected by FA composition of the rearing substrate [14] or the stage in the life cycle of the insect [50]. HI larvae has been associated with high SFA content (ranging usually between 40–80%) and low PUFA content, i.e., around 10–20% of total FA [51] or even less, in larvae [52,53]. Thus, the lower extent of oxidation observed in muscle from HI-fed chicken could be associated with the reduced content of the easily oxidized PUFA and the elevated content of the (slowly oxidized) SFA in HI treatment. In support of this, Schiavone et al. [18] reported that 50% or 100% substitution of soybean oil with HI larvae increased the ratio of SFA of broiler chicken breast proportionally to the level of substitution and to the detriment of the PUFA fraction. The diminishing effect of HI larvae (meal or fat) on PUFA content was confirmed in Schiavone et al. [53] and Cullere et al. [54]. In addition, Schiavone et al. [53] noted that the defatting process not only enhances the protein content but may also reduce the risk of lipid oxidation in meat. Finally, volatile analysis showed that hexanal, a compound considered as an oxidation marker [55], was detected in lower levels under the HI diet than under the other two diets in both trials, which could imply less oxidation. Although the fatty acid profile of breast meat in our studies was not analyzed, a recent study by Altmann et al. [20] using the same animals as in Trial 1 reported a higher content of SFA in thigh muscle from chickens fed with the HI diet. It should be noted that lipid content between breast and thigh muscle in poultry meat differs [56] as well as the fatty acid composition [57]. As a consequence, the results of Altmann et al. [20] can only be seen as indicative.

On the other hand, the higher levels of the two thiols in HI-fed chickens compared with the other two diets in Trial 2 could reflect differences in the free amino acid content among the three treatments, especially considering that thiol formation is linked with Strecker degradation of amino acids (e.g., cysteine) when reacting with secondary lipid oxidation compounds [58]. Differences in thiamine content may also play a role, as thiamine degradation is another pathway for thiol formation [59].

It appears that several factors could be involved in the extent of lipid oxidation when feeds of high nutritional value, like microalga or insects, are included in poultry diets. These factors may be related to the type of microalga [24] or insect species [60], the level of supplementation, the stage of inclusion in the basal diets (starter, grower, finisher phase), the type and level of antioxidants (e.g., β-carotene, tocopherol, carotenoids) that microalga contain and may affect the oxidative stability [61], the type of substrate on which larvae were reared [14], the use of defatted biomasses (algae or insect) [62,63], the defatting method [64], or the interaction with other factors, like amino acid supplementation [24,65].

Overall, the results of the two trials indicate that alternative protein sources may affect the aromatic profile of chicken meat and this could impact sensory perception. For example, hexanal (detected in the highest abundance in both trials) is associated with a "green/grassy" odor [66] and has been identified as one of the most important odor active compounds in chicken breast on the basis of its odor activity value [67]. 2-Heptenal, heptanal, octanal, nonanal, and decanal (described as "oily/fatty", "fatty/roasty/citrus", "fatty/sweet", "roasted/meaty/fatty", and "sweet/fruity", respectively [66,68]) due to their low odor threshold values [69] and the relatively high level at which they were detected in C and SP diets, mainly in Trial 2, could potentially affect aroma or flavor perception in chickens fed these diets. Although the significant alcohols have been previously reported in chicken aroma studies, only 1-octen-3-ol, with the characteristic "mushroom" odor [68], has a low odor threshold (1 ng/g) compared to the other alcohols [69,70] and was detected in significant quantities in both trials. Hydrocarbons are derived from lipid oxidation and are compounds that showed significant variations among treatments. However, these are generally not considered to impact the flavor of lipid-based foods [71]. Thiols are significant volatile compounds in meat aroma [72]. The higher levels of the two thiols in HI-fed chicken compared with the other two diets in Trial 2 could signify differences in the aromatic quality of the three diets. As meat flavor is a combination of volatile aroma and non-volatile taste compounds, further analytical research (e.g., defining free amino acid content; performing gas chromatography-olfactometry analysis) in combination with sensory evaluation would be required to identify compounds that could influence eating quality and possibly consumer preference.

LDA indicated the potential of the volatile analysis/profile to discriminate the three dietary treatments in both trials. Compounds that had higher impact on the group separation were mainly those with higher concentration, rather than compounds that were only detected in traces. The content of linoleic acid—the precursor of most of these compounds—seemed to play a significant role in the characterization and identification of these treatments.

## 5. Conclusions

This is the first study to report differences in the volatile profile of chicken meat after dietary replacement of soybean meal with alternative protein sources (i.e., microalga or partially defatted larvae meal). The differences detected were mainly in the lipid oxidation-deriving compounds that could play a significant role in the development of the typical aroma of chicken meat. Multivariate analysis confirmed that the dietary treatments led to a discriminatory volatile profile in both trials. Considering that both microalga and insects could stand as sustainable options in animal feeding in the years to come, future research should focus on identifying the type of biomasses and the proper inclusion level in order to improve chicken meat quality.

**Supplementary Materials:** The following are available online at http://www.mdpi.com/2304-8158/9/9/1235/s1, Table S1: Linear retention indices (LRIs) and method of identification of the main compounds detected in chicken meat (Trial 1 and Trial 2); Table S2: Factor loadings of the 18 most discriminant compounds of Trial 1 that lead to a confusion matrix with a Cohen's kappa coefficient value of 1.0, i.e., clear separation of the three groups, Table S3: Factor loadings of the 14 most discriminant compounds of Trial 2 that lead to a confusion matrix with a Cohen's kappa coefficient value of 1.0, i.e., clear separation of the three groups.

**Author Contributions:** Conceptualization, B.A.A. and D.M.; methodology, V.G.; statistical analysis, A.O.S. and V.G.; validation, V.G. and M.C.; formal analysis, V.G.; investigation, V.G.; resources, D.M.; data curation, V.G.; writing—original draft preparation, V.G.; writing—review and editing, V.G.; M.C.; B.A.A., A.O.S., and D.M.; visualization, V.G., M.C., B.A.A. and D.M.; supervision, V.G., M.C., and D.M.; project administration, B.A.A. and D.M.; funding acquisition, D.M. All authors have read and agreed to the published version of the manuscript.

**Funding:** This research was funded by the initiative Niedersächsisches Vorab by the Ministry for Science and Culture, Lower Saxony, Germany (#ZN 3041).

**Acknowledgments:** We would like to thank Carmen Neumann, Susanne Rothstein, and Frank Liebert from the Division Animal Nutrition Physiology, University of Göttingen, Germany, for their contributions towards the animal nutrition and animal rearing. We also would like to thank Ruth Wigger for technical maintenance of the GC–MS and Elisa Oertel for assisting in data preprocessing (peak alignment) in the two trials. This study would not be possible without the many technical staff at the Department of Animal Sciences, University of Göttingen, who provided support during animal slaughtering.

**Conflicts of Interest:** The authors declare no conflict of interest.

## References

1. Phillips, S.M.; Chevalier, S.; Leidy, H.J. Protein "requirements" beyond the RDA: Implications for optimizing health. *Appl. Physiol. Nutr. Metab.* **2016**, *41*, 565–572. [CrossRef] [PubMed]

2.	Kralik, G.; Kralik, Z.; Grčević, M.; Hanžek, D. Quality of Chicken Meat. *Anim. Husb. Nutr.* **2018**, *63*. [CrossRef]

3.	Mir, N.A.; Rafiq, A.; Kumar, F.; Singh, V.; Shukla, V. Determinants of broiler chicken meat quality and factors affecting them: A review. *J. Food Sci. Technol.* **2017**, *54*, 2997–3009. [CrossRef]

4.	Bou, R.; Guardiola, F.; Barroeta, A.; Codony, R. Effect of dietary fat sources and zinc and selenium supplements on the composition and consumer acceptability of chicken meat. *Poult. Sci.* **2005**, *84*, 1129–1140. [CrossRef] [PubMed]

5.	Kareem, K.Y.; Abdulla, N.R.; Foo, H.L.; Zamri, A.N.M.; Shazali, N.; Loh, T.C.; Alshelmani, M.I. Effect of feeding larvae meal in the diets on growth performance, nutrient digestibility and meat quality in broiler chicken. *Indian J. Anim. Res.* **2018**, *88*, 1180–1185.

6.	Gaylord, T.; Barrows, F.; Rawles, S. Apparent amino acid availability from feedstuffs in extruded diets for rainbow trout Oncorhynchus mykiss. *Aquac. Nutr.* **2010**, *16*, 400–406. [CrossRef]

7.	Hanson, M.N.; Persia, M.E. Effects of Dietary Soy Inclusion on Broiler Chick Performance and Metabolizable Energy. *Iowa State Univ. Anim. Ind. Rep.* **2014**. [CrossRef]

8.	Veldkamp, T.; van Duinkerken, G.; van Huis, A.; Lakemond, C.M.M.; Bosch, G.; van Boekel, M.A.J.S. *Insects as a Sustainable Feed Ingredient in Pig and Poultry Diets—A Feasibility Study*; Wageningen UR Livestock Research: Lelystad, The Netherlands, 2012.

9.	OECD/FAO. *OECD-FAO Agricultural Outlook 2020–2029*; OECD: Paris, France, 2020. [CrossRef]

10.	Kristensen, L.; Støier, S.; Würtz, J.; Hinrichsen, L. Trends in meat science and technology: The future looks bright, but the journey will be long. *Meat Sci.* **2014**, *98*, 322–329. [CrossRef]

11.	Finley, J.W. Evolution and future needs of food chemistry in a changing world. *J. Agric. Food Chem.* **2020**, in press. [CrossRef]

12.	Akhtar, Y.; Isman, M.B. Insects as an Alternative Protein Source. In *Proteins in Food Processing*, 2nd ed.; Yada, Y.R., Ed.; Woodhead Publishing: Cambridge, UK, 2017; pp. 263–288.

13.	Madeira, M.S.; Cardoso, C.; Lopes, P.A.; Coelho, D.; Afonso, C.; Bandarra, N.M.; Prates, J.A.M. Microalgae as feed ingredients for livestock production and meat quality: A review. *Livest. Sci.* **2017**, *205*, 111–121. [CrossRef]

14.	Barragan-Fonseca, K.B.; Dicke, M.; van Loon, J.J. Nutritional value of the black soldier fly (*Hermetia illucens* L.) and its suitability as animal feed—A review. *J. Insects Food Feed.* **2017**, *3*, 105–120. [CrossRef]

15.	Józefiak, D.; Józefiak, A.; Kierończyk, B.; Rawski, M.; Świątkiewicz, S.; Długosz, J.; Engberg, R.M. Insects—A Natural Nutrient Source for Poultry—A Review. *Ann. Anim. Sci.* **2016**, *16*, 297–313. [CrossRef]

16.	Müller, A.; Wolf, D.; Gutzeit, H.O. The black soldier fly, Hermetia illucens—A promising source for sustainable production of proteins, lipids and bioactive substances. *Z. Naturforsch. C Biosci.* **2017**, *72*, 351–363. [CrossRef] [PubMed]

17.	Gasco, L.; Biasato, I.; Dabbou, S.; Schiavone, A.; Gai, F. Animals fed insect-based diets: State-of-the-art on digestibility, performance and product quality. *Animals* **2019**, *9*, 170. [CrossRef]

18.	Schiavone, A.; Cullere, M.; De Marco, M.; Meneguz, M.; Biasato, I.; Bergagna, S.; Dezzutto, D.; Gai, F.; Dabbou, S.; Gasco, L. Partial or total replacement of soybean oil by black soldier fly larvae (*Hermetia illucens* L.) fat in broiler diets: Effect on growth performances, feed-choice, blood traits, carcass characteristics and meat quality. *Ital. J. Anim. Sci.* **2017**, *16*, 93–100. [CrossRef]

19.	Cullere, M.; Tasoniero, G.; Giaccone, V.; Acuti, G.; Marangon, A.; Dalle Zotte, A. Black soldier fly as dietary protein source for broiler quails: Meat proximate composition, fatty acid and amino acid profile, oxidative status and sensory traits. *Animals* **2018**, *12*, 640–647. [CrossRef]

20.	Altmann, B.; Wigger, R.; Ciulu, M.; Mörlein, D. The effect of insect or microalga alternative protein feeds on broiler meat quality. *J. Sci. Food Agric.* **2020**. [CrossRef]

21.	Falquet, J.; Hurni, J.P. *The Nutritional Aspects of Spirulina*; Antenna Foundation: Geneva, Switzerland, 2017.

22.	Jung, F.; Krüger-Genge, A.; Waldeck, P.; Küpper, J.-H. Spirulina platensis, a super food? *J. Cell. Biotechnol.* **2019**, *5*, 43–54. [CrossRef]

23.	Khan, Z.; Bhadouria, P.; Bisen, P. Nutritional and therapeutic potential of Spirulina. *Curr. Pharm. Biotechnol.* **2005**, *6*, 373–379. [CrossRef]

24.	El-Bahr, S.; Shousha, S.; Shehab, A.; Khattab, W.; Ahmed-Farid, O.; Sabike, I.; El-Garhy, O.; Albokhadaim, I.; Albosadah, K. Effect of Dietary Microalgae on Growth Performance, Profiles of Amino and Fatty Acids, Antioxidant Status, and Meat Quality of Broiler Chickens. *Animals* **2020**, *10*, 761. [CrossRef]

25. El-Hady, A.A.; El-Ghalid, O. Spirulina platensis Algae (SPA): A novel poultry feed additive Effect of SPA supplementation in broiler chicken diets on productive performance, lipid profile and calcium-phosphorus metabolism. In Proceedings of the VI Mediterranean Poultry Summit, Torino, Italy, 18–20 June 2018.

26. Venkataraman, L.V.; Somasekaran, T.; Becker, E.W. Replacement Value of blue-green alga (spirulina platensis) for fishmeal and a vitamin-mineral premix for broiler chicks. *Br. Poult. Sci.* **1994**, *35*, 373–381. [CrossRef] [PubMed]

27. Vasta, V.; Priolo, A. Ruminant fat volatiles as affected by diet. A review. *Meat Sci.* **2006**, *73*, 218–228. [CrossRef]

28. Calkins, C.R.; Hodgen, J.M. A fresh look at meat flavor. *Meat Sci.* **2007**, *77*, 63–80. [CrossRef] [PubMed]

29. Flores, M. The Eating Quality of Meat: III—Flavor. In *Lawrie's Meat Science*, 8th ed.; Toldrá, F., Ed.; Woodhead Publishing Series in Food Science; Technology and Nutrition: Cambridge, UK, 2017; pp. 383–417. [CrossRef]

30. Monahan, F.J.; Schmidt, O.; Moloney, A.P. Meat provenance: Authentication of geographical origin and dietary background of meat. *Meat Sci.* **2018**, *144*, 2–14. [CrossRef] [PubMed]

31. Neumann, C.; Velten, S.; Liebert, F. The Graded Inclusion of Algae (*Spirulina platensis*) or Insect (*Hermetia illucens*) Meal as a Soybean Meal Substitute in Meat Type Chicken Diets Impacts on Growth, Nutrient Deposition and Dietary Protein Quality Depending on the Extent of Amino Acid Supplementation. *Open J. Anim. Sci.* **2018**, *8*, 163–183. [CrossRef]

32. Wang, C.; Zhang, W.; Li, H.; Mao, J.; Guo, C.; Ding, R.; Wang, Y.; Fang, L.; Chen, Z.; Yang, G. Analysis of volatile compounds in pears by HS-SPME-GCxGC-TOFMS. *Molecules* **2019**, *24*, 1795. [CrossRef]

33. R Core Team. *R A Language and Environment for Statistical Computing*; R Foundation for Statistical Computing: Vienna, Austria, 2019.

34. Ottensmann, M.; Stoffel, M.A.; Nichols, H.J.; Hoffman, J.I. GCalignR: An R package for aligning gas-chromatography data for ecological and evolutionary studies. *PLoS ONE* **2018**, *13*. [CrossRef]

35. Robinson, D. Fuzzyjoin: Join Tables Together on Inexact Matching. R Package Version 0.1.6. 2020. Available online: https://cran.r-project.org/web/packages/fuzzyjoin/index.html (accessed on 4 September 2020).

36. Venables, W.N.; Ripley, B.D. *Modern Applied Statistics with S*, 4th ed.; Springer: New York, NY, USA, 2002.

37. Revelle, W. *Procedures for Psychological, Psychometric, and Personality Research*; American Chemical Society: Washington, DC, USA, 2019.

38. Cortinas Hernández, L.; Barroeta, A.C.; Villaverde Haro, C.; Galobarti Cots, J.; Guardiola Ibarz, F.; Baucells Sánchez, M.D. Influence of the Dietary Polyunsaturation Level on Chicken Meat Quality: Lipid Oxidation. *Poult. Sci.* **2005**, *84*, 48–55. [CrossRef]

39. Cullere, M.; Woods, M.J.; van Emmenes, L.; Pieterse, E.; Hoffman, L.C.; Dalle Zotte, A. Hermetia illucens Larvae Reared on Different Substrates in Broiler Quail Diets: Effect on Physicochemical and Sensory Quality of the Quail Meat. *Animals* **2019**, *9*, 525. [CrossRef]

40. Lopez-Ferrer, S.; Baucells, M.D.; Barroeta, A.; Grashorn, M. n-3 enrichment of chicken meat. 1. Use of very long-chain fatty acids in chicken diets and their influence on meat quality: Fish oil. *Poult. Sci.* **2001**, *80*, 741–752. [CrossRef]

41. Jayasena, D.D.; Ahn, D.U.; Nam, K.C.; Jo, C. Flavour Chemistry of Chicken Meat: A Review. *Asian-Australas J. Anim. Sci.* **2013**, *26*, 732–742. [CrossRef] [PubMed]

42. Elmore, J.S.; Cooper, S.L.; Enser, M.; Mottram, D.S.; Sinclair, L.A.; Wilkinson, R.G.; Wood, J.D. Dietary manipulation of fatty acid composition in lamb meat and its effect on the volatile aroma compounds of grilled lamb. *Meat Sci.* **2005**, *69*, 233–242. [CrossRef]

43. Vandamme, J.; Nikiforov, A.; De Roose, M.; Leys, C.; De Cooman, L.; Van Durme, J. Controlled accelerated oxidation of oleic acid using a DBD plasma: Determination of volatile oxidation compounds. *Food Res. Int.* **2016**, *79*, 54–63. [CrossRef]

44. Raes, K.; De Smet, S.; Demeyer, D. Effect of dietary fatty acids on incorporation of long chain polyunsaturated fatty acids and conjugated linoleic acid in lamb, beef and pork meat: A review. *Anim. Feed. Sci. Technol.* **2004**, *113*, 199–221. [CrossRef]

45. Diraman, H.; Koru, E.; Dibeklioglu, H. Fatty acid profile of Spirulina platensis used as a food supplement. *Isr. J. Aquacult. Bamidgeh* **2009**, *91*, 134–142.

46. Hudson, B.J.; Karis, I.G. The lipids of the alga Spirulina. *J. Sci. Food Agric.* **1974**, *25*, 759–763. [CrossRef] [PubMed]

47. Azman, M.; Konar, V.; Seven, P. Effects of different dietary fat sources on growth performances and carcass fatty acid composition of broiler chickens. *Rev. Med. Vet.* **2004**, *155*, 278–286.

48. Waldroup, P.; Waldroup, A. Fatty acid effect on carcass: The influence of various blends of dietary fats added to corn-soybean meal based diets on the fatty acid composition of broilers. *Int. J. Poult. Sci.* **2005**, *4*, 123–132. [CrossRef]

49. Bonos, E.; Kasapidou, E.; Kargopoulos, A.; Karampampas, A.; Christaki, E.; Florou-Paneri, P.; Nikolakakis, I. Spirulina as a functional ingredient in broiler chicken diets. *S. Afr. J. Anim. Sci.* **2016**, *46*, 94. [CrossRef]

50. Hadžić, D. Using of Black Soldier Fly (Hermetia Illucens) Larvae Meal in Fish Nutrition. In *Answers for Forthcoming Challenges in Modern Agriculture*; Brka, M.O.-M.E., Karić, L., Falan, V., Toroman, A., Eds.; Springer: Cham, Switzerland, 2020; pp. 132–140.

51. Ewald, N. Fatty Acid Composition of Black Soldier Fly Larvae. Master's Thesis, Swedish University of Agricultural Sciences, Uppsala, Sweden, 2019.

52. Ushakova, N.; Brodskii, E.; Kovalenko, A.; Bastrakov, A.; Kozlova, A.; Pavlov, D. Characteristics of lipid fractions of larvae of the black soldier fly Hermetia illucens. *Dokl. Biochem. Biophys.* **2016**, *468*, 209–212. [CrossRef]

53. Schiavone, A.; Dabbou, S.; Petracci, M.; Zampiga, M.; Sirri, F.; Biasato, I.; Gai, F.; Gasco, L. Black soldier fly defatted meal as a dietary protein source for broiler chickens: Effects on carcass traits, breast meat quality and safety. *Animal* **2019**, *13*, 2397–2405. [CrossRef] [PubMed]

54. Cullere, M.; Schiavone, A.; Dabbou, S.; Gasco, L.; Dalle Zotte, A. Meat quality and sensory traits of finisher broiler chickens fed with black soldier fly (*Hermetia Illucens* L.) larvae fat as alternative fat source. *Animals* **2019**, *9*, 140. [CrossRef] [PubMed]

55. Shahidi, F.; Pegg, R.B. Hexanal as an indicator of the flavor deterioration of meat and meat products. In *Lipids in Food Flavours*; Ho, C.T., Hartman, T.G., Eds.; ACS Publications: Denver, CO, USA, 1994; pp. 256–279.

56. Barroeta, A. Nutritive value of poultry meat: Relationship between vitamin E and PUFA. *World Poult. Sci. J.* **2007**, *63*, 277–284. [CrossRef]

57. Schneiderová, D.; Zelenka, J.; Mrkvicová, E. Poultry meat production as a functional food with a voluntary n-6 and n-3 polyunsaturated fatty acids ratio. *Czech. J. Anim. Sci.* **2007**, *52*, 203–213. [CrossRef]

58. Whitfield, F.B.; Mottram, D.S. Volatiles from interactions of Maillard reactions and lipids. *Crit. Rev. Food Sci. Nutr.* **1992**, *31*, 1–58. [CrossRef]

59. Belitz, H.D.; Grosch, W.; Schieberle, P. 5. Aroma Substances. In *Food Chemistry*; Springer: Berlin/Heidelberg, Germany, 2004; pp. 319–376.

60. Vrabec, V.; Kulma, M.; Cocan, D. Insects as an Alternative Protein Source for Animal Feeding: A Short Review about Chemical Composition. *Bul. Univ. Agric. Sci. Vet. Med. Cluj-Napoca Anim. Sci. Biotechnol.* **2014**, *72*, 116–126. [CrossRef]

61. Pestana, J.; Puerta, B.; Santos, H.; Madeira, M.; Alfaia, C.; Lopes, P.; Pinto, R.; Lemos, J.; Fontes, C.; Lordelo, M. Impact of dietary incorporation of Spirulina (Arthrospira platensis) and exogenous enzymes on broiler performance, carcass traits, and meat quality. *Poult. Sci.* **2020**, *99*, 2519–2532. [CrossRef]

62. Schiavone, A.; De Marco, M.; Martínez, S.; Dabbou, S.; Renna, M.; Madrid, J.; Hernandez, F.; Rotolo, L.; Costa, P.; Gai, F. Nutritional value of a partially defatted and a highly defatted black soldier fly larvae (Hermetia illucens L.) meal for broiler chickens: Apparent nutrient digestibility, apparent metabolizable energy and apparent ileal amino acid digestibility. *J. Anim. Sci. Biotech.* **2017**, *8*, 51. [CrossRef]

63. Gatrell, S.K.; Kim, J.; Derksen, T.J.; O'Neil, E.V.; Lei, X.G. Creating $\omega$-3 fatty-acid-enriched chicken using defatted green microalgal biomass. *J. Agric. Food Chem.* **2015**, *63*, 9315–9322. [CrossRef]

64. Laroche, M.; Perreault, V.; Marciniak, A.; Gravel, A.; Chamberland, J.; Doyen, A. Comparison of Conventional and Sustainable Lipid Extraction Methods for the Production of Oil and Protein Isolate from Edible Insect Meal. *Foods* **2019**, *8*, 572. [CrossRef]

65. Moghadam, M.B.; Shehab, A.; Cherian, G. Methionine supplementation augments tissue n-3 fatty acid and tocopherol content in broiler birds fed flaxseed. *Anim. Feed. Sci. Technol.* **2017**, *228*, 149–158. [CrossRef]

66. Zhang, M.; Chen, X.; Hayat, K.; Duhoranimana, E.; Zhang, X.; Xia, S.; Yu, J.; Xing, F. Characterization of odor-active compounds of chicken broth and improved flavor by thermal modulation in electrical stewpots. *Food Res. Int.* **2018**, *109*, 72–81. [CrossRef] [PubMed]

67. Ayseli, M.T.; Filik, G.; Selli, S. Evaluation of volatile compounds in chicken breast meat using simultaneous distillation and extraction with odour activity value. *J. Food Nutr. Res.* **2014**, *53*, 137–142.

68. Ba, H.V.; Hwang, I.; Jeong, D.; Touseef, A. Principle of Meat Aroma Flavors and Future Prospect. In *Latest Research into Quality Control*; Akyar, I., Ed.; InTech: Rijeka, Croatia, 2012; pp. 146–176.

69. Qi, J.; Wang, H.; Zhou, G.; Xu, X.; Li, X.; Bai, Y.; Yu, X. Evaluation of the taste-active and volatile compounds in stewed meat from the Chinese yellow-feather chicken breed. *Int. J. Food Prop.* **2018**, *20*, S2579–S2595. [CrossRef]

70. Gkarane, V.; Brunton, N.P.; Harrison, S.M.; Gravador, R.S.; Allen, P.; Claffey, N.A.; Diskin, M.G.; Fahey, A.G.; Farmer, L.J.; Moloney, A.P. Volatile Profile of Grilled Lamb as Affected by Castration and Age at Slaughter in Two Breeds. *J. Food Sci.* **2018**, *83*, 2466–2477. [CrossRef] [PubMed]

71. Ho, C.-T.; Chen, Q. Lipids in food flavors. An overview. In *Lipids in Food Flavors*; Ho, C.-T., Hartman, T.G., Eds.; American Chemical Society: Washington, DC, USA, 1994; pp. 1–14.

72. Mottram, D.S. Flavour formation in meat and meat products: A review. *Food Chem.* **1998**, *62*, 415–424. [CrossRef]

 *foods*

*Article*

# Effects of Photoperiod Regime on Meat Quality, Oxidative Stability, and Metabolites of Postmortem Broiler Fillet (*M. Pectoralis major*) Muscles

Jacob R. Tuell [1], Jun-Young Park [1,2], Weichao Wang [3], Bruce Cooper [4], Tiago Sobreira [4], Heng-Wei Cheng [5] and Yuan H. Brad Kim [1,*]

1   Meat Science and Muscle Biology Laboratory, Department of Animal Sciences, Purdue University, West Lafayette, IN 47907, USA; tuell@purdue.edu (J.R.T.); park1126@purdue.edu (J.-Y.P.)
2   Division of Applied Life Sciences (BK 21 Plus), Gyeongsang National University, Gyeongsangnam-do 52828, Korea
3   Department of Animal Sciences, Purdue University, West Lafayette, IN 47907, USA; wang2077@purdue.edu
4   Bindley Bioscience Center, Purdue University, West Lafayette, IN 47907, USA; brcooper@purdue.edu (B.C.); sobreira@purdue.edu (T.S.)
5   Livestock Behavior Research Unit, USDA-Agricultural Research Service, West Lafayette, IN 47907, USA; Heng-Wei.Cheng@ARS.USDA.GOV
*   Correspondence: bradkim@purdue.edu; Tel.: +1-765-496-1631

Received: 3 February 2020; Accepted: 14 February 2020; Published: 19 February 2020

**Abstract:** The objective of this study was to evaluate the effects of photoperiod on meat quality, oxidative stability, and metabolites of broiler fillet (*M. Pectoralis major*) muscles. A total of 432 broilers was split among 4 photoperiod treatments [hours light(L):dark(D)]: 20L:4D, 18L:6D, 16L:8D, and 12L:12D. At 42 days, a total of 48 broilers (12 broilers/treatment) was randomly selected and harvested. At 1 day postmortem, fillet muscles were dissected and displayed for 7 days. No considerable impacts of photoperiods on general carcass and meat quality attributes, such as carcass weight, yield, pH, water-holding capacity, and shear force, were found ($p > 0.05$). However, color and oxidative stability were influenced by the photoperiod, where muscles from 20L:4D appeared lighter and more discolored, coupled with higher lipid oxidation ($p < 0.05$) and protein denaturation ($p = 0.058$) compared to 12L:12D. The UPLC–MS metabolomics identified that 20 metabolites were different between the 20L:4D and 12L:12D groups, and 15 were tentatively identified. In general, lower aromatic amino acids/dipeptides, and higher oxidized glutathione and guanine/methylated guanosine were observed in 20L:4D. These results suggest that a photoperiod would result in no considerable impact on initial meat quality, but extended photoperiods might negatively impact oxidative stability through an alteration of the muscle metabolites.

**Keywords:** antioxidative status; aromatic amino acids; broiler; lighting program; meat quality; metabolite profiling

## 1. Introduction

Broiler chicken production plays a key role in supplying consumers with high quality protein, as evidenced by the steady increase in US broiler chicken production from 50.4 billion pounds in 2008 [1] to 56.8 billion pounds in 2018 [2]. While genetic improvement plays the most critical factor in meeting this continuously rising demand [3,4], environmental factors such as rearing temperature, nutrition, and lighting schedule/intensity must be managed to fully capitalize on the genetic potential of modern broiler chickens [5–7].

The poultry industry has traditionally reared broilers under long photoperiods to maximize growth performance [7,8]. Several studies have identified continuous or near-continuous photoperiod regimes as positively impacting breast meat yield, feed consumption and conversion, and growth rates [9–11]. However, there is growing evidence that long photoperiods might negatively impact broiler health and welfare. For example, rapid growth rates in broiler chickens have been implicated in causing skeletal deformities, metabolic disorders, and increased mortality [7,12]. The increased breast meat yield commonly observed in long photoperiod regimes has often been shown to be inversely related to the yield of the thigh and drum [10,11,13], contributing to an increased frequency of leg abnormalities with impaired walking ability [9,14,15]. These detriments to skeletal health associated with long photoperiods might also associate with the disruption of the normal diurnal rhythm, which plays a critical role in regulating bone modeling/remodeling [16].

Several studies have investigated the impacts of photoperiod on broiler performance and carcass traits, but few studies have examined its potential impacts on meat quality. Previously, Li et al. [17] found that breast meat from broilers reared under a 12L:12D photoperiod has lower malondialdehyde (MDA, a secondary lipid oxidation product) concentrations compared to 23L:1D controls. Guob et al. [18] corroborated this finding, reporting higher blood serum MDA levels in broilers reared at a longer photoperiod, indicative of increased oxidative stress. It has been well-established that chronic stress is detrimental to broiler meat quality and oxidative stability [19–21]. However, to our knowledge, no studies have evaluated the impact of photoperiod on meat quality and oxidative stability of broiler fillet (*M. Pectoralis major*) muscles during aerobic display. Aerobic packaging (using oxygen permeable polyvinylchloride film/overwrap with a polystyrene foam tray) is the most common method of packaging for fresh broiler meat products in the US [22], despite often exhibiting a shorter shelf-life [23,24]. As such, aerobic display storage could potentially exacerbate any oxidative defect already present in fresh broiler products. The aim of this study was to evaluate the effects of photoperiod on meat quality, oxidative stability, and metabolites of postmortem broiler fillet (*M. Pectoralis major*) muscles.

## 2. Materials and Methods

All animal use and procedures were approved by the Purdue Animal Care and Use Committee (1712001657).

### 2.1. Photoperiod Treatments

Ross 308 broiler chicks ($n$ = 432) at 1 day of age were weighed in groups ($n$ = 18/group) and allocated among 24 pens (110 cm × 110 cm) for equal distribution of weight across the pens. The pens were randomly assigned to one of four photoperiod treatment rooms at the Poultry Unit of the Animal Sciences Research and Education Center at Purdue University. Lighting schedule regimens were performed as follows: [hours light(L):dark(D)] 20L:4D, 18L:6D, 16L:8D, and 12L:12D. For all treatments, the birds were provided with constant (24L:0D) lighting at 30 lux at 1 day of age, reduced to 23L:1D from day 2 to day 7. After this, the photoperiods were adjusted in gradual increments until reaching the final expected photo schedule at day 14, which were maintained until 42 days of age. Brooder temperature was 34 °C until day 3, after which temperature was gradually reduced until 21–24 °C was reached and maintained until 42 days of age.

All broilers were provided a starter diet with 23.43% crude protein (CP) and 3,050 kcal metabolizable energy (ME)/kg from day 1 to day 14, a grower diet with 22.81% CP and 3,150 kcal ME/kg from day 15 to day 28, and a finisher diet with 19.17% CP and 3,200 kcal ME/kg from day 29 until day 42. Food and water was provided ad libitum throughout the course of the study.

### 2.2. Harvest and Sample Preparation

At day 42, 2 broilers per pen ($n$ = 12/treatment; a total of 48 broilers) were randomly selected, transported approximately 30 min, and harvested under the standard procedures. The hot carcass weight (HCW) was recorded as the weight of the carcass following plucking, evisceration, and removal of the head and feet, prior to carcass chilling. The carcasses were chilled in a commercial air cooler with an ambient temperature of 2 °C. The chilled carcass weight (CCW) was recorded as the weight of the carcass after 24 h of chilling. Cooler shrink was calculated as the percent difference between the HCW and CCW. Broiler fillet (*M. Pectoralis major*) muscles, with *M. Pectoralis minor* and skin removed, were dissected from both sides of each carcass at 1 day postmortem and weighed. Fillet yield was calculated as the weight of the fillet muscles as a percentage of CCW.

Approximately 30 g of each left side fillet muscle was collected for drip loss measurement. The remaining left side muscles were frozen and stored at −80 °C as 1 day postmortem samples until later analyses. The right side fillets were displayed for 7 days at 2 °C under 1450 lux fluorescent lighting to mimic the retail store conditions. The samples were displayed bone-side up on polystyrene foam trays with soaking pads, overwrapped with a commercial polyvinyl chloride film. After aerobic display storage, the samples were frozen and stored at −80 °C as 7 day displayed samples. Prior to analysis, the samples were frozen in liquid nitrogen and pulverized to form a homogenous powder by using a commercial blender (Waring Products, Inc., Stamford, CT, USA).

*2.3. pH Measurement*

The pH of the fillet muscles was measured at 1 d postmortem, following carcass chilling. Values were obtained from the left fillet muscles in duplicates by using a pH probe (Hanna Instrument, Inc., Warner, NH, USA) calibrated with pH 4 and 7 buffers prior to the analysis.

*2.4. Water-Holding Capacity (WHC)*

Prior to any measure of WHC, the samples were gently blotted with a paper towel to remove excess moisture from the muscle surface. Drip loss was measured in accordance with the method published by Honikel [25], with some modifications. At 1 day postmortem, approximately 30 grams of the caudal portion of the left side fillet muscles without skin and visible connective tissue was suspended with netting in an airtight container, at 2 °C for 48 h. The drip loss was expressed as the percentage difference between the weight of the sample prior to and after hanging storage. Display weight loss was determined as the percent difference between the initial and final weights of the samples before and after 7 days of aerobic display storage. Freezing/thawing loss was assessed as the percent difference between weights prior to and after freezing/thawing of 1 day postmortem samples at −80 °C and 24 h of thawing at 2 °C. Cooking loss was measured by cooking samples in a water-impermeable plastic bag submerged in an 80 °C water bath. Cooking temperature was monitored using a thermocouple (Type-T, Omega Engineering, Stamford, CT, USA) connected to a data logger (Madge Tech, Inc., Warner, NH, USA). After 71 °C was reached in the geometrical center of the fillet, the samples were immediately submerged in an ice water bath to halt the cooking process. Cooking loss was expressed as the percent change between the initial and final weight of the samples.

*2.5. Instrumental Tenderness*

The samples used for cooking loss were chilled at 4 °C for 16 h prior to the measurement of shear force. Six slices (1 cm × 1 cm) per sample were taken in a direction parallel to that of the muscle fibers. Slices were sheared perpendicularly to the fiber direction using a Warner-Bratzler type V-shaped blade attached to a TA-XT Plus Texture Analyser (Stable Micro System Ltd., UK) at 2 mm/sec. Peak shear force from the cores in Newtons was determined, and the mean values of the replicates were used for statistical analysis.

*2.6. Proximate Composition*

Proximate composition of the fillet muscles at 1 day postmortem was analyzed according to the AOAC methods [26]. Moisture was determined in triplicates at 100 °C using the oven air-drying method. Percent nitrogen was measured in duplicates following the Dumas combustion method (Leco, St. Joseph, MI, USA), and concentration of crude protein was determined by multiplying percent nitrogen by 6.25. Crude ash was measured in duplicates by combusting the dried samples in a 580 °C muffle furnace. Crude lipid was determined as follows, by the formula:

$$100\% - [\%\ \text{moisture} + \%\ \text{crude protein (wet matter basis)} + \%\ \text{crude ash (wet matter basis)}] \quad (1)$$

*2.7. Instrumental Color Attributes*

Instrumental color was assessed daily on 7 day displayed samples. Commission internationale de l'éclairage (CIE) L*, a*, and b* values were obtained in triplicates from three randomly selected locations per fillet using a CR-400 Chroma Meter (Konica Minolta, Chiyoda, Tokyo, Japan) equipped with a CIE standard illuminant $D_{65}$. CIE a* and b* data were used to determine the hue angle (discoloration) and chroma (saturation) values based on the American Meat Science Association meat color measurement guidelines [27].

*2.8. Oxidative Stability and Transmission Value*

Lipid oxidation was assessed using the 2-thiobarbituric acid reactive substances (TBARS) assay to determine the formation of MDA according to the method published by Buege and Aust [28] with modifications for broiler meat by Kim et al. [29]. Values of both day 1 and day 7 samples were obtained in duplicates by measuring absorbance of the obtained supernatant at 531 nm using a microplate spectrophotometer (Epoch, Biotek Instruments, Inc., Winooski, VT, USA) and multiplying by 5.54 to calculate the TBARS value. The values were expressed as milligrams MDA per kilogram of fillet muscle.

Protein oxidation, by measuring loss of thiol groups on day 1 and day 7 of aerobic display storage, was assessed in duplicates following the method published by Berardo et al. [30]. Absorbance of the sample filtrate was measured at 412 nm using a microplate spectrophotometer (Epoch, Biotek Instruments, Inc., Winooski, VT, USA), and the values of the thiol content presented in nanomoles thiol groups per milligram protein were determined using the Lambert–Beer formula ($\varepsilon_{412} = 14{,}000\ \text{M}^{-1}\ \text{cm}^{-1}$). Protein concentration of the sample filtrate was assessed using a bovine serum albumin standard curve.

Transmission values, a determinant of protein denaturation, of day 1 postmortem samples were assessed in duplicates using the method of Ockerman and Cahill [31] with modifications as described by Kim et al. [32]. Briefly, turbidity of the mixed sample, 1 milliliter of sample filtrate mixed with 5 milliliters of 0.1 M citric acid in 0.2 M sodium phosphate buffer (pH 4.6), was measured at 600 nm wavelength using a spectrophotometer (UV-1600PC, VWR International, LLC, Radnor, PA, USA).

*2.9. Fatty Acid Profile*

Intramuscular lipids were extracted from the fillet muscles in duplicates using the method of Folch et al. [33] with modifications by Shin and Ajuwon [34]. Briefly, fatty acid methyl esters (FAME) were prepared by a trans-esterification reaction, after which FAMEs were extracted in hexane. FAMEs were analyzed on a gas chromatograph (Varian CP 3900 with CP-8400 autosampler, Agilent, Santa Clara, CA, USA) equipped with a 105 m Rtx-2330 fused silica capillary GC column (10729, Restek, Bellefonte, PA, USA). Fatty acids were identified by comparison of retention times with the known standards (Supelco 37 components FAME Mix, Sigma Aldrich, St. Louis, MO, USA). Detected fatty acids were expressed in grams per 100 grams of intramuscular lipid.

*2.10. UPLC–MS Metabolite Profiling*

For metabolomics data, one sample per pen of the two extreme treatments (20L:4D and 12L:12D) was randomly selected and analyzed. Sample extraction and removal of protein was performed in accordance to the method published by Bligh and Dyer [35]. In brief, 100 mg of fillet muscle tissue powder was vortexed with 300 μL of chloroform and 300 μL of methanol, after which 300 μL of distilled water was added and centrifuged for 10 min at 16,000× *g*. The upper phase, containing the polar metabolites, was transferred to a microcentrifuge tube, evaporated to dryness with a vacuum concentrator, and reconstituted in 60 μL of HPLC diluent (95% water and 5% acetonitrile, containing 0.1% formic acid). Samples were sonicated, centrifuged, and transferred to HPLC vials.

UPLC–MS was performed using an Agilent 1290 Infinity II UPLC system (Agilent Technologies, Palo Alto, CA, USA) equipped with Waters Acquity HSS T3 separation column (2.1 mm × 100 mm, 1.8 μm) and HSS T3 guard column (2.1 mm × 5 mm, 1.8 μm) (Waters, Milford, MA, USA). A gradient of water and acetonitrile was used. Following chromatographic separation, an Agilent 6545 quadrupole time-of-flight (Q-TOF) mass spectrometer was employed. Mass data were collected and analyzed with Agilent MassHunter B.06 software from an m/z of 70–1,000. Agilent Reference Mass Correction Solution (G1969-85001) was infused to improve mass accuracy. Agilent ProFinder B.10 was employed for peak deconvolution and alignment. Peaks were annotated using the HMDB (www.hmdb.ca) metabolite database with a mass error ≤ 10 ppm.

### 2.11. Statistical Analysis

The experimental design was a randomized complete block with the photoperiod treatment (20L:4D, 18L:6D, 16L:8D, and 12L:12D) as the fixed effect and the pen ($n$ = 6/treatment) as the experimental unit. Individual broilers and their interactions with the fixed effect were considered as a random effect. Data including the aerobic display storage period (i.e., color and oxidative stability) were considered as a split plot design, where the photoperiod treatment was a whole plot and the display duration was a sub-plot. Data were analyzed using the PROC MIXED and PROC GLIMMIX procedures of SAS 9.4 (SAS Institute Inc., Cary, NC, USA) with the PDIFF option for separation of the least square means ($p < 0.05$). Trends were defined as ($0.10 > p \geq 0.05$). For metabolomics data, one sample from each photoperiod treatment lacked adequate correlation with others within the respective treatment and was omitted from the analysis, resulting in 5 replicates per treatment. Metabolomics data were analyzed using the unpaired $t$-test significance analysis, and significant metabolites ($p < 0.05$) were used for principal component analysis (PCA) modeling.

## 3. Results

### 3.1. Carcass and Meat Quality

Effects of photoperiod on carcass traits including HCW, CCW, cooler shrink, fillet weight, and fillet yield are presented in Table 1. Overall, the photoperiod did not result in any considerable impacts on general carcass quality attributes ($p > 0.05$). However, there was a tendency for carcasses from 20L:4D to lose more weight, as shown by the greater cooler shrink percentage compared to other shorter photoperiod treatments ($p = 0.070$).

**Table 1.** Effect of photoperiod on broiler carcass characteristics ($n$ = 6/treatment).

| Trait | 20L:4D | 18L:6D | 16L:8D | 12L:12D | SEM | Significance of $p$-Value |
|---|---|---|---|---|---|---|
| Hot carcass weight (kg) | 2.28 | 2.42 | 2.22 | 2.38 | 0.07 | 0.199 |
| Chilled carcass weight (kg) | 2.18 | 2.33 | 2.16 | 2.30 | 0.06 | 0.182 |
| Cooler shrink (%) | 4.7 | 3.7 | 3.2 | 3.3 | 0.4 | 0.070 |
| Fillet weight (g) | 547.7 | 575.0 | 537.8 | 593.4 | 21.4 | 0.269 |
| Fillet yield (%) | 25.2 | 24.8 | 25.0 | 25.8 | 0.6 | 0.681 |

Similarly, no significant effects of photoperiod on pH, WHC, and instrumental tenderness of broiler fillet muscles were found (Table 2, $p > 0.05$). There was a trend of fillets from 16L:8D showing higher freezing/thawing loss compared to others ($p = 0.098$).

Proximate composition of the fillet muscles was unaffected by the treatments as well (Table 3, $p > 0.05$).

**Table 2.** Effect of photoperiod on pH, water-holding capacity, and instrumental shear force of broiler fillet (*M. Pectoralis major*) muscles (*n* = 6/treatment).

| Trait | | 20L:4D | 18L:6D | 16L:8D | 12L:12D | SEM | Significance of *p*-Value |
|---|---|---|---|---|---|---|---|
| pH (24 h) | | 5.93 | 5.93 | 5.91 | 5.95 | 0.03 | 0.797 |
| Water-holding capacity (%) | Drip loss | 2.9 | 3.7 | 4.5 | 3.3 | 0.8 | 0.534 |
| | Freezing/thawing loss | 2.9 | 3.1 | 4.5 | 2.9 | 0.5 | 0.098 |
| | Display weight loss | 2.9 | 2.7 | 3.1 | 3.2 | 0.3 | 0.773 |
| | Cooking loss | 11.5 | 12.2 | 12.3 | 12.4 | 0.7 | 0.808 |
| Shear force (*N*) | | 22.3 | 17.4 | 21.6 | 24.8 | 2.4 | 0.228 |

**Table 3.** Effect of photoperiod on proximate composition (wet-matter basis) of broiler fillet (*M. Pectoralis major*) muscles (*n* = 6/treatment).

| Trait | 20L:4D | 18L:6D | 16L:8D | 12L:12D | SEM | Significance of *p*-Value |
|---|---|---|---|---|---|---|
| Moisture (%) | 74.6 | 75.0 | 74.4 | 74.2 | 0.3 | 0.350 |
| Protein (%) | 22.0 | 21.9 | 22.3 | 22.1 | 0.3 | 0.914 |
| Lipid (%) | 1.9 | 1.5 | 1.6 | 2.1 | 0.2 | 0.133 |
| Ash (%) | 1.6 | 1.6 | 1.7 | 1.7 | 0.1 | 0.190 |

### 3.2. Color, Oxidative Stability, and Fatty Acid Profile

There were significant interactions between the photoperiod treatment and aerobic display storage on CIE L* (lightness), CIE a* (redness), CIE b* (yellowness), and hue angle (discoloration) values of the fillet muscles (Table 4, $p < 0.05$). At day 1 of display, the muscles from 20L:4D exhibited a lighter color than the muscles from 18L:6D and 12L:12D ($p < 0.05$) but not 16L:8D ($p > 0.05$). At day 2 of display and forward, 18L:6D fillets had lower CIE L* values compared to the 20L:4D only ($p < 0.05$), while 16L:8D and 12L:12D were intermediates ($p > 0.05$). For CIE a* values, the fillets from 16L:8D were higher than other treatments on day 1 of display ($p < 0.05$), although both 18L:6D and 12L:12D were redder in color than 20L:4D ($p < 0.05$). By day 7 of display, 16L:8D fillets maintained higher redness than 20L:4D ($p < 0.05$), but were not different from 18L:6D and 12L:12D groups ($p > 0.05$). For CIE b*, there were no differences across treatments from day 1 to day 5 of display ($p > 0.05$). However, on day 6 and day 7, the fillets from 18L:6D were less yellow in color than 16L:8D ($p < 0.05$) only.

A significant interaction between the photoperiod treatment and the display time on hue angle values was found (Table 4). The highest hue angle values (indication of discoloration) were observed in 20L:4D compared to other photoperiod treatments on day 1 and day 2 of display ($p < 0.05$). This could be attributed to the lower CIE a* values observed in 20L:4D during the same display duration, coupled with numerically higher CIE b* values. From day 3 and onwards, the differences across treatments on each respective display day were less pronounced, although 18L:6D maintained a lower hue angle than 20L:4D ($p < 0.05$) but was not different compared to 16L:8D and 12L:12D treatments at any point of the display ($p > 0.05$). Chroma values were affected by display storage duration only ($p < 0.05$, data not shown), regardless of the photoperiod group ($p > 0.05$). Overall, chroma values exhibited a similar pattern as CIE b* values, increasing from day 1 to day 4 of display before decreasing by day 7.

Significant main effects of the photoperiod and the display period were observed for the TBARS values (Table 5, $p < 0.05$). Higher concentration of MDA was found in the fillet muscle samples from 20L:4D (0.50 mg MDA/kg fillet) and 18L:6D (0.52 mg MDA/kg fillet), compared to the 12L:12D (0.37 mg MDA/kg fillet) ($p < 0.05$), while 16L:8D (0.42 mg MDA/kg fillet) was intermediate ($p > 0.05$). As expected, MDA accumulated in the displayed samples from day 1 to day 7 ($p < 0.05$). There was no significant interaction between the photoperiod and the display observed in the TBARS values. There was, however, an interaction between the photoperiod and the display period for protein oxidation, as assessed by the content of the thiol groups ($p < 0.05$). A detectable loss in thiol groups was found in 20L:4D from day 1 to day 7 only ($p < 0.05$).

**Table 4.** Effect of photoperiod on $D_{65}$ instrumental color attributes [CIE L* (lightness), CIE a* (redness), CIE b* (yellowness), hue angle (discoloration), and chroma (color intensity)] of broiler filet (*M. Pectoralis major*) muscles during 7 days of aerobic display ($n$ = 6/treatment).

| Trait | $D$ [2] | $P$ [1] | | | | SEM | Significance of $p$-Value | | |
|---|---|---|---|---|---|---|---|---|---|
| | | 20L:4D | 18L:6D | 16L:8D | 12L:12D | | $P$ | $D$ | $P \times D$ |
| CIE L* | 1 d | 49.8 [abc] | 46.5 [ij] | 48.6 [bcdefg] | 47.7 [fghij] | | | | |
| | 2 d | 50.8 [a] | 48.8 [bcdef] | 49.7 [abcd] | 49.5 [abcde] | | | | |
| | 3 d | 50.8 [a] | 48.8 [bcdef] | 50.0 [ab] | 49.5 [abcde] | | | | |
| | 4 d | 49.6 [abcde] | 47.2 [ghij] | 49.0 [bcdef] | 48.6 [bcdefg] | 0.6 | 0.037 | <0.001 | 0.035 |
| | 5 d | 49.2 [bcdef] | 46.8 [ij] | 48.9 [bcdef] | 48.2 [cdefghi] | | | | |
| | 6 d | 49.1 [bcdef] | 46.5 [i] | 48.5 [cdefgh] | 48.1 [efghij] | | | | |
| | 7 d | 49.3 [bcdef] | 46.9 [hij] | 49.1 [bcdef] | 48.1 [defghij] | | | | |
| CIE a* | 1 d | 2.9 [efghi] | 3.6 [bcd] | 4.4 [a] | 3.7 [bc] | | | | |
| | 2 d | 3.0 [efgh] | 3.4 [cde] | 4.0 [b] | 3.5 [bcde] | | | | |
| | 3 d | 3.0 [efgi] | 3.4 [cdef] | 3.6 [bcd] | 3.4 [cdef] | | | | |
| | 4 d | 2.7 [ghijk] | 3.2 [cdefg] | 3.5 [bcde] | 3.1 [defg] | 0.2 | 0.018 | <0.001 | 0.010 |
| | 5 d | 2.4 [ijk] | 2.9 [fghi] | 3.0 [efgh] | 3.0 [efghi] | | | | |
| | 6 d | 2.3 [ijk] | 2.7 [ghijk] | 3.2 [cdefg] | 2.5 [hijk] | | | | |
| | 7 d | 2.3 [k] | 2.8 [ghijk] | 2.8 [fghij] | 2.5 [hijk] | | | | |
| CIE b* | 1 d | 7.0 [abcdefgh] | 6.2 [h] | 7.0 [abcdefgh] | 6.5 [efgh] | | | | |
| | 2 d | 6.9 [abcdefgh] | 6.3 [gh] | 6.8 [defgh] | 6.2 [h] | | | | |
| | 3 d | 6.8 [cdefgh] | 6.2 [h] | 6.9 [bcdefgh] | 6.4 [gh] | | | | |
| | 4 d | 7.7 [abcd] | 7.0 [abcdefgh] | 7.9 [a] | 7.3 [abcdefg] | 0.4 | 0.347 | <0.001 | 0.026 |
| | 5 d | 7.5 [abcde] | 6.9 [bcdefgh] | 7.8 [abc] | 7.4 [abcdef] | | | | |
| | 6 d | 7.5 [abcdef] | 6.8 [defgh] | 7.9 [ab] | 7.2 [abcdefgh] | | | | |
| | 7 d | 7.0 [abcdefgh] | 6.4 [fgh] | 7.6 [abcd] | 7.2 [abcdefgh] | | | | |
| Hue angle | 1 d | 67.0 [cdef] | 59.5 [ij] | 57.4 [j] | 60.7 [hij] | | | | |
| | 2 d | 66.5 [cdefg] | 61.1 [hij] | 59.0 [ij] | 60.9 [hij] | | | | |
| | 3 d | 66.4 [defg] | 61.2 [hij] | 62.1 [ghi] | 62.5 [ghi] | | | | |
| | 4 d | 70.6 [abc] | 64.9 [fgh] | 65.8 [efg] | 67.2 [bcdef] | 1.6 | 0.049 | <0.001 | 0.007 |
| | 5 d | 71.5 [ab] | 66.5 [cdefg] | 68.4 [abcdef] | 68.1 [abcdef] | | | | |
| | 6 d | 72.4 [a] | 67.6 [bcdef] | 67.8 [bcdef] | 70.3 [abcde] | | | | |
| | 7 d | 71.7 [ab] | 66.4 [cdefg] | 69.3 [abcdef] | 70.4 [abcd] | | | | |
| Chroma | 1 d | 7.6 | 7.3 | 8.4 | 7.5 | | | | |
| | 2 d | 7.6 | 7.2 | 7.9 | 7.3 | | | | |
| | 3 d | 7.5 | 7.1 | 7.8 | 7.3 | | | | |
| | 4 d | 8.2 | 7.8 | 8.7 | 8.0 | 0.3 | 0.309 | <0.001 | 0.208 |
| | 5 d | 7.9 | 7.6 | 8.4 | 8.0 | | | | |
| | 6 d | 7.8 | 7.4 | 8.5 | 7.7 | | | | |
| | 7 d | 7.4 | 7.1 | 8.1 | 7.7 | | | | |

[1] Photoperiod effect. [2] Display period effect. [a–k] Means lacking a common superscript within a color attribute differ due to the interaction of photoperiod treatment and display period ($p$ < 0.05).

**Table 5.** Effect of photoperiod on the 2-thiobarbituric acid reactive substance values, thiol content, and transmission value of broiler fillet (*M. Pectoralis major*) muscles, during aerobic display storage (*n* = 6/treatment).

| Trait | | $P$[1] | | | | | $D$[2] | | | Significance of *p*-Value | | |
|---|---|---|---|---|---|---|---|---|---|---|---|---|
| | D | 20L:4D | 18L:6D | 16L:8D | 12L:12D | SEM | 1d | 7d | SEM | P | D | P × D |
| TBARS[3] | - | 0.50 [a] | 0.52 [a] | 0.42 [ab] | 0.37 [b] | 0.04 | 0.38 [y] | 0.53 [x] | 0.02 | 0.038 | <0.001 | 0.312 |
| Transmission value (%) | - | 53.2 | 37.1 | 48.8 | 33.9 | 5.4 | - | - | - | 0.058 | - | - |
| Thiol content[4] | 1d | 28.9 [A] | 27.0 [AB] | 27.4 [AB] | 26.8 [AB] | 1.0 | 27.5 | 26.8 | 0.5 | 0.970 | 0.114 | 0.025 |
| | 7d | 25.9 [B] | 27.6 [AB] | 27.0 [AB] | 26.8 [AB] | | | | | | | |

[1] Photoperiod effect. [2] Display period effect. [3] 2-thiobarbituric acid reactive substances values expressed as milligrams MDA per kilogram fillet muscle. [4] Thiol content expressed as nanomoles thiol groups per milligrams protein. [a,b] Means lacking a common superscript within a row differ due to photoperiod treatment (*p* < 0.05). [x,y] Means lacking a common superscript within a row differ due to display period (*p* < 0.05). [A,B] Means lacking a common superscript differ due to the interaction of the photoperiod treatment and the display period (*p* < 0.05).

In terms of protein denaturation, there was a strong trend that the fillet muscle samples from 20L:4D and 16L:8D had a higher transmission than the samples from 18L:6D and 12L:12D groups ($p = 0.058$), indicating a higher degree of denaturation in the sarcoplasmic protein in those samples.

Most detected fatty acids were not different across the photoperiod treatments (Table 6, $p > 0.05$). Higher polyunsaturated fatty acids (PUFA) were found in lower photoperiod groups (16L:8D and 12L:12D) compared to 20L:4D ($p < 0.05$). Of the PUFA, higher omega-3 fatty acids were found in 12L:12D compared to the 20L:4D ($p < 0.05$), while higher omega-6 fatty acids were found in both 16L:8D and 12L:12D compared to 20L:4D ($p < 0.05$). The differences in omega-3 and omega-6 fatty acid contents were not pronounced enough to cause a significant difference in the omega-6:omega-3 ratio ($p = 0.114$), nor the ratio of unsaturated to saturated fatty acids ($p = 0.588$). There was, however, a tendency for higher palmitic (C16:0) acid in 20L:4D compared to other photoperiod groups ($p = 0.082$).

**Table 6.** Effect of photoperiod on the fatty acid profile (grams per 100 grams of intramuscular lipid) of broiler fillet (*M. Pectoralis major*) muscles ($n = 6$/treatment).

| Fatty acid | 20L:4D | 18L:6D | 16L:8D | 12L:12D | SEM | Significance of *p*-Value |
|---|---|---|---|---|---|---|
| C14:0 | 0.37 | 0.33 | 0.35 | 0.34 | 0.01 | 0.328 |
| C14:1 | 0.09 | 0.08 | 0.07 | 0.06 | 0.01 | 0.108 |
| C16:0 | 19.9 | 19.1 | 19.2 | 18.9 | 0.3 | 0.082 |
| C16:1 | 2.91 | 2.61 | 2.36 | 2.62 | 0.17 | 0.181 |
| C18:0 | 7.58 | 7.93 | 7.73 | 7.72 | 0.30 | 0.872 |
| C18:1(n9) | 24.0 | 23.3 | 22.6 | 22.8 | 0.7 | 0.474 |
| C18:2(n6) | 25.6 | 26.5 | 27.2 | 27.1 | 0.6 | 0.234 |
| C18:3(n3) | 2.17 | 2.28 | 2.37 | 2.35 | 0.11 | 0.577 |
| C18:3(n6) | 0.32 | 0.32 | 0.31 | 0.31 | 0.01 | 0.961 |
| C20:0 | 0.06 | 0.04 | 0.05 | 0.05 | 0.01 | 0.496 |
| C20:1(n9) | 0.19 | 0.19 | 0.18 | 0.17 | 0.01 | 0.776 |
| C20:3(n3) | 0.09 | 0.09 | 0.08 | 0.12 | 0.02 | 0.503 |
| C20:3(n6) | 1.12 | 1.23 | 1.35 | 1.17 | 0.27 | 0.935 |
| C20:4(n6) | 3.68 | 4.05 | 3.98 | 4.05 | 0.32 | 0.823 |
| C20:5(n3) | 0.26 | 0.25 | 0.23 | 0.25 | 0.02 | 0.873 |
| C22:1(n9) | 0.02 | 0.05 | 0.03 | 0.01 | 0.02 | 0.504 |
| C22:6(n3) | 0.52 | 0.59 | 0.57 | 0.67 | 0.06 | 0.359 |
| SFA [1] | 27.9 | 27.5 | 27.4 | 27.2 | 0.3 | 0.463 |
| MUFA [2] | 27.2 | 26.3 | 25.3 | 25.6 | 0.8 | 0.346 |
| PUFA [3] | 33.7 [b] | 35.3 [ab] | 36.1 [a] | 36.1 [a] | 0.6 | 0.032 |
| Total UFA [4] | 61.0 | 61.6 | 61.4 | 61.7 | 0.8 | 0.917 |
| n3 | 3.04 [b] | 3.21 [ab] | 3.24 [ab] | 3.38 [a] | 0.07 | 0.022 |
| n6 | 30.7 [b] | 32.1 [ab] | 32.9 [a] | 32.7 [a] | 0.5 | 0.037 |
| n6:n3 | 10.1 | 10.0 | 10.2 | 9.7 | 0.1 | 0.114 |
| UFA:SFA | 2.18 | 2.25 | 2.25 | 2.28 | 0.05 | 0.588 |

[1] Saturated fatty acids. [2] Monounsaturated fatty acids. [3] Polyunsaturated fatty acids. [4] Unsaturated fatty acids. [a,b] Means lacking a common superscript within a row differ due to the photoperiod treatment ($p < 0.05$).

### 3.3. Metabolite Profiling

Metabolite profiling was conducted for the samples from the two extreme treatments (20L:4D and 12L:12D) in order to obtain insight into the biological and biochemical processes that might be differently affected by the photoperiod treatments. Untargeted metabolite profiling detected 1472 metabolites in the fillet muscle samples from the 20L:4D and 12L:12D treatments. PCA was performed to discriminate between the 20L:4D and 12L:12D treatments based on the 20 significant metabolites with $p < 0.05$ (Figure 1). PC1 was shown to explain 31.65% of the total variance and PC2 as 17.02%, leading to a total PC of 48.67%.

Of the 20 metabolites found to be significantly impacted by the photoperiod treatment, 15 were tentatively identified using the HMDB metabolite database with a mass error of $\leq 10$ ppm (Table 7). Overall, the muscles from 12L:12D were found to be higher in amino acids/dipeptides of aromatic amino acids (tyrosine, tryptophan, phenylalanine) with leucine/isoleucine, as well as piperidine. Samples from 20L:4D were found to have a higher oxidized glutathione, methylated histidine, and guanine and methylated/demethylated guanosine ($p < 0.05$).

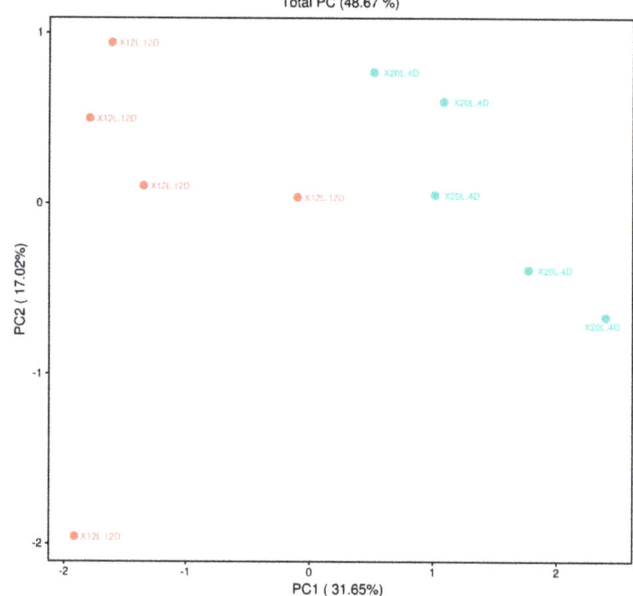



**Figure 1.** PCA modeling based on the 20 metabolites present within the broiler fillet (*M. Pectoralis major*) muscles were found to be different ($p < 0.05$) between the 20L:4D and 12L:12D photoperiod treatment groups ($n$ = 5/treatment).

**Table 7.** Metabolites significantly different between the 20L:4D and 12L:12D photoperiod treatments ($n$ = 5/treatment).

| Component Name [1] | Mass (Da) | Formula | Δppm | FC [2] 20L:4D/12L:12D | *p*-Value |
|---|---|---|---|---|---|
| *Amino acids, peptides, and analogues* | | | | | |
| L-Phenylalanine | 165.0790 | $C_9H_{11}NO_2$ | 0 | −0.8094 | 0.041 |
| Tryptophan-Isoleucine/Leucine | 317.1758 | $C_{17}H_{23}N_3O_3$ | 6 | −0.7910 | 0.029 |
| N-Acryloylglycine | 129.0423 | $C_5H_7NO_3$ | 2 | −0.7386 | 0.036 |
| Alanine-Isoleucine/Leucine | 202.1317 | $C_9H_{18}N_2O_3$ | 0 | −0.6628 | 0.025 |
| Tyrosine-Isoleucine/Leucine | 294.1583 | $C_{15}H_{22}N_2O_4$ | 1 | −0.5159 | 0.043 |
| Tyrosine-Isoleucine/Leucine | 294.1579 | $C_{15}H_{22}N_2O_4$ | 0 | −0.3290 | 0.004 |
| 1-Methylhistidine/3-Methylhistidine | 169.0849 | $C_7H_{11}N_3O_2$ | 1 | 0.6057 | 0.017 |
| Oxidized glutathione | 612.1534 | $C_{20}H_{32}N_6O_{12}S_2$ | 2 | 0.8536 | 0.011 |
| *Nucleosides, nucleotides, and analogues* | | | | | |
| 1,7-Dimethylguanosine | 311.1226 | $C_{12}H_{17}N_5O_5$ | 1 | 0.3058 | 0.007 |
| 1-Methylguanosine/2-Methylguanosine | 297.1091 | $C_{11}H_{15}N_5O_5$ | 6 | 0.4195 | 0.009 |
| Guanine | 151.0509 | $C_5H_5N_5O$ | 10 | 0.4677 | 0.017 |
| *Others* | | | | | |
| Piperidine | 85.0889 | $C_5H_{11}N$ | 3 | −0.9575 | 0.041 |
| Tyramine | 137.0840 | $C_8H_{11}NO$ | 0 | 0.2922 | 0.017 |
| 4′-O-Methyldelphinidin 3-O-beta-D-glucoside | 479.1165 | $C_{22}H_{23}O_{12}$ | 5 | 0.4325 | 0.019 |
| 1H-Indole-3-carboxaldehyde | 145.0524 | $C_9H_7NO$ | 3 | 1.095 | 0.020 |

[1] Compounds were tentatively identified using the HMDB (www.hmdb.ca) metabolite databases with a mass error ≤ 10 ppm. [2] Fold change from 20L:4D to 12L:12D. Positive fold change indicates higher levels in the 20L:4D group and lower levels in the 12L:12D group. Negative fold change indicates lower levels in the 20L:4D group and higher levels in the 12L:12D group.

## 4. Discussion

Previous studies evaluating the influence of photoperiod on broiler meat production have primarily focused on growth performance and yield [7,8]. Several studies have reported a positive impact of the length of photoperiods on growth performance during the starter phase and breast meat yield [10,11,13]. The present study observed no impact of photoperiod on HCW and CCW, agreeing with previous literatures which showed that compensatory gain occurring in the finishing stage minimizes the differences in market weight [11,13,17]. In addition, the present study found no impact of the photoperiod on the weight and yield of the fillet muscles, in disagreement to previous findings [10,11,13,17], except that the percentage of cooler shrink was linearly receded with a shorter photoperiod. In particular, there was a tendency ($p = 0.070$) for 20L:4D carcasses to lose more moisture during the carcass chilling process compared to the 16L:8D and 12L:12D groups. The difference in the current study might be caused by multiple factors, such as the rearing conditions (pen size, group density, and room temperature) or combination of test factors (light intensity, feed energy, and nutrient density). These findings do indicate, however, that the photoperiod-associated differences in meat quality cannot be attributed to weight differences of the fillet muscles.

In addition, aside from a trend of higher freezing/thawing loss in 16L:8D ($p = 0.098$), there was no other impact of photoperiod on WHC. Similarly, other studies have found no impact of photoperiod on WHC [17,36]. No differences in shear force values were also observed across treatments, and, to our knowledge, studies measuring instrumental tenderness of broiler fillet muscles associated with photoperiod effects are unavailable in the current literature. Li et al. [17] reported a higher percentage meat protein in broiler fillet muscles from 12L:12D compared to 20L:4D and 23L:1D groups. Proximate composition including crude protein concentration was unaffected in the present study. This could be explained in part by the lack of a 23L:1D group in the present study, as the meat protein values reported by Li et al. [17] were 0.67% different between the 23L:1D and 12L:12D groups but only 0.32% different between the 12L:12D and 20L:4D groups, The current and previous findings support the postulation that the photoperiod regimes of the present study are unlikely to have any considerable impacts on the composition and general meat quality attributes of the broiler fillet muscles.

In terms of oxidative stability during chilling storage/display times, photoperiod treatments had multiple main effects and two-way interactions on instrumental color and oxidative stability attributes. The fillet muscles from the 20L:4D and 18L:6D groups had higher MDA contents than the 12L:12D group. This finding is interesting considering the lower PUFA content in 20L:4D compared to 12L:12D, as well as 16L:8D, as it has been well-established that PUFA is much more susceptible to lipid oxidation [37]. However, it has been demonstrated that broiler thigh muscle is less susceptible to lipid oxidation during refrigerated storage than breast meat, despite higher free long-chain PUFA content [38]. Numerous factors can promote lipid oxidation in fresh broiler muscles including microbial growth, enzymatic activity, exposure to light and oxygen, and others [39]. As all broilers were treated in the same manner during and following harvest, it is likely differences in oxidative stability can be attributed to pre-harvest factors. The samples from 20L:4D showed a significant decrease in thiol contents from day 1 to day 7, while the samples from other groups had no difference in the thiol contents during display. The finding of longer photoperiod being pro-oxidative to broiler muscles has been corroborated by Li et al. [17] who observed a higher MDA in the fillet muscles from 23L:1D compared to 12L:12D. In addition, Guob et al. [18] reported that a 12L:12D treatment decreased the serum MDA content compared to 20L:4D. The findings were further supported by a reported trend of greater activity of superoxide dismutase, a well-known indicator of antioxidative capacity, in 16L:8D and 12L:12D treatments compared to 23L:1D and 20L:4D treatments [18]. Although antioxidant indices were not assessed in this study, current and previous data do support a positive impact of shorter photoperiod on oxidative stability. The photoperiod effect on oxidative reactions and antioxidative capacity of broiler meat would warrant further research.

The ratio of oxidized (GSSG) to reduced (GSH) glutathione, known as the glutathione redox ratio, has been used as a reliable marker of oxidative stress, based on the role of GSH in protecting against free radicals [40,41]. In the present study, UPLC–MS metabolomics tentatively identified GSSG as higher in the fillet muscles of 20L:4D compared to 12L:12D. Previously, Guob et al. [18] assessed activity of serum GSH peroxidase and found no significant relationship between its activity and photoperiod. However, Asensi et al. [40] reported a positive relationship between the glutathione redox ratio and buildup of lactate/pyruvate through exhaustive physical exercise. Thus, it is reasonable to postulate that postmortem muscle metabolism might have been altered between the 20L:4D and 12L:12D groups. The problem of pale, soft, exudative (PSE) meat in broilers has been well-documented, with findings of higher muscle CIE L*, poorer WHC, and poorer protein functionality [42,43]. The PSE condition arises from rapid glycolysis during early postmortem, causing rapid muscle acidification when the carcass is not chilled, resulting in a denaturation of muscle proteins. Given the higher CIE L* in 20L:4D early in display coupled with trends of higher moisture loss during carcass chilling ($p = 0.070$) and a great transmission value ($p = 0.058$), it would be reasonable to postulate that longer photoperiods might contribute to the PSE-like condition in broiler fillet muscles. However, given the lack of differences in pH, other measures of WHC, and metabolites related to glycolytic pathways, the hypothesis would need to be tested in further studies.

In this study, several tentatively identified metabolites with important biological functions, including guanine, methylated guanosine, and dimethylguanosine, were upregulated in 20L:4D fillet muscles compared to 12L:12D. It has been implicated that hypermethylation of purine bases is a biological marker of disrupting tumor-suppressor genes and inactivating DNA repair genes [44]. In fact, Asensi et al. [40] identified a positive relationship between glutathione redox ratio and oxidative damage of DNA. Adding to this, the current results indicated that there was an upregulation of methylhistidine in 20L:4D. A positive relationship between skeletal muscle mass [45] and turnover of myofibrillar protein [46,47] with methylhistidine has been demonstrated in humans. As broilers grow, both an increase in absolute rates of breast muscle protein synthesis and degradation is observed, leading to an overall net increase in protein deposition [48]. Thus, there is some evidence to suggest 20L:4D treatment might alter muscle metabolome in a way that would support the rapid deposition of fillet muscle tissue.

L-Phenylalanine, tryptophan-leucine/isoleucine dipeptide, and tyrosine-leucine/isoleucine dipeptide were tentatively identified and found to be upregulated in the fillet muscles from 12L:12D broilers. L-Phenylalanine and L-tyrosine are known to be the precursors for catecholamine neurotransmitters, including dopamine, epinephrine, and norepinephrine [49]. Particularly, L-tyrosine has a positive impact on reducing levels of stress hormones [50] and ameliorates negative effects of sleep deprivation [51,52], and L-tryptophan has been well-established as a precursor for melatonin and serotonin [53,54]. Melatonin has a key role in regulation of circadian rhythm, and its production is suppressed by light exposure [55,56]. Given this relationship of these amino acids to stress and diurnal rhythm, the present study provides some evidence for a mechanism of extreme lengths of photoperiods on increasing oxidative stress.

## 5. Conclusions

Photoperiod treatments had minimal impacts on the carcass and meat quality traits of broiler fillet muscles. However, color and oxidative stability were influenced by the current photoperiod treatments and aerobic display storage. In general, the 20L:4D treatment fillets appeared to be lighter and more discolored, distinct from other photoperiod treatments during early display. This was coupled with a higher lipid oxidation in 20L:4D and 18L:6D treatments compared to 12L:12D. Metabolomic analyses indicated that compared to 12L:12D, the 20L:4D group exhibited a downregulation of aromatic amino acids, known to be related to neurotransmitter production, and an upregulation of oxidized glutathione, a biomarker of oxidative stress. These findings support a potential mechanism for the generation of long photoperiod-associated oxidative defects. For practical implications, the results of this study could provide valuable information and practical insights for the poultry industry to develop some pre- and post-harvest strategies for minimizing any quality defects of fresh meat products from broilers exposed to extended photoperiod environments.

**Author Contributions:** Conceptualization, H.-W.C. and Y.H.B.K.; Methodology, J.R.T., T.S., W.W., H.-W.C., and Y.H.B.K.; Software, J.R.T., T.S., and B.C.; Validation, H.-W.C. and Y.H.B.K.; Formal Analysis, J.R.T. and T.S.; Investigation, J.R.T., J.-Y.P., W.W., and B.C.; Resources, W.W. and H.-W.C.; Data Curation, J.R.T., T.S., and B.C.; Writing—Original Draft Preparation, J.R.T.; Writing—Review and Editing, J.R.T., H.-W.C., and Y.H.B.K.; Visualization, J.R.T. and T.S.; Supervision, H.-W.C. and Y.H.B.K.; Project Administration, H.-W.C. and Y.H.B.K.; Funding Acquisition, H.-W.C. and Y.H.B.K. All authors have read and agreed to the published version of the manuscript.

**Funding:** This research received no external funding.

**Acknowledgments:** Authors would like to thank Erin Will of the Purdue Meat Science and Muscle Biology Laboratory, Jason Fields of the Animal Sciences Research and Education Center Poultry Unit at Purdue University, and Blaine Brown and Gary Waters of the Purdue Boilermaker Butcher Block for their assistance in sample collection and processing. This research did not receive external funding. Specification of commercial products or trade names does not imply recommendation or endorsement by the USDA. The USDA is an equal opportunity provider and employer.

**Conflicts of Interest:** The authors declare no conflict of interest.

# References

1. *National Agricultural Statistics Service Poultry-Production and Value 2008 Summary*; United States Department of Agriculture: Washington, DC, USA, 2009.

2. *National Agricultural Statistics Service Poultry-Production and Value 2018 Summary*; United States Department of Agriculture: Washington, DC, USA, 2019.

3. Zuidhof, M.J.; Schneider, B.L.; Carney, V.L.; Korver, D.R.; Robinson, F.E. Growth, efficiency, and yield of commercial broilers from 1957, 1978, and 2005. *Poult. Sci.* **2014**, *93*, 2970–2982. [CrossRef] [PubMed]

4. Havenstein, G.; Ferket, P.; Qureshi, M. Carcass composition and yield of 1957 versus 2001 broilers when fed representative 1957 and 2001 broiler diets. *Poult. Sci.* **2003**, *82*, 1509–1518. [CrossRef] [PubMed]

5. Howlider, M.A.R.; Rose, S.P. Temperature and the growth of broilers. *World's Poult. Sci. J.* **1987**, *43*, 228–237. [CrossRef]

6. Zubair, A.K.; Leeson, S. Compensatory growth in the broiler chicken: A review. *World's Poult. Sci. J.* **1996**, *52*, 189–201. [CrossRef]

7. Olanrewaju, H.A.; Thaxton, J.P.; Dozier III, W.A.; Purswell, J.; Roush, W.B.; Branton, S.L. A Review of Lighting Programs for Broiler Production. *Int. J. Poult. Sci.* **2006**, *5*, 301–308.

8. Olanrewaju, H.A.; Miller, W.W.; Maslin, W.R.; Collier, S.D.; Purswell, J.L.; Branton, S.L. Interactive effects of light-sources, photoperiod, and strains on growth performance, carcass characteristics, and health indices of broilers grown to heavy weights. *Poult. Sci.* **2019**, *98*, 6232–6240. [CrossRef]

9. Classen, H.L.; Riddell, C.; Robinson, F.E. Effects of increasing photoperiod length on performance and health of broiler chickens. *Br. Poult. Sci.* **1991**, *32*, 21–29. [CrossRef]

10. Lien, R.J.; Hess, J.B.; McKee, S.R.; Bilgili, S.F.; Townsend, J.C. Effect of Light Intensity and Photoperiod on Live Performance, Heterophil-to-Lymphocyte Ratio, and Processing Yields of Broilers. *Poult. Sci.* **2007**, *86*, 1287–1293. [CrossRef]

11. Lien, R.J.; Hooie, L.B.; Hess, J.B. Influence of long-bright and increasing-dim photoperiods on live and processing performance of two broiler strains. *Poult. Sci.* **2009**, *88*, 896–903. [CrossRef]

12. Julian, R.J. Production and growth related disorders and other metabolic diseases of poultry—A review. *Vet. J.* **2005**, *169*, 350–369. [CrossRef]

13. Downs, K.M.; Lien, R.J.; Hess, J.B.; Bilgili, S.F.; Dozier, W.A. The Effects of Photoperiod Length, Light Intensity, and Feed Energy on Growth Responses and Meat Yield of Broilers. *J. Appl. Poult. Res.* **2006**, *15*, 406–416. [CrossRef]

14. Classen, H.L.; Riddell, C. Photoperiodic Effects on Performance and Leg Abnormalities in Broiler Chickens. *Poult. Sci.* **1989**, *68*, 873–879. [CrossRef] [PubMed]

15. Sanotra, G.S.; Lund, J.D.; Vestergaard, K.S. Influence of light-dark schedules and stocking density on behaviour, risk of leg problems and occurrence of chronic fear in broilers. *Br. Poult. Sci.* **2002**, *43*, 344–354. [CrossRef] [PubMed]

16. Elefteriou, F. Regulation of bone remodeling by the central and peripheral nervous system. *Arch. Biochem. Biophys.* **2008**, *473*, 231–236. [CrossRef]

17. Li, W.; Guo, Y.; Chen, J.; Wang, R.; He, Y.; Su, D. Influence of Lighting Schedule and Nutrient Density in Broiler Chickens: Effect on Growth Performance, Carcass Traits and Meat Quality. *Asian Australas. J. Anim. Sci.* **2010**, *23*, 1510–1518. [CrossRef]

18. Guob, Y.L.; Li, W.B.; Chen, J.L. Influence of nutrient density and lighting regime in broiler chickens: Effect on antioxidant status and immune functiona. *Br. Poult. Sci.* **2010**, *51*, 222–228. [CrossRef]

19. Lu, Q.; Wen, J.; Zhang, H. Effect of Chronic Heat Exposure on Fat Deposition and Meat Quality in Two Genetic Types of Chicken. *Poult. Sci.* **2007**, *86*, 1059–1064. [CrossRef]

20. Ali, M.S.; Kang, G.-H.; Joo, S.T. A Review: Influences of Pre-slaughter Stress on Poultry Meat Quality. *Asian Australas. J. Anim. Sci.* **2008**, *21*, 912–916. [CrossRef]

21. Zhang, Z.Y.; Jia, G.Q.; Zuo, J.J.; Zhang, Y.; Lei, J.; Ren, L.; Feng, D.Y. Effects of constant and cyclic heat stress on muscle metabolism and meat quality of broiler breast fillet and thigh meat. *Poult. Sci.* **2012**, *91*, 2931–2937. [CrossRef]

22. Dawson, P.L.; Owens, C.M.; Alvarado, C.; Sams, A.R. Packaging. In *Poultry Meat Processing*; Taylor & Francis: Boca Raton, FL, USA, 2010; pp. 101–123.

23. Kraft, A.A.; Reddy, K.V.; Hasiak, R.J.; Lind, K.D.; Galloway, D.E. Microbiological Quality of Vacuum Packaged Poultry With or Without Chlorine Treatment. *J. Food Sci.* **1982**, *47*, 380–385. [CrossRef]

24. Jiménez, S.M.; Salsi, M.S.; Tiburzi, M.C.; Rafaghelli, R.C.; Tessi, M.A.; Coutaz, V.R. Spoilage microflora in fresh chicken breast stored at 4 °C: Influence of packaging methods. *J. Appl. Microbiol.* **1997**, *83*, 613–618.

25. Honikel, K.O. Reference methods for the assessment of physical characteristics of meat. *Meat Sci.* **1998**, *49*, 447–457. [CrossRef]

26. *AOAC Official Methods of Analysis of AOAC*; Association of Official Analytical Chemists: Gaithersburg MD, USA, 2006.

27. *AMSA Meat Color Measurement Guidelines*; American Meat Science Association: Champaign, IL, USA, 2012; pp. 1–17.

28. Buege, J.A.; Aust, S.D. Microsomal lipid peroxidation. In *Methods in Enzymology*; Elsevier: Amsterdam, The Netherlands, 1978; Volume 52, pp. 302–310. ISBN 978-0-12-181952-1.

29. Kim, H.-W.; Kim, J.-H.; Yan, F.; Cheng, H.; Brad Kim, Y.H. Effects of heat stress and probiotic supplementation on protein functionality and oxidative stability of ground chicken leg meat during display storage: Heat stress and probiotic feeding on meat quality of ground chicken thigh. *J. Sci. Food Agric.* **2017**, *97*, 5343–5351. [CrossRef] [PubMed]

30. Berardo, A.; Claeys, E.; Vossen, E.; Leroy, F.; De Smet, S. Protein oxidation affects proteolysis in a meat model system. *Meat Sci.* **2015**, *106*, 78–84. [CrossRef] [PubMed]

31. Ockerman, H.W.; Cahill, V.R. Water extractability of muscle protein and factors which affect this procedure as a method of determining pork quality. *J. Anim. Sci.* **1968**, *27*, 31–38. [CrossRef]

32. Kim, Y.H.; Huff-Lonergan, E.; Sebranek, J.G.; Lonergan, S.M. High-oxygen modified atmosphere packaging system induces lipid and myoglobin oxidation and protein polymerization. *Meat Sci.* **2010**, *85*, 759–767. [CrossRef]

33. Folch, J.; Lees, M.; Sloane Stanley, G.H. A simple method for the isolation and purification of total lipides from animal tissues. *J. Biol. Chem.* **1957**, *226*, 497–509.

34. Shin, S.; Ajuwon, K. Effects of Diets Differing in Composition of 18-C Fatty Acids on Adipose Tissue Thermogenic Gene Expression in Mice Fed High-Fat Diets. *Nutrients* **2018**, *10*, 256. [CrossRef]

35. Bligh, E.G.; Dyer, W.J. A Rapid Method of Total Lipid Extraction and Purification. *Can. J. Biochem. Physiol.* **1959**, *37*, 911–917. [CrossRef]

36. Fidan, E.D.; Nazlıgül, A.; Türkyılmaz, M.K.; Aypak, S.Ü.; Kilimci, F.S.; Karaarslan, S.; Kaya, M. Effect of photoperiod length and light intensity on some welfare criteria, carcass, and meat quality characteristics in broilers. *R. Bras. Zootec.* **2017**, *46*, 202–210. [CrossRef]

37. Shahidi, F.; Zhong, Y. Lipid oxidation and improving the oxidative stability. *Chem. Soc. Rev.* **2010**, *39*, 4067. [CrossRef]

38. Alasnier, C.; Meynier, A.; Viau, M.; Gandemer, G. Hydrolytic and Oxidative Changes in the Lipids of Chicken Breast and Thigh Muscles During Refrigerated Storage. *J. Food Sci.* **2000**, *65*, 9–14. [CrossRef]

39. Huis in't Veld, J.H.J. Microbial and biochemical spoilage of foods: An overview. *Int. J. Food Microbiol.* **1996**, *33*, 1–18. [CrossRef]

40. Asensi, M.; Sastre, J.; Pallardo, F.V.; Lloret, A.; Lehner, M.; Garcia-de-la Asuncion, J.; Viña, J. Ratio of reduced to oxidized glutathione as indicator of oxidative stress status and DNA damage. In *Methods in Enzymology*; Elsevier: Amsterdam, The Netherlands, 1999; Volume 299, pp. 267–276. ISBN 978-0-12-182200-2.

41. Vina, J. *Handbook of Glutathione, Metabolism and Physiological Functions*; CRC Press Inc.: Boca Raton, FL, USA, 1988.

42. Van Laack, R.L.J.M.; Liu, C.-H.; Smith, M.O.; Loveday, H.D. Characteristics of Pale, Soft, Exudative Broiler Breast Meat. *Poult. Sci.* **2000**, *79*, 1057–1061. [CrossRef]

43. Barbut, S. Colour measurements for evaluating the pale soft exudative (PSE) occurrence in turkey meat. *Food Res. Int.* **1993**, *26*, 39–43. [CrossRef]

44. Baylin, S.B.; Herman, J.G. DNA hypermethylation in tumorigenesis: Epigenetics joins genetics. *Trends Genet.* **2000**, *16*, 168–174. [CrossRef]

45. Wang, Z.; Deurenberg, P.; Matthews, D.E.; Heymsfield, S.B. Urinary 3-Methylhistidine Excretion: Association With Total Body Skeletal Muscle Mass by Computerized Axial Tomography. *JPEN J. Parenter Enteral Nutr.* **1998**, *22*, 82–86. [CrossRef]

46. McKeran, R.O.; Halliday, D.; Purkiss, P. Comparison of Human Myofibrillar Protein Catabolic Rate Derived from 3-Methylhistidine Excretion with Synthetic Rate from Muscle Biopsies during l-[α-15N]Lysine Infusion. *Clin. Sci.* **1978**, *54*, 471–475. [CrossRef]

47. Garlick, P.J.; McNurlan, M.A.; Bark, T.; Lang, C.H.; Gelato, M.C. Hormonal Regulation of Protein Metabolism in Relation to Nutrition and Disease. *J. Nutr.* **1998**, *128*, 356S–359S. [CrossRef]

48. Kang, C.W.; Sunde, M.L.; Swick, R.W. Growth and Protein Turnover in the Skeletal Muscles of Broiler Chicks. *Poult. Sci.* **1985**, *64*, 370–379. [CrossRef]

49. Fernstrom, J.D.; Fernstrom, M.H. Tyrosine, Phenylalanine, and Catecholamine Synthesis and Function in the Brain. *J. Nutr.* **2007**, *137*, 1539S–1547S. [CrossRef]

50. Reinstein, D.K.; Lehnert, H.; Wurtman, R.J. Dietary tyrosine suppresses the rise in plasma corticosterone following acute stress in rats. *Life Sci.* **1985**, *37*, 2157–2163. [CrossRef]

51. Neri, D.F.; Wiegmann, D.; Stanny, R.R.; Shappell, S.A.; McCardie, A.; McKay, D.L. The effects of tyrosine on cognitive performance during extended wakefulness. *Aviat. Space Environ. Med.* **1995**, *66*, 313–319. [PubMed]

52. Magill, R.A.; Waters, W.F.; Bray, G.A.; Volaufova, J.; Smith, S.R.; Lieberman, H.R.; McNevin, N.; Ryan, D.H. Effects of Tyrosine, Phentermine, Caffeine d-amphetamine, and Placebo on Cognitive and Motor Performance Deficits During Sleep Deprivation. *Nutr. Neurosci.* **2003**, *6*, 237–246. [CrossRef]

53. Hardeland, R.; Pandi-Perumal, S.R.; Cardinali, D.P. Melatonin. *Int. J. Biochem. Cell Biol.* **2006**, *38*, 313–316. [CrossRef]

54. Lucki, I. The spectrum of behaviors influenced by serotonin. *Biol. Psychiatry* **1998**, *44*, 151–162. [CrossRef]

55. Reiter, R.J. The melatonin rhythm: Both a clock and a calendar. *Experientia* **1993**, *49*, 654–664. [CrossRef]

56. Talpur, H.; Chandio, I.; Brohi, R.; Worku, T.; Rehman, Z.; Bhattarai, D.; Ullah, F.; JiaJia, L.; Yang, L. Research progress on the role of melatonin and its receptors in animal reproduction: A comprehensive review. *Reprod. Dom. Anim.* **2018**, *53*, 831–849. [CrossRef]

 *foods*

*Article*

# Effect of Rearing System on the Straight and Branched Fatty Acids of Goat Milk and Meat of Suckling Kids

Guillermo Ripoll [1,2,*], María Jesús Alcalde [3], Anastasio Argüello [4], María de Guía Córdoba [5] and Begoña Panea [1,2]

[1] Instituto Agroalimentario de Aragón–IA2–(CITA-Universidad de Zaragoza), C/Miguel Servet, 177, 50013 Zaragoza, Spain; bpanea@aragon.es

[2] Animal Production and Health Department, Centro de Investigación y Tecnología Agroalimentaria de Aragón, Avda. Montañana, 930, 50059 Zaragoza, Spain

[3] Department of Agroforesty Science, Universidad de Sevilla, Crta. Utrera, 41013 Sevilla, Spain; aldea@us.es

[4] Department of Animal Pathology, Animal Production and Science and Technology of Foods, Universidad de Las Palmas de Gran Canaria, 35416 Las Palmas, Spain; tacho@ulpgc.es

[5] Nutrición y Bromatología, Instituto Universitario de Investigación de Recursos Agrarios (INURA), Nutrición y Bromatología, Escuela de Ingeniería Agrarias, Universidad de Extremadura, Avda. Adolfo Suarez s/n, 06007 Badajoz, Spain; mdeguia@unex.es

* Correspondence: gripoll@aragon.es; Tel.: +34-976-716-452

Received: 26 February 2020; Accepted: 8 April 2020; Published: 9 April 2020

**Abstract:** Goat meat is considered healthy because it has fewer calories and fat than meat from other traditional meat species. It is also rich in branched chain fatty acids that have health advantages when consumed. We studied the effects of maternal milk and milk replacers fed to suckling kids of four breeds on the straight and branched fatty acid compositions of their muscle. In addition, the proximal and fatty acid compositions of colostrum and milk were studied. Goat colostrum had more protein and fat and less lactose than milk. Goat milk is an important source of healthy fatty acids such as C18:1 c9 and C18:2 n–6. Suckling kid meat was also an important source of C18:1c9. Dairy goat breeds had higher percentages of *trans* monounsaturated fatty acids (MUFAs) and most of the C18:1 isomers but lower amounts of total MUFAs than meat breeds. However, these dairy kids had meat with a lower percentage of conjugated linoleic acid (CLA) than meat kids. The meat of kids fed natural milk had higher amounts of CLA and branched chain fatty acids (BCFAs) and lower amounts of n–6 fatty acids than kids fed milk replacers. Both milk and meat are a source of linoleic, α-linolenic, docosahexaenoic, eicosapentaenoic and arachidonic fatty acids, which are essential fatty acids and healthy long-chain fatty acids.

**Keywords:** goat; milk; BCFA; replacer; methyl; colostrum

## 1. Introduction

Approximately 119,000 tons of caprine meat were produced in Europe in 2017 [1]. However, Mediterranean goat farms are mainly focused on dairy production, including cheese and milk products, because they have higher prices than cow milk [2]. In addition, goat milk is also generating great interest for human consumption due to its nutritional advantages and lack of allergenicity compared to cow milk [3]. Accordingly, Europe produces 2,824,715 tons of goat milk, and 45% is produced in South Europe [1]. Although most income per goat on the dairy farm comes from the sale of milk, 20% of the total income comes from the sale of kids [4]. These kids are weaned very early and reared with milk replacers. These milk replacers are specifically formulated for kids, resulting in good daily weight gain. Moreover, this meat is perceived by consumers to be high-quality meat, with most kids being slaughtered at the very light carcass weight of 5–7 kg [5]. However, some farmers, especially those who use autochthonous breeds, believe that early separation decreases both milk yield and growth of the kids. In addition, they also believe that meat from kids reared with milk replacers is tough [6] and, as a consequence, they are opposed to this practice. This belief could be explained because most of the kid meat with high pH, which may induce tough meat, comes from kids raised on milk replacers [7]. On the other hand, the meat of kids reared with milk replacers was preferred by consumers according to visual appraisal, and consequently, as the purchase intention of these consumers was high [8,9]. Another advantage of meat from suckling kids fed natural milk is that it has a high percentage of hexanal, which has been positively related to both the flavor and overall acceptability [10].

On the other hand, goat meat is considered healthy because it has fewer calories and fat than meat from other meat species, such as pork or lamb [11] and because its fatty acid profile complies with the recommendation of the World Health Organization, which states that trans fatty acid intake should not reach more than 5% [12]. In addition, goat meat is rich in branched chain fatty acids (BCFAs), and BCFAs are of interest for two main reasons: first, BCFAs, particularly short chain BCFAs, have an impact on the characteristic flavor of meat and dairy products [13–16]. Second, some authors have described health advantages with BCFA consumption. Additionally, the intake of BCFAs is related to the correct function of the newborn gut [17] and the induction of apoptosis in breast cancer cells [18]. Considering these benefits, BCFAs could be considered bioactive compounds and deserve deep insight. BCFAs are mainly saturated fatty acids with at least one methyl branch on the carbon chain. Often, the branch is close to the end of the chain, producing the *iso-* and *anteiso-* isomers when the methyl branch is on the penultimate or antepenultimate carbon atoms. BCFAs are synthetized by bacteria as a main component of the bacterial membrane. Hence, BCFAs are found in meat and milk from ruminants because of rumen bacterial activity. In general, adipose tissues of goats are richer in these fatty acids than those of other ruminants [19]. However, very light suckling kids are functionally non-ruminants [20], and the presence of BCFAs in their meat probably originates mainly from maternal milk. In addition, kids fed milk replacers from cow milk do not consume many BCFAs. Setting aside the origins of the BCFAs in meat, we studied the effects of feeding suckling kids with maternal milk or milk replacers on the straight and BCFAs of their muscles. In addition, the proximal and fatty acid compositions of colostrum and milk of four breeds were studied.

## 2. Materials and Methods

### 2.1. Animals

All procedures were conducted according to the guidelines of Directive 2010/63/EU on the protection of animals used for experimental and other scientific purposes [21].

One hundred and twenty-four suckling male kids of 4 goat breeds (Cabra del Guadarrama, GU; Palmera, PL; Retinta, RE; Tinerfeña, TI) were evenly reared at two (PA and TI) or three farms (GU and RE) per breed in their respective local areas. Therefore, 15, 15, 15 and 16 kids of GU, PL, RE and TI, respectively, were fed milk replacers (MR), and 16, 16, 15 and 16 kids of GU, PL, RE and TI, respectively, were fed natural milk (NM). Animals were selected to be as unrelated as possible to ensure that the full range of genetic diversity was present within the breeds used in the study. Animals were all born from a single parturition. Kids in the MR rearing systems were fed colostrum for the first two days and had free access to milk replacer 24 h a day, which was suckled from a teat connected to a unit for feeding a liquid diet. The goats grazed the natural resources of the area during the day and were supplemented with hay of the pastures and similar commercial concentrates between breeds.

### 2.2. Milk Sampling

Commercial milk replacers were reconstituted at 17% (w/v) and given warm (40 °C). The main ingredients were skimmed milk (≈ 60%) and whey. The chemical composition (on an as-dry matter basis) of milk replacers was as follows: total fat 25% ± 0.6, crude protein 24% ± 0.5, crude cellulose 0.1% ± 0.0, ash 7% ± 0.6, Ca 0.8% ± 0.1, Na 0.5% ± 0.2, P 0.7% ± 0.0, Fe 36 mg/kg ± 4.0, Cu 3 mg/kg

± 1.7, Zn 52 mg/kg ± 18.8, Mn 42 mg/kg ± 14.4, I 0.22 mg/kg ± 0.06, Se 0.1 mg/kg ± 0.06 and BHT 65 ppm ± 30. Kids in the NM rearing system suckled directly from dams with no additional feedstuff. At night, they were housed with their dams in a stable. Kids in both rearing systems had no access to concentrates, hay, forages or other supplements.

The natural milk of dams was collected from 10:00 h to 11:00 h in the morning at 1, 10 and 30 d of lactation. Two 50 mL Falcon tubes were filled with milk and three drops of Azidiol (Panreac Applichem, Barcelona, Spain). No oxytocin was used. The chemical composition (protein, fat and lactose) of the milk was determined by using a DMA2001 Milk Analyzer (Miris Inc., Uppsala, Sweden). A subsample of natural milk at 1 and 30 d of lactation was freeze-dried and stored at −80 °C until fatty acid analysis.

### 2.3. Carcass Sampling

The 124 kids were slaughtered at a live weight of 8.4 kg ± 0.12 kg. Standard commercial procedures according to the European normative of protection of animals at the time of killing [22] were followed. Head-only electrical stunning was applied (1.00 A) to the kids, which were then exsanguinated and dressed. Thereafter, the hot carcasses, including the heads and kidneys, were weighed to achieve a hot carcass weight (HCW) of 5.0 kg ± 0.10 kg. Afterwards, carcasses were hung by the Achilles tendon and chilled for 24 h at 4 °C. After carcass chilling, the *longissimus thoracis* muscle of the left half of the carcasses was extracted, vacuum packed and frozen at −20 °C until fatty acid composition analyses.

### 2.4. Fatty Acid Analyses

The fatty acid methyl esthers (FAMEs) from lyophilized milk fat samples were prepared by direct transesterification using KOH in methanol (2 N) and extracted with hexane [23]. Fat depots for fatty acid profile analysis were processed according to Folch et al. [24].

The determination of FAME was carried out using a Bruker 436 Scion gas chromatograph (Bruker, Billerica, MA, USA) equipped with a cyanopropyl capillary column (BR-2560, 100 m × 0.25 mm ID × 0.20 μm thickness, Bruker, Billerica, MA, USA), a flame ionization detector and Compass CDS software. Fatty acid quantification was performed as described in the UNE-EN 12966-4 Official Method (2015). The identification was performed using the GLC 538 and GLC 463 standard references (Nu-Chek-Prep Inc., Elysian, Minnesota, USA). Fatty acid contents are expressed as a percentage of the total amount of identified fatty acids. After individual fatty acid determinations, the total contents of saturated fatty acids (SFAs), monounsaturated fatty acids (MUFAs), polyunsaturated fatty acids (PUFAs), PUFA n–6 and PUFA n–3 were calculated. The PUFA n–6/n–3 ratios were also calculated. Desirable fatty acids of the meats were calculated as MUFA + PUFA + C18:0 [25]. In addition, the sums of conjugated linoleic acid (CLA), *iso-* and *anteiso-* BCFA, *cis-* and *trans-* MUFA were also calculated.

### 2.5. Statistical Analysis

All statistics were calculated using the XLSTAT statistical package v.3.05 (Addinsoft, New York, NY, USA). The proximal composition and fatty acid composition of milk were analyzed using the MIXED procedure for repeated measures. The factors included were breed as between-subject fixed effects, time as within-subject effects and random animal effects as subjects (experimental units). The lowest Akaike Information Criterion (AIC) was used to choose the matrix of the error structure. Least square means were estimated, and differences were tested with a t-test at the 0.05 level.

The fatty acid composition of meat was analyzed using the ANCOVA procedure with the breed (B) and the rearing system (RS) as fixed effects and the hot carcass weight (HCW) as a covariate. The HCW was used as covariate to avoid the influence of weight differences on the fatty acid composition. The least square means were adjusted for an HCW of 5.02 kg. Differences between means were tested with Duncan's test at a 0.05 level of significance.

Two principal component analyses were performed with the main fatty acid groups of both milk and meat. The variables included in these principal component analyses were SFA, MUFA, PUFA, n–3, n–6, n–6:n–3 ratio, ΣCLA, Σ*iso*-BCFA, Σ*anteiso*-BCFA, ΣBCFA, *cis*-MUFA, and Σ*trans*-MUFA.

## 3. Results

### 3.1. Milk

The chemical composition of natural milk through the first 30 days of lactation is shown in Figure 1. The milk protein percentage was significantly affected by breed and time of lactation ($P < 0.0001$). However, the milk fat and lactose percentages were affected by the interaction between breed and time of lactation ($P < 0.0001$).

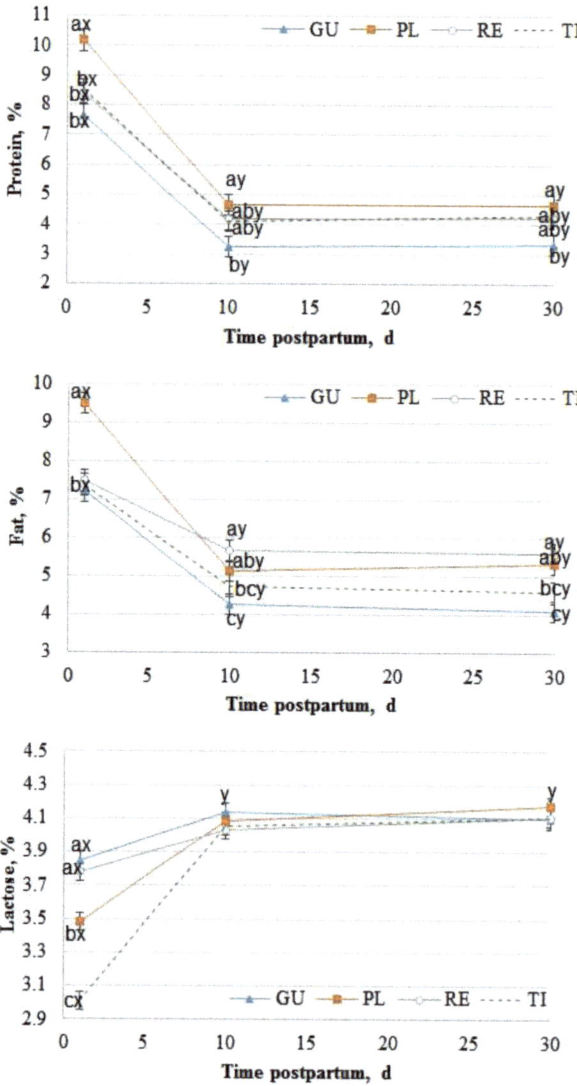

**Figure 1.** Chemical composition of natural milk at 1, 10 and 30 days of lactation. GU, Cabra del Guadarrama; PL, Palmera; RE, Retinta; TI, Tinerfeña. Different superscripts (a, b, c) indicate significant differences within a time of lactation ($P < 0.05$). Different superscripts (x,y) indicate significant differences within a breed ($P < 0.05$).

The protein content was higher on the first day of lactation (colostrum) than at the other studied times, especially for the Palmera breed, which presented higher values than the other breeds ($P < 0.05$). Thereafter, the protein percentage decreased significantly ($P < 0.05$) for all the breeds from the 1st to the 10th day of lactation and remained steady from the 10th to the 30th day ($P > 0.05$). Similarly, the fat percentage decreased from the 1st day to the 10th day of lactation, with Palmera presenting a more acute decrease from the 1st day to the 10th day than the other breeds. Conversely, for protein and fat, the lactose percentage increased significantly from the 1st day to the 10th day ($P < 0.05$) and remained constant thereafter, without differences among breeds ($P > 0.05$). The fatty acid composition is shown as g/100 g of FAMES in Tables 1–4. There were 17 straight SFAs in the milk (Table 1), with C14:0, C16:0 and C18:0 being the most abundant, followed by C10:0 and C12:0. The less abundant fatty acids were mainly odd fatty acids such as C7:0, C9:0, C13:0, C22:0 and C23:0. All SFAs were affected by the interaction of breed and time of lactation ($P < 0.01$), except C17:0.

**Table 1.** Individual straight saturated fatty acids of goat milk at 1 and 30 days of lactation (g/100 g of fatty acid methyl esthers (FAMEs)).

| Time (T) | 1 d of Lactation | | | | 30 d of Lactation | | | | | | | |
|---|---|---|---|---|---|---|---|---|---|---|---|---|
| Breed (B) | GU [†] | PL | RE | TI | GU | PL | RE | TI | s.e. | B | T | B × T |
| C4:0 | 1.48[c] | 1.06[d] | 1.72[b] | 1.27[cd] | 2.03[a] | 1.62[bc] | 1.77[ab] | 2.02[a] | 0.103 | 0.005 | 0.0001 | 0.002 |
| C6:0 | 1.44[d] | 1.01[e] | 1.79[c] | 1.21[de] | 2.37[ab] | 2.22[ab] | 2.20[b] | 2.54[a] | 0.110 | 0.042 | 0.0001 | 0.0001 |
| C7:0 | 0.012[bc] | 0.010[bc] | 0.012[bc] | 0.005[c] | 0.051[a] | 0.048[a] | 0.023[b] | 0.062[a] | 0.006 | 0.018 | 0.0001 | 0.001 |
| C8:0 | 1.57[c] | 1.08[d] | 2.07[b] | 1.18[d] | 2.85[a] | 2.94[a] | 2.78[a] | 2.86[a] | 0.135 | 0.005 | 0.0001 | 0.0001 |
| C9:0 | 0.035[cd] | 0.031[cd] | 0.040[cd] | 0.025[d] | 0.112[b] | 0.130[ab] | 0.068[c] | 0.146[a] | 0.012 | 0.031 | 0.0001 | 0.0001 |
| C10:0 | 4.76[d] | 3.42[d] | 6.35[c] | 3.51[d] | 8.75[b] | 11.18[a] | 9.53[ab] | 10.13[a] | 0.483 | 0.020 | 0.0001 | 0.0001 |
| C11:0 | 0.125[e] | 0.091[e] | 0.176[d] | 0.080[e] | 0.273[bc] | 0.376[a] | 0.252[c] | 0.314[ab] | 0.020 | 0.40 | 0.0001 | 0.0001 |
| C12:0 | 2.48[de] | 1.85[ef] | 2.92[d] | 1.73[f] | 3.52[c] | 4.98[a] | 4.18[ab] | 4.00[bc] | 0.241 | 0.007 | 0.0001 | 0.0001 |
| C13:0 | 0.087[c] | 0.059[cd] | 0.077[c] | 0.041[d] | 0.153[ab] | 0.193[a] | 0.131[b] | 0.157[ab] | 0.013 | 0.12 | 0.0001 | 0.007 |
| C14:0 | 11.29[ab] | 9.90[gh] | 11.84[a] | 9.50[h] | 10.06[gh] | 12.52[a] | 11.76[a] | 12.07[a] | 0.694 | 0.23 | 0.059 | 0.007 |
| C15:0 | 0.593[b] | 0.530[b] | 0.618[b] | 0.521[b] | 0.635[b] | 0.945[a] | 0.818[a] | 0.954[a] | 0.049 | 0.012 | 0.0001 | 0.0001 |
| C16:0 | 24.37[b] | 27.88[a] | 24.47[b] | 26.86[a] | 19.62[c] | 27.83[a] | 23.85[b] | 28.24[a] | 0.718 | 0.0001 | 0.057 | 0.0001 |
| C17:0 | 0.804[bcd] | 0.955[ab] | 0.879[abc] | 0.965[a] | 0.540[e] | 0.746[cd] | 0.708[d] | 0.698[d] | 0.047 | 0.0001 | 0.0001 | 0.60 |
| C18:0 | 13.30[b] | 12.31[bcd] | 10.28[de] | 12.91[bc] | 17.64[a] | 9.25[e] | 12.29[bcd] | 10.87[cde] | 0.663 | 0.0001 | 0.52 | 0.0001 |
| C20:0 | 0.212[b] | 0.190[bcd] | 0.171[cd] | 0.221[b] | 0.338[a] | 0.150[cd] | 0.191[bc] | 0.140[d] | 0.015 | 0.0001 | 0.56 | 0.0001 |
| C22:0 | 0.050[bc] | 0.047[bcd] | 0.048[bcd] | 0.057[bc] | 0.074[a] | 0.037[cd] | 0.060[b] | 0.034[d] | 0.005 | 0.001 | 0.83 | 0.0001 |
| C23:0 | 0.006[c] | 0.008[c] | 0.015[b] | 0.017[b] | 0.006[c] | 0.008[c] | 0.022[a] | 0.014[b] | 0.002 | 0.0001 | 0.47 | 0.002 |

[†] GU, del Guadarrama; PL, Palmera; RE, Retinta; TI, Tinerfeña; s.e., standard error. Different superscripts in the same row indicate significant differences ($P \leq 0.05$).

**Table 2.** Individual monounsaturated fatty acids of goat milk at 1 and 30 days of lactation (g/100 g of FAMEs).

| Time (T) | 1 d of Lactation | | | | 30 d of Lactation | | | | | | | |
|---|---|---|---|---|---|---|---|---|---|---|---|---|
| Breed (B) | GU [†] | PL | RE | TI | GU | PL | RE | TI | s.e. | B | T | B × T |
| C12:1 | 0.050 | 0.032 | 0.034 | 0.023 | 0.037 | 0.050 | 0.047 | 0.035 | 0.015 | 0.47 | 0.31 | 0.38 |
| C14:1c9 | 0.078[bc] | 0.066[bcd] | 0.103[a] | 0.062[cd] | 0.055[d] | 0.084[abc] | 0.084[ab] | 0.063[cd] | 0.007 | 0.0001 | 0.242 | 0.025 |
| C15:1 | 0.011 | 0.004 | 0.005 | 0.005 | 0.012 | 0.005 | 0.005 | 0.005[a] | 0.005 | 0.33 | 0.942 | 0.99 |
| C16:1t9 | 0.361[cd] | 0.445[ab] | 0.487[a] | 0.444[ab] | 0.272[e] | 0.377[cd] | 0.404[bc] | 0.336[d] | 0.049 | 0.0001 | 0.0001 | 0.79 |
| C16:1c7 | 0.338[abc] | 0.358[ab] | 0.344[abc] | 0.385[a] | 0.293[bcd] | 0.264[d] | 0.274[cd] | 0.237[d] | 0.022 | 0.99 | 0.0001 | 0.061 |
| C16:1c9 | 1.185[b] | 1.487[a] | 1.495[a] | 1.588[a] | 0.727[d] | 1.070[bc] | 1.056[bc] | 0.965[c] | 0.063 | 0.0001 | 0.0001 | 0.323 |
| C17:1c9 | 0.337[b] | 0.353[b] | 0.537[a] | 0.512[a] | 0.142[d] | 0.205[cd] | 0.290[bc] | 0.252[bcd] | 0.037 | 0.0001 | 0.0001 | 0.50 |
| C18:1t11 | 0.939[c] | 1.257[a] | 0.683[c] | 0.717[c] | 1.157[ah] | 0.674[c] | 0.017[c] | 0.901[bc] | 0.104 | 0.024 | 0.86 | 0.004 |
| C18:1c9 | 27.217[b] | 29.100[ab] | 27.559[ab] | 30.274[a] | 22.953[c] | 17.745[de] | 21.396[cd] | 17.713[e] | 1.186 | 0.58 | 0.0001 | 0.0001 |
| C18:1t15 | 0.125[bc] | 0.120[bc] | 0.128[ba] | 0.121[ba] | 0.156[ab] | 0.089[c] | 0.171[a] | 0.116[c] | 0.011 | 0.000 | 0.22 | 0.007 |
| C18:1c11 | 0.173[b] | 0.123[bc] | 0.117[c] | 0.114[c] | 0.239[a] | 0.142[bc] | 0.163[b] | 0.136[bc] | 0.015 | 0.0001 | 0.001 | 0.32 |
| C18:1c12 | 0.139[ab] | 0.139[ab] | 0.074[e] | 0.082[c] | 0.159[a] | 0.105[bc] | 0.121[b] | 0.079[c] | 0.010 | 0.0001 | 0.33 | 0.005 |
| C18:1c13 | 0.047 | 0.055 | 0.037 | 0.041 | 0.047 | 0.046 | 0.057 | 0.039 | 0.006 | 0.41 | 0.631 | 0.11 |
| C18:1t16 | 0.146[ab] | 0.139[ab] | 0.134[b] | 0.130[b] | 0.153[ab] | 0.116[bc] | 0.168[a] | 0.100[c] | 0.011 | 0.001 | 0.69 | 0.010 |
| C18:1c15 | 0.066[ab] | 0.058[ab] | 0.060[ab] | 0.059[ab] | 0.059[ab] | 0.048[b] | 0.070[a] | 0.050[b] | 0.005 | 0.038 | 0.23 | 0.12 |
| C20:1n–9 | 0.007 | 0.002 | 0.002 | 0.001 | 0.001 | 0.001 | 0.002 | 0.001[a] | 0.003 | 0.49 | 0.30 | 0.47 |
| C22:1 | 0.012[d] | 0.015[cd] | 0.022[ab] | 0.026[a] | 0.003[e] | 0.007[e] | 0.019[bc] | 0.015[cd] | 0.002 | 0.0001 | 0.0001 | 0.065 |

[†] GU, del Guadarrama; PL, Palmera; RE, Retinta; TI, Tinerfeña; s.e., standard error. Different superscripts in the same row indicate significant differences ($P \leq 0.05$).

**Table 3.** Individual polyunsaturated fatty acids of goat milk at 1 and 30 days of lactation (g/100 g of FAMEs).

| Time (T) | 1 d of Lactation | | | | 30 d of Lactation | | | | | | | |
|---|---|---|---|---|---|---|---|---|---|---|---|---|
| Breed (B) | GU[†] | PL | RE | TI | GU | PL | RE | TI | s.e. | B | T | B × T |
| C18:2 n–6 t9,12 | 0.122[abc] | 0.111[bcd] | 0.107[bcd] | 0.097[cd] | 0.130[ab] | 0.087[d] | 0.149[a] | 0.083[d] | 0.009 | 0.0001 | 0.67 | 0.002 |
| C18:2 n–6 | 3.37[a] | 3.31[a] | 2.13[c] | 2.77[b] | 2.62[b] | 1.88[cd] | 1.78[d] | 2.00[cd] | 0.120 | 0.0001 | 0.0001 | 0.003 |
| C18:2 c9, t11 | 0.221[cd] | 0.464[a] | 0.307[b] | 0.254[bc] | 0.176[d] | 0.248[bcd] | 0.293[bc] | 0.236[bcd] | 0.027 | 0.0001 | 0.0001 | 0.015 |
| C18:2 t9, c11 | 0.042[b] | 0.052[ab] | 0.051[ab] | 0.047[b] | 0.041[b] | 0.048[b] | 0.059[a] | 0.045[b] | 0.003 | 0.0001 | 0.94 | 0.20 |
| C18:2 t10, c12 | 0.019[ab] | 0.022[ab] | 0.019[ab] | 0.019[ab] | 0.017[ab] | 0.013[b] | 0.023[a] | 0.017[ab] | 0.002 | 0.35 | 0.17 | 0.065 |
| C18:3 n–6 | 0.031[a] | 0.020[ab] | 0.012[b] | 0.021[ab] | 0.009[b] | 0.012[ab] | 0.005[b] | 0.012[b] | 0.005 | 0.11 | 0.005 | 0.30 |
| C18:3 n–3 | 0.418[a] | 0.245[bcd] | 0.350[ab] | 0.265[bc] | 0.383[a] | 0.223[cd] | 0.420[a] | 0.171[d] | 0.031 | 0.0001 | 0.365 | 0.030 |
| C20:2 n–6 | 0.027[b] | 0.041[a] | 0.018[c] | 0.029[b] | 0.010[d] | 0.013[cd] | 0.012[d] | 0.011[d] | 0.002 | 0.0001 | 0.0001 | 0.0001 |
| C20:3 n–9 | 0.032[b] | 0.036[ab] | 0.026[cd] | 0.042[a] | 0.014[f] | 0.018[ef] | 0.021[de] | 0.028[bc] | 0.002 | 0.0001 | 0.0001 | 0.016 |
| C20:3 n–6 | 0.001[c] | 0.002[c] | 0.006[b] | 0.009[a] | 0.001[c] | 0.003[c] | 0.008[ab] | 0.008[ab] | 0.001 | 0.0001 | 0.63 | 0.43 |
| C20:4 n–6 | 0.521[a] | 0.483[a] | 0.356[b] | 0.526[a] | 0.164[c] | 0.190[c] | 0.170[c] | 0.215[c] | 0.029 | 0.001 | 0.0001 | 0.008 |
| C20:5 n–3 | 0.096[a] | 0.071[b] | 0.075[b] | 0.078[b] | 0.048[c] | 0.041[c] | 0.066[b] | 0.036[c] | 0.006 | 0.006 | 0.0001 | 0.001 |
| C22:3 n–3 | 0.013[b] | 0.018[ab] | 0.015[b] | 0.020[a] | 0.004[d] | 0.012[bc] | 0.022[a] | 0.006[cd] | 0.002 | 0.0001 | 0.0001 | 0.0001 |
| C22:4 n–6 | 0.084[a] | 0.109[a] | 0.045[b] | 0.089[a] | 0.028[bc] | 0.026[bc] | 0.018[c] | 0.025[c] | 0.008 | 0.0001 | 0.0001 | 0.007 |
| C22:5 n–3 | 0.314[a] | 0.158[c] | 0.201[bc] | 0.227[b] | 0.074[d] | 0.051[d] | 0.088[d] | 0.053[d] | 0.017 | 0.0001 | 0.0001 | 0.0001 |
| C22:6 n–3 | 0.054[a] | 0.016[c] | 0.033[b] | 0.015[c] | 0.014[c] | 0.005[c] | 0.018[c] | 0.006[c] | 0.004 | 0.0001 | 0.0001 | 0.0001 |

[†] GU, del Guadarrama; PL, Palmera; RE, Retinta; TI, Tinerfeña; s.e., standard error. Different superscripts in the same row indicate significant differences ($P \leq 0.05$).

**Table 4.** Main groups of straight and branched fatty acids of goat milk at 1 and 30 days of lactation (g/100 g of FAMEs).

| Time (T) | 1 d of Lactation | | | | 30 d of Lactation | | | | | | | |
|---|---|---|---|---|---|---|---|---|---|---|---|---|
| Breed (B) | GU[†] | PL | RE | TI | GU | PL | RE | TI | s.e. | B | T | B × T |
| SFA | 63.48[c] | 61.03[c] | 64.37[c] | 60.85[c] | 69.79[b] | 76.07[a] | 71.64[b] | 75.96[a] | 1.384 | 0.40 | 0.0001 | 0.0001 |
| ΣBCFA | 0.678[b] | 0.475[c] | 0.730[ab] | 0.564[c] | 0.569[c] | 0.696[ab] | 0.814[a] | 0.575[c] | 0.035 | 0.0001 | 0.046 | 0.0001 |
| Σiso-BCFA | 0.404[a] | 0.238[c] | 0.423[a] | 0.320[b] | 0.311[b] | 0.305[bc] | 0.407[a] | 0.279[bc] | 0.018 | 0.0001 | 0.120 | 0.002 |
| Σanteiso-BCFA | 0.274[b] | 0.237[b] | 0.307[b] | 0.244[b] | 0.258[b] | 0.391[a] | 0.407[a] | 0.296[b] | 0.021 | 0.0001 | 0.0001 | 0.001 |
| MUFA | 31.13[b] | 33.75[ab] | 31.82[ab] | 34.58[a] | 26.47[c] | 21.03[d] | 25.14[cd] | 21.04[d] | 1.292 | 0.72 | 0.0001 | 0.0001 |
| Σcis-MUFA | 29.66[b] | 31.79[ab] | 30.39[ab] | 33.17[a] | 24.73[c] | 19.77[de] | 23.58[cd] | 19.59[e] | 1.269 | 0.74 | 0.0001 | 0.001 |
| Σtrans-MUFA | 1.47[bc] | 1.96[a] | 1.43[bc] | 1.41[bc] | 1.74[ab] | 1.26[c] | 1.56[abc] | 1.45[bc] | 0.122 | 0.34 | 0.45 | 0.009 |
| ΣCLA | 0.283[de] | 0.538[a] | 0.377[b] | 0.320[bcd] | 0.235[e] | 0.309[bcde] | 0.376[bc] | 0.298[cde] | 0.030 | 0.0001 | 0.001 | 0.013 |
| PUFA | 5.36[a] | 5.16[a] | 3.75[c] | 4.51[b] | 3.73[c] | 2.86[d] | 3.15[d] | 2.95[d] | 0.175 | 0.0001 | 0.0001 | 0.000 |
| n–6 | 4.15[a] | 4.080[a] | 2.68[cd] | 3.54[b] | 2.96[c] | 2.21[de] | 2.14[e] | 2.35[de] | 0.139 | 0.0001 | 0.0001 | 0.001 |
| n–3 | 0.895[a] | 0.508[cd] | 0.674[b] | 0.605[bc] | 0.523[c] | 0.331[de] | 0.613[bc] | 0.273[e] | 0.044 | 0.0001 | 0.0001 | 0.0001 |
| n–6:n–3 | 4.78[c] | 8.14[a] | 4.08[cd] | 5.95[b] | 5.80[b] | 6.69[b] | 3.80[d] | 8.95[a] | 0.360 | 0.0001 | 0.031 | 0.0001 |

[†] GU, del Guadarrama; PL, Palmera; RE, Retinta; TI, Tinerfeña; s.e., standard error. FA. Fatty acids; SFA, saturated fatty acids; BCFA, branched chain fatty acids; MUFA, monounsaturated fatty acids; CLA, conjugated linoleic acid; PUFA, polyunsaturated fatty acids. Different superscripts in the same row indicate significant differences ($P \leq 0.05$).

Independent of the breed, the percentages of C6:0, C8:0, C10:0, C11:0, C12:0 and C13:0 increased from the 1st day to the 30th day of lactation. In the same way, the percentages of C17:0 decreased over time independent of the breed. Nevertheless, the percentages of C4:0, C7:0 and C9:0 increased with lactation time in all breeds except RE, which remained constant, and C15:0 increased in all breeds except GU, which remained constant. In contrast, C14:0 percentages increased only in TI, C16:0 percentages decreased only in GU, and C23:0 percentages increased only in RE. Finally, C20:0 and C22:0 increased over time in GU and decreased in TI without changes in PL and RE, and C18:0 increased in GU, decreased in PL and remained constant in RE and TI. The 17 detected MUFAs are shown in Table 2. C18:1 c9 had the highest percentage and was affected by the magnitude of the interaction between breed and time of lactation ($P = 0.001$). The percentage of C18:1 c9 decreased for all breeds, but the decrease in PL and TI was twofold compared to the decrease in GU and RE. C18:1t11 remained constant with the time of lactation in RE and TI but increased in GU and decreased in PL. The following most abundant fatty acids were the isomers of C16:1. Those isomers, in general, decreased with the time of lactation, although the C16:1c7 of GU and RE remained constant ($P > 0.05$). The time of lactation did not influence the percentages of C18:1t15, C18:1c11 and C18:1c12 of PL and TI, but the percentages of these fatty acids of RE increased with the time of lactation ($P < 0.05$).

Most of the 16 detected PUFAs (Table 3) were significantly affected by the interaction between breed and time of lactation. The most predominant PUFA was C18:2 n–6, which decreased from the 1st day to the 30th day of lactation for all breeds, although PL registered the most pronounced decrease. The FAs showed the following order in terms of their quantities: C18:2 c9, t11, C18:3 n–3 and C20:4 n–6. These FAs also decreased from the 1st day to the 30th day of lactation for all breeds but in different amounts. FA C22:5 n–3 was also relatively abundant in milk at one day of lactation, but its amount was dramatically reduced at 30 days.

Table 4 shows the main groups of straight and branched fatty acids of milk at the 1st day and the 30th day of lactation. All the studied groups were affected by the interaction between breed and time of lactation ($P < 0.005$). Independent of the breed, the SFA content increased as lactation time increased ($P > 0.05$). At 30 days, PL and TI had higher SFA contents than GU and RE ($P < 0.05$). RE and TI did not show changes in the contents of ΣBCFA with time ($P > 0.05$), but GU showed increases, while PL showed decreases with time ($P < 0.05$). The content of Σ*iso*-BCFA was affected by lactation time only in GU, showing decreases over time. On the first day of lactation, there were no differences between breeds for the contents of Σ*anteiso*-BCFA, whereas in PL and RE, it increased from the 1st day to the 30th day ($P < 0.05$), and no changes over time were observed in GU and TI. For all breeds, MUFA and Σ*cis*-MUFA decreased with the time of lactation ($P < 0.05$), with the decrease being more intense for PL and TI than GU and RE. Nevertheless, the Σ*trans*-MUFA content was unaffected by time in GU, RE and TI ($P > 0.05$), but it decreased in PL. The ΣCLA content changed, showing decreases, with time only in Pl ($P < 0.05$). For all the breeds, the total PUFA content decreased with the time of lactation ($P < 0.05$), but this decrease was smaller in RE than in the other breeds. The n–6:n–3 ratio decreased with time of lactation ($P < 0.05$), except in RE, in which no changes were observed ($P < 0.05$).

The biplot of the principal component analysis is shown in Figure 2. The first two principal components summarized approximately 66% of the variation in the data. Milks at the 1st day and the 30th day were clearly discriminated by the variables used in the first principal component. Therefore, milk on the first day was positively related to Σ*cis*-MUFA, ΣMUFA and ΣPUFA, while milk at 30 days was correlated with ΣSFA.

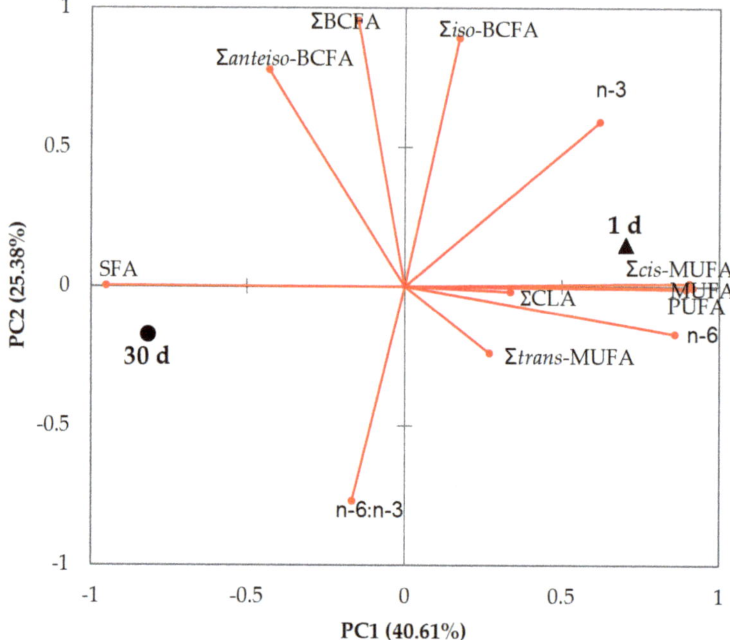



**Figure 2.** Principal component analysis of the main groups and branched fatty acids of goat milk at 1 and 30 days of lactation. SFA, saturated fatty acids; BCFA, branched chain fatty acids; MUFA, monounsaturated fatty acids; CLA, conjugated linoleic acid; PUFA, polyunsaturated fatty acids.

*3.2. Meat*

The 15 detected straight SFAs are shown in Table 5. The predominant SFAs were C14:0 and C16:0, and C18:0. C9:0, C11:0, C12:0, C14:0, C16:0, C18:0 and C22:0 were affected by the interaction between breed and rearing system ($P < 0.01$), whereas C6:0 and C8:0 were not affected by any of the study effects ($P > 0.05$). In general terms, meat from suckling kids fed natural milk had higher values of C15:0, C16:0, C17:0 and C18:0 than meat from kids fed milk replacers ($P < 0.05$), and the effect of rearing system was more noticeable for TI, in which 9 of the 15 acids were affected, than in the other breeds. The most abundant MUFA (Table 6) was C18:1c9, which was affected by the interaction between breed and rearing system ($P < 0.001$). GU fed natural milk had higher contents and RE lower contents of this FA than kids fed milk replacers ($P < 0.05$), whereas the rearing system did not affect PL and TI ($P > 0.05$). There were 16 detected PUFAs, as shown in Table 7. The most abundant PUFA was C18:2 n–6, which was affected by the interaction between rearing system and breed ($P < 0.01$). Meat from PL, RE and TI kids fed natural milk had a lower content of C18:2 n–6 than meat from kids fed milk replacers, without differences between rearing systems ($P > 0.05$). The following fatty acid in quantity was C20:4 n–6. This fatty acid was only affected by breed ($P = 0.0001$), having higher values in GU than the rest of the breeds, especially when reared with milk replacers.

**Table 5.** Individual straight saturated fatty acids of suckling kid meat (g/100 g of FAMEs).

| RS [+] | Milk Replacer | | | | Natural Milk | | | | | | | |
|---|---|---|---|---|---|---|---|---|---|---|---|---|
| Breed (B) | GU | PL | RE | TI | GU | PL | RE | TI | s.e. | B | RS | B × RS |
| C6:0 | 0.002[a] | 0.005[a] | 0.002[a] | 0.002[a] | 0.002[a] | 0.003[a] | 0.004[a] | 0.003[a] | 0.001 | 0.49 | 0.72 | 0.17 |
| C8:0 | 0.004[b] | 0.011[a] | 0.005[ab] | 0.007[ab] | 0.006[ab] | 0.007[ab] | 0.006[ab] | 0.009[ab] | 0.002 | 0.401 | 0.83 | 0.30 |
| C9:0 | 0.013[b] | 0.077[a] | 0.009[b] | 0.001[b] | 0.030[ab] | 0.048[a] | 0.025[ab] | 0.049[a] | 0.012 | 0.008 | 0.16 | 0.008 |
| C10:0 | 0.08[d] | 0.154[cd] | 0.142[cd] | 0.196[cd] | 0.239[c] | 0.511[a] | 0.274[bc] | 0.407[ab] | 0.048 | 0.034 | 0.0001 | 0.10 |
| C11:0 | 0.013[c] | 0.066[ab] | 0.012[c] | 0.011[c] | 0.045[abc] | 0.073[ab] | 0.036[bc] | 0.080[a] | 0.012 | 0.07 | 0.001 | 0.036 |
| C12:0 | 3.1[a] | 4.25[a] | 0.36[c] | 3.75[a] | 0.79[bc] | 1.12[bc] | 0.90[bc] | 1.56[b] | 0.360 | 0.0001 | 0.0001 | 0.0001 |
| C13:0 | 0.04[de] | 0.091[abc] | 0.021[e] | 0.065[cd] | 0.067[bcd] | 0.093[ab] | 0.054[cd] | 0.107[a] | 0.010 | 0.0001 | 0.003 | 0.16 |
| C14:0 | 12.13[a] | 10.78[abc] | 7.66[d] | 11.65[ab] | 8.98[cd] | 9.26[cd] | 9.79[bc] | 8.61[cd] | 0.641 | 0.013 | 0.006 | 0.0001 |
| C15:0 | 0.245[c] | 0.403[bc] | 0.199[c] | 0.452[b] | 0.425[b] | 0.634[a] | 0.455[b] | 0.715[a] | 0.054 | 0.0001 | 0.0001 | 0.78 |
| C16:0 | 22.60[cd] | 25.12[b] | 21.12[d] | 24.73[b] | 21.21[d] | 27.49[a] | 23.99[bc] | 27.64[a] | 0.697 | 0.0001 | 0.003 | 0.001 |
| C17:0 | 0.44[d] | 0.562[bcd] | 0.530[cd] | 0.691[bc] | 0.693[bc] | 1.021[a] | 0.794[b] | 0.977[a] | 0.062 | 0.002 | 0.0001 | 0.30 |
| C18:0 | 10.3[c] | 9.48[c] | 13.15[ab] | 9.79[c] | 14.07[a] | 13.57[a] | 13.09[ab] | 11.875[b] | 0.640 | 0.017 | 0.0001 | 0.003 |
| C20:0 | 0.062[ab] | 0.050[b] | 0.066[ab] | 0.061[ab] | 0.081[a] | 0.063[ab] | 0.071[ab] | 0.067[ab] | 0.006 | 0.34 | 0.028 | 0.54 |
| C22:0 | 0.05[a] | 0.018[b] | 0.020[b] | 0.011[b] | 0.010[b] | 0.018[b] | 0.023[b] | 0.014[b] | 0.007 | 0.035 | 0.07 | 0.000 |
| C23:0 | 0.205[a] | 0.072[bc] | 0.057[c] | 0.043[c] | 0.173[ab] | 0.049[c] | 0.098[bc] | 0.047[c] | 0.023 | 0.0001 | 0.90 | 0.31 |

[+] RS, Rearing system; GU, del Guadarrama; PL, Palmera; RE, Retinta; TI, Tinerfeña; s.e., standard error. Different superscripts in the same row indicate significant differences (*P* ≤ 0.05). The least square means were adjusted for a hot carcass weight (HCW) of 5.02 kg.

**Table 6.** Individual monounsaturated fatty acids of suckling kid meat (g/100 g of FAMEs).

| RS [+] | Milk Replacer | | | | Natural Milk | | | | | | | |
|---|---|---|---|---|---|---|---|---|---|---|---|---|
| Breed (B) | GU | PL | RE | TI | GU | PL | RE | TI | s.e. | B | RS | B × RS |
| C14:1c9 | 0.229[bc] | 0.266[ab] | 0.063g | 0.276[a] | 0.107[f] | 0.175[de] | 0.143[e] | 0.200[cd] | 0.015 | 0.0001 | 0.0001 | 0.0001 |
| C15:1 | 0.002[e] | 0.013[a] | 0.001[b] | 0.003[b] | 0.003[b] | 0.003[b] | 0.004[b] | 0.004[b] | 0.002 | 0.14 | 0.36 | 0.009 |
| C16:1c7 | 1.92[e] | 0.28[b] | 0.85[b] | 0.34[b] | 1.40[ab] | 0.31[b] | 0.82[b] | 0.31[b] | 0.317 | 0.001 | 0.59 | 0.71 |
| C16:1c9 | 2.86[e] | 2.37[ab] | 2.55[a] | 2.64[a] | 2.10[b] | 2.42[ab] | 2.58[a] | 2.65[a] | 0.132 | 0.18 | 0.1 | 0.001 |
| C17:1c9 | 0.147[e] | 0.324[cde] | 0.238[de] | 0.425[bc] | 0.182[e] | 0.537[ab] | 0.379[bcd] | 0.626[a] | 0.054 | 0.0001 | 0.001 | 0.22 |
| C18:1c11 | 0.079[d] | 0.125[bcd] | 0.088[cd] | 0.127[bc] | 0.191[ab] | 0.192[a] | 0.150[ab] | 0.135[b] | 0.014 | 0.12 | 0.0001 | 0.0001 |
| C18:1c12 | 0.042[d] | 0.097[cd] | 0.069[d] | 0.162[b] | 0.145[bc] | 0.228[a] | 0.154[bc] | 0.164[b] | 0.018 | 0.005 | 0.0001 | 0.0001 |
| C18:1c13 | 0.010 | 0.035[b] | 0.021[bc] | 0.040[b] | 0.026[bc] | 0.065[a] | 0.036[b] | 0.035[b] | 0.006 | 0.001 | 0.004 | 0.017 |
| C18:1c15 | 0.026[e] | 0.069[ab] | 0.016[c] | 0.075[ab] | 0.032[c] | 0.080[a] | 0.052[b] | 0.053[b] | 0.007 | 0.0001 | 0.86 | 0.0001 |
| C18:1t11 | 0.349[e] | 0.680[abc] | 0.303[c] | 0.929[ab] | 0.651[abc] | 1.207[a] | 0.592[bc] | 1.049[ab] | 0.169 | 0.031 | 0.022 | 0.63 |
| C18:1t15 | 0.131[e] | 0.137[a] | 0.134[a] | 0.141[a] | 0.173[a] | 0.159[a] | 0.139[a] | 0.141[a] | 0.014 | 0.55 | 0.12 | 0.26 |
| C18:1t16 | 0.018[e] | 0.079[bc] | 0.035[c] | 0.120[b] | 0.134[b] | 0.216[a] | 0.125[b] | 0.136[b] | 0.021 | 0.054 | 0.0001 | 0.006 |
| C18:1c9 | 30.51[c] | 33.78[b] | 41.53[a] | 35.65[b] | 36.30[b] | 33.61[b] | 36.477[b] | 35.661[b] | 1.088 | 0.0001 | 0.86 | 0.0001 |
| C20:1n–9 | 0.002[e] | 0.003[a] | 0.002[a] | 0.002[a] | 0.004[a] | 0.001[a] | 0.003[a] | 0.004[a] | 0.001 | 0.81 | 0.71 | 0.32 |
| C22:1 | 0.010[a] | 0.003 [b] | 0.010 [ab] | 0.006 [b] | 0.010 [ab] | 0.004 [b] | 0.007 [ab] | 0.004 [b] | 0.002 | 0.010 | 0.44 | 0.71 |

[+] RS, Rearing system; GU, del Guadarrama; PL, Palmera; RE, Retinta; TI, Tinerfeña; s.e., standard error. Different superscripts in the same row indicate significant differences (*P* ≤ 0.05).

**Table 7.** Individual polyunsaturated fatty acids of suckling kid meat (g/100 g of FAMEs).

| RS [†] | Milk Replacer | | | | Natural Milk | | | | | | | |
|---|---|---|---|---|---|---|---|---|---|---|---|---|
| Breed (B) | GU | PL | RE | TI | GU | PL | RE | TI | s.e. | B | RS | B × RS |
| C18:2c9,t11 | 0.174$^{cd}$ | 0.264$^{bcd}$ | 0.097$^{d}$ | 0.343$^{b}$ | 0.159$^{d}$ | 0.500$^{a}$ | 0.270$^{bc}$ | 0.353$^{b}$ | 0.046 | 0.002 | 0.006 | 0.007 |
| C18:2 n–6 | 7.275$^{a}$ | 6.048$^{ab}$ | 7.428$^{a}$ | 4.465$^{b}$ | 6.339$^{a}$ | 2.765$^{c}$ | 4.283$^{bc}$ | 3.188$^{c}$ | 0.468 | 0.0001 | 0.0001 | 0.007 |
| C18:2n–6t9,12 | 0.024$^{c}$ | 0.088$^{b}$ | 0.023$^{c}$ | 0.118$^{b}$ | 0.110$^{b}$ | 0.173$^{a}$ | 0.108$^{b}$ | 0.115$^{b}$ | 0.016 | 0.010 | 0.0001 | 0.001 |
| C18:2t10,c12 | 0.027$^{a}$ | 0.026$^{ab}$ | 0.017$^{b}$ | 0.025$^{ab}$ | 0.016$^{b}$ | 0.021$^{ab}$ | 0.021$^{ab}$ | 0.027$^{a}$ | 0.003 | 0.177 | 0.218 | 0.005 |
| C18:2t9,c11 | 0.042$^{a}$ | 0.014$^{b}$ | 0.023$^{b}$ | 0.014$^{b}$ | 0.026$^{b}$ | 0.011$^{b}$ | 0.016$^{b}$ | 0.014$^{b}$ | 0.004 | 0.0001 | 0.077 | 0.152 |
| C18:3 n–3 | 0.372$^{c}$ | 0.230$^{d}$ | 0.764$^{a}$ | 0.227$^{d}$ | 0.435$^{c}$ | 0.191$^{d}$ | 0.529$^{b}$ | 0.193$^{d}$ | 0.033 | 0.0001 | 0.019 | 0.0001 |
| C18:3 n–6 | 0.035$^{a}$ | 0.023$^{bc}$ | 0.018$^{bc}$ | 0.025$^{b}$ | 0.024$^{bc}$ | 0.020$^{bc}$ | 0.014$^{c}$ | 0.023$^{bc}$ | 0.004 | 0.000 | 0.087 | 0.494 |
| C20:2 n–6 | 0.125$^{b}$ | 0.062$^{c}$ | 0.182$^{a}$ | 0.020$^{d}$ | 0.073$^{c}$ | 0.022$^{d}$ | 0.059$^{c}$ | 0.021$^{d}$ | 0.009 | 0.0001 | 0.0001 | 0.0001 |
| C20:3 n–6 | 0.003$^{abc}$ | 0.002$^{bcd}$ | 0.005$^{a}$ | 0.001$^{d}$ | 0.004$^{ab}$ | 0.001$^{d}$ | 0.003$^{abcd}$ | 0.002$^{cd}$ | 0.001 | 0.033 | 0.447 | 0.303 |
| C20:3 n–9 | 0.252$^{a}$ | 0.163$^{b}$ | 0.093$^{bcd}$ | 0.074$^{d}$ | 0.151$^{bc}$ | 0.078$^{cd}$ | 0.083$^{bcd}$ | 0.073$^{d}$ | 0.022 | 0.0001 | 0.005 | 0.013 |
| C20:4 n–6 | 3.23$^{a}$ | 1.95$^{bc}$ | 0.93$^{c}$ | 0.87$^{c}$ | 2.38$^{b}$ | 1.15$^{c}$ | 1.25$^{c}$ | 0.94$^{c}$ | 0.321 | 0.0001 | 0.215 | 0.097 |
| C20:5 n–3 | 0.234$^{a}$ | 0.140$^{ab}$ | 0.090$^{b}$ | 0.078$^{b}$ | 0.169$^{ab}$ | 0.107$^{b}$ | 0.179$^{ab}$ | 0.091$^{b}$ | 0.029 | 0.001 | 0.954 | 0.020 |
| C22:3 n–3 | 0.161$^{a}$ | 0.0001$^{b}$ | 0.036$^{b}$ | 0.0001$^{b}$ | 0.075$^{b}$ | 0.0001$^{b}$ | 0.044$^{b}$ | 0.001$^{b}$ | 0.025 | 0.0001 | 0.386 | 0.071 |
| C22:4 n–6 | 0.463$^{a}$ | 0.264$^{b}$ | 0.149$^{b}$ | 0.167$^{b}$ | 0.227$^{b}$ | 0.178$^{b}$ | 0.135$^{b}$ | 0.171$^{b}$ | 0.042 | 0.0001 | 0.013 | 0.003 |
| C22:5 n–3 | 0.598$^{a}$ | 0.336$^{bc}$ | 0.235$^{c}$ | 0.197$^{c}$ | 0.527$^{ab}$ | 0.293$^{c}$ | 0.410$^{bc}$ | 0.252$^{c}$ | 0.061 | 0.0001 | 0.543 | 0.117 |
| C22:6 n–3 | 0.093$^{a}$ | 0.045$^{bc}$ | 0.044$^{c}$ | 0.029$^{c}$ | 0.109$^{a}$ | 0.040$^{c}$ | 0.091$^{ab}$ | 0.037$^{c}$ | 0.011 | 0.0001 | 0.072 | 0.127 |

[†] RS, Rearing system; GU, del Guadarrama; PL, Palmera; RE, Retinta; TI, Tinerfeña; s.e., standard error. Different superscripts in the same row indicate significant differences ($P \leq 0.05$). The least square means were adjusted for a hot carcass weight (HCW) of 5.02 kg.

The main groups of straight and branched fatty acids are shown in Table 8. SFAs of meat from GU, PL and TI were similar in both rearing systems ($P > 0.05$), but RE had a higher ΣSFA content when fed natural milk ($P < 0.05$). The *iso-*, *anteiso-* and total BCFAs were affected by the interaction of the principal effects ($P < 0.01$). Hence, for PL and RE, the total BCFA amount was higher when animals were fed natural milk, whereas no effect of the rearing system was observed for GU and TI. All the breeds presented higher *iso*-BCFA amounts when fed natural milk except TI, for which no rearing system effects were observed, and regarding *anteiso*, only RE was affected by the rearing system, with higher amounts when animals were fed natural milk. ($P < 0.05$). The rearing system did not affect the ΣMUFAs of GU, PL and TI ($P > 0.05$). However, the meat of RE had lower ΣMUFAs when kids were fed natural milk ($P < 0.05$). The ΣCLA was higher in PL and RE when kids were fed natural milk ($P < 0.05$) but remained constant in meat of GU and TI ($P < 0.05$). PUFAs were affected by breed and rearing system ($P < 0.001$). Therefore, in general terms, meat from suckling kids fed natural milk had lower amounts of PUFAs. Finally, the desirable fatty acids were affected by the interaction between rearing system and breed ($P < 0.001$). Therefore, RE had the most desirable fatty acids when fed milk replacers but this index diminished when they were fed natural milk. Conversely, GU had the highest desirable fatty acid index when fed natural milk. PL and TI were not influenced by the rearing system ($P > 0.05$).

Figure 3 shows the principal component analysis of the main groups and branched fatty acids as a function of the rearing system. The two first principal components summarized almost 63% of the variation in the data (Figure 3). Feeding suckling kids NM was positively related to ΣBCFA, ΣCLA, Σ*iso*-BCFA and Σtrans-MUFA, and negatively related to n–6:n–3 in their meat. However, feeding suckling kids MR was related to n–6 and PUFAs in their meat.

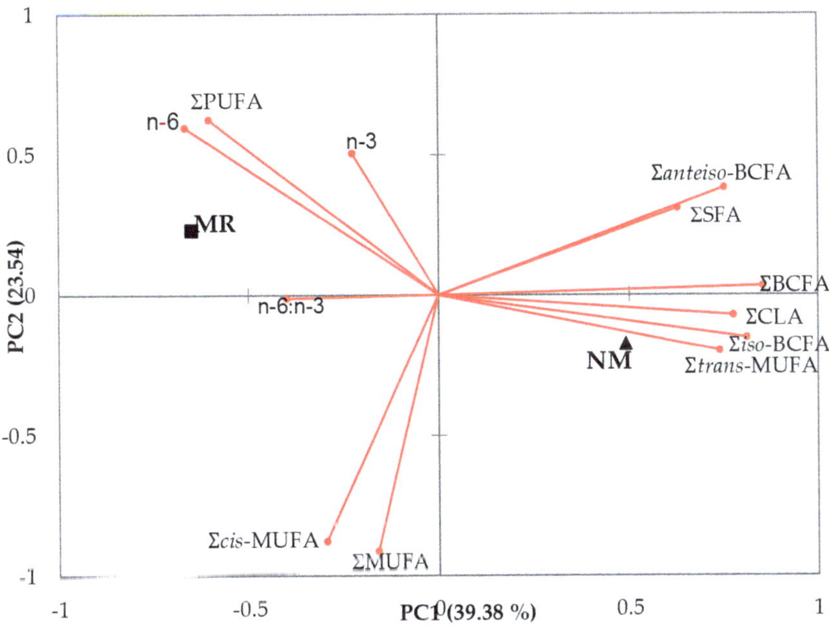


**Figure 3.** Principal component analysis of the main groups and branched fatty acids of meat of kids reared with milk replacers (MR) or natural milk (NM). SFA, saturated fatty acids; BCFA, branched chain fatty acids; MUFA, monounsaturated fatty acids; CLA, conjugated linoleic acid; PUFA, polyunsaturated fatty acids.

**Table 8.** Main groups of straight and branched fatty acids of suckling kid meat (g/100 g of FAMEs).

| RS [†] | Milk Replacer | | | | Natural Milk | | | | | | | |
|---|---|---|---|---|---|---|---|---|---|---|---|---|
| Breed (B) | GU | PL | RE | TI | GU | PL | RE | TI | s.e. | B | RS | B × RS |
| ΣSFA | 49.87[bc] | 51.62[ab] | 43.68[d] | 51.96[ab] | 47.34[c] | 54.58[a] | 50.21[b] | 52.69[ab] | 1.216 | 0.0001 | 0.047 | 0.001 |
| ΣBCFA | 0.41[c] | 0.46[bc] | 0.31[d] | 0.49[abc] | 0.48[abc] | 0.60[a] | 0.58[ab] | 0.51[ab] | 0.035 | 0.30 | 0.0001 | 0.001 |
| Σiso-BCFA | 0.23[cd] | 0.27[bc] | 0.21[d] | 0.31[b] | 0.31[b] | 0.39[a] | 0.39[a] | 0.33[ab] | 0.024 | 0.22 | 0.0001 | 0.004 |
| Σanteiso-BCFA | 0.18[a] | 0.19[a] | 0.11[b] | 0.18[a] | 0.17[a] | 0.21[a] | 0.19[a] | 0.18[a] | 0.015 | 0.073 | 0.06 | 0.001 |
| ΣMUFA | 36.48[c] | 38.49[bc] | 46.03[a] | 41.20[b] | 41.66[b] | 39.65[b] | 42.01[b] | 41.57[b] | 1.126 | 0.0001 | 0.45 | 0.0001 |
| Σcis-MUFA | 35.84[c] | 37.36[bc] | 45.43[a] | 39.74[b] | 40.50[b] | 37.63[bc] | 40.80[b] | 39.85[b] | 1.131 | 0.0001 | 0.90 | 0.000 |
| Σtrans-MUFA | 0.65[c] | 1.13[bc] | 0.59[c] | 1.47[b] | 1.16[bc] | 2.02[a] | 1.21[bc] | 1.72[ab] | 0.212 | 0.017 | 0.001 | 0.42 |
| ΣCLA | 0.24[cd] | 0.30[bcd] | 0.14[d] | 0.38[b] | 0.20[cd] | 0.53[a] | 0.31[bc] | 0.39[b] | 0.047 | 0.005 | 0.014 | 0.005 |
| ΣPUFA | 13.11[a] | 9.64[bc] | 10.13[b] | 6.66[d] | 10.83[b] | 5.55[d] | 7.94[cd] | 5.50[d] | 0.844 | 0.0001 | 0.0001 | 0.31 |
| n–6 | 11.16[a] | 8.43[bc] | 8.73[b] | 5.67[d] | 9.16[b] | 4.31[d] | 5.85[cd] | 4.46[d] | 0.744 | 0.0001 | 0.0001 | 0.18 |
| n–3 | 1.46[a] | 0.74[b] | 1.17[a] | 0.53[b] | 1.32[a] | 0.63[b] | 1.25[a] | 0.57[b] | 0.117 | 0.0001 | 0.73 | 0.62 |
| n–6:n–3 | 7.88[b] | 13.00[a] | 7.46[b] | 12.08[a] | 6.88[b] | 7.82[b] | 4.78[c] | 8.88[b] | 0.708 | 0.0001 | 0.0001 | 0.02 |
| Desirable FA | 59.97[cd] | 58.51[cd] | 69.24[a] | 58.20[d] | 65.62[b] | 59.28[cd] | 61.80[c] | 59.28[cd] | 0.970 | 0.0001 | 0.98 | 0.0001 |

[†] RS, Rearing system; GU, del Guadarrama; PL, Palmera; RE, Retinta; TI, Tinerfeña; s.e., standard error. FA. Fatty acids; SFA, saturated fatty acids; BCFA, branched chain fatty acids; MUFA, monounsaturated fatty acids; CLA, conjugated linoleic acid; PUFA, polyunsaturated fatty acids. Different superscripts in the same row indicate significant differences ($P \leq 0.05$). The least square means were adjusted for a hot carcass weight (HCW) of 5.02 kg.

## 4. Discussion

### 4.1. Milk

It is well reported that milk composition changes in the first few days of lactation and that goat colostrum has more protein and fat and less lactose than milk [26–29]. In agreement with this, in the current experiment, protein and fat percentages decreased and the lactose percentage increased from partum to 10 d of lactation. However, once colostrum production ceased, the milk composition was almost constant over time because the mammary gland developed a high tolerance to external factors, mainly diet [30,31], over the course of evolution to preserve functions and ensure the survival of the newborn ruminants [32]. In addition, this resilience is extendable to the fatty acid composition [33]. Current results for the proximate composition of colostrum are in accordance with those reported in the Majorera breed, with values of protein ranging from 7%–10%, fat ranging from 8%–9% and lactose ranging from 2%–4% [34,35]. Similar results for protein, fat and lactose (6.2%, 7.4% and 4.1%, respectively) were also reported for Murciano-Granadina goats [36]. However, low values of protein and fat were described in Tinerfeña colostrum [37]. Regarding the milk composition, we found higher protein and fat percentages but lower lactose percentages than the milk of Majorera and Payoya at comparable times of lactation reported by other authors [26,27,38]. However, the milk of Sarda had similar percentages of protein and fat but higher lactose content than that of other breeds [39]. Because goat milk has lower percentages of lactose than cow milk, it can be an alternative to people with lactose-related health problems.

Literature comparing fatty acids of goat colostrum and milk is scarce. Lou et al. [40] reported higher values of SFAs and lower values of MUFAs in milk than in colostrum of Laoshan breeds, which agreed with our results. Additionally, these latter authors [39] reported that the SFA proportion in colostrum ranged from 70% to 80%, in agreement with our results, while Marziali et al. [41] reported that the SFA proportion in colostrum of Murciano-Granadina ranged from 48% to 58%. On the other hand, it can be seen in Table 4 that this decrease in the total MUFA amount was mainly due to a decrease in the cis-MUFA series, whereas the change was less noticeable for trans-MUFAs. Similarly, there was a decrease in the total PUFA amount, whereas the n6 amount was lower in milk than in colostrum for all the studied breeds. However, the n3 amount only decreased from colostrum to milk in GU and TI. Therefore, both the relative proportions of different groups of fatty acids as well as the kind of fatty acids differ between colostrum and milk, which could be reflected in the kid meat composition. Ruminant milk is the greatest source of CLA [30], and C18:1c9 is one of the most important fatty acids detected both in milk and colostrum, as reported by Marziali, Guerra, Cerdán-Garcia, Segura-Carretero, Caboni and Verardo [41]. As reported by previous researchers [42,43], cis- and trans- C18:1 acids are often detected in goat milk, with a preeminent occurrence of C18:1 t11. However, from a sensory point of view, C8:0, C10:0 and 4-methyloctanoic fatty acids were the most influential in the flavor of milk and more predominant in caprine than in bovine or ovine milk [40,44]. Monomethyl-branched substitutions on short-chain fatty acids (C4-C6) are found only in goat milk and are implicated in goat-like flavors [45]. Regarding branched fatty acids, LeDoux, Rouzeau, Bas and Sauvant [42] also reported values of approximately 2.2% in goat milk, whereas Massoart Leën et al. [46] found that branched chain fatty acids comprised 2% of goat milk and 3% of cow milk. These values are much higher than those found in the present study. BCFAs in milk tat are derived from the incorporation of BCFAs of rumen bacterial lipids, whereas endogenous synthesis is limited [47]. The discrepancies in the results of this study and previous results could be derived from several factors, such as the diets used, and deserves more attention in the future. In literature, several fatty acids have been widely studied for several reasons. For example, linoleic and alpha linolenic acid are the only essential fatty acids [48], and some long-chain fatty acids (docosahexaenoic, eicosapentaenoic and arachidonic) are precursors of bioactive molecules such as prostaglandins, thromboxanes, leukotrienes and others [49]. In addition, as pointed out by Vlaeminck, Fievez, Cabrita, Fonseca and Dewhurst [47], CLA is generally a lesser component of milk fat than BCFA, but both have similar potential activity against cancer.

Although milk and dairy products have a high SFA content, it was concluded that the effects on human cardiovascular health are neutral or even positive [50]. It is possible that BCFAs play an important role in health benefits and must be investigated deeply.

*4.2. Meat*

The rumen of goat kids is fully functional only at approximately 56 days of age [51]. Therefore, ruminal biohydrogenation is limited or nonexistent until this time [52], so the fatty acid content of the intramuscular fat of kids should be influenced largely by dietary fatty acids (milk). Because 6%–20% of the de novo fatty acids synthetized arise mainly in the mammary gland and adipose tissue, the fatty acids of intramuscular fat should be influenced largely by the diets of the dams [52,53]. The endogenous synthesis of MUFA is specifically catalyzed by the $\Delta$-9 desaturase enzyme. It has been reported that mammary desaturase activity is higher in dairy animals than in meat animals [54–56]. However, in the present study, PL and TI, which are both dairy breeds, had higher percentages of *trans* MUFAs and most of the C18:1 isomers but lower amounts of the total MUFAs than RE (meat) and GU (mixed purpose). It was reported that meat kids show a higher percentage of CLA than dairy kids [57]. In the current results, the most abundant C18:1 isomers were C18:1c9 and C18:1 t11, in agreement with Adeyemi et al. [58]. These fatty acids are important because C18:1c9 decreases the blood cholesterol content [59], while an anti-atherogenic effect of 18:1 c9t11 has been presumed [60].

It is well known that the n–6:n–3 ratio is an indicator of the role of fatty acids in coronary heart diseases. Although the effects of the ratio on human health are less consistent than expected [61], it is accepted that this ratio must be lower than four [62]. Current results for kid meat varied between 4.78 and 13.00, in agreement with those of other authors [57]. The intramuscular fat from dairy animals, such as PL and TI, had a more unfavorable n–6:n–3 ratio than other breeds [55]. Although some authors did not find a relationship between slightly different milks and intramuscular fatty acids [63], suckling kid goats have a less healthy n–6:n–3 ratio than adult goats fed forages [58].

Several authors reported that fatty acid profiles in the fat of suckling kids usually reflect the fatty acid profiles of the suckled meat [64,65]. Although the lack of the chemical and fatty acid composition of milk replacers is a potential limitation of this study, we want to draw attention to the fact that commercial milk replacers for kids used were very similar between breeds. In addition, the chemical composition of the milk replacer is the same for all kids while the composition of the fat of the kids varies. Tsiplakou et al. [66] reported that the FA profile of goat kids' muscles reflected that of suckled milk (natural or artificial) and the FA of muscle might help us to discover when artificially reared goat kids are sold as naturally suckled kids. According to De Palo et al. [67], artificial feeding could increase the amount of unsaturated fatty acids to improve its profitability of milk. Other authors reported that the muscle FA profile of goat kids is healthier when fed natural milk that when fed milk replacers [66], although this is related with the composition of milk replacer. Hence, the modification of the fatty acid composition of milk replacers can be a useful tool to improve the quality of the intermuscular fat of suckling kids. Supplementation of milk replacers with docohexanoic acid increased the DHA concentration and the n–6/n–3 ratios were reduced in tissues of goats kids. However, some fatty acids of muscle are less prone to be modified than others. Joy et al. [68] reported similar conclusions in suckling lambs. Therefore, CLA and n–6/n–3 of natural milk and muscles had high correlations while SFA and PUFA n–3 had low correlations. Other authors found that same positive relationship between linolenic acid in milk and meat and the lack of relationship between C18:2 n–6 in milk and meat [69]. Other authors found that same positive relationship between linolenic acid in milk and meat and the lack of relationship between C18:2 n–6 in milk and meat [69]. However, to modify the fatty acids of the suckled natural milk is mandatory to modify the goats' diets. Hence, the feeding affected to the PUFA, CLA c9,t11, PUFA/SFA, PUFA n–3 and PUFA n–6/n–3 of sheep milk, revealing higher CLA, PUFA/SFA, PUFA n–3 and PUFA n–6/n–3 in milk and suckling lamb meat when grazing pastures instead of being hay fed. Sanz-Sampelayo et al. [70] investigated the possibility of improving the composition of goat meat, in terms of the fatty acid composition, using concentrates supplemented with polyunsaturated fatty acids. These authors reported that milk contained fat with a lower content of saturated fatty acids and a higher content of n–3 PUFA, trans-C18: 1 and CLA. The intramuscular fat presented of the suckling kids had a higher proportion of n–3 PUFA, trans C18: 1 and CLA, while that of n–6 PUFA remained unchanged. In consequence, it is demonstrated that important fatty acids of milk can be improved by modifying the goats' diets and the modified milk can improve the fatty acid composition of the intramuscular fat of the suckling kids

## 5. Conclusions

This study confirmed that a great change in composition occurs from colostrum to milk. Once colostrum production has finished, milk composition is almost constant over time, demonstrating that the mammary gland has developed a high tolerance to external factors, such as diet. Goat colostrum has more protein and fat and less lactose than milk. The amount of lactose is actually lower in goat milk than in cow milk, which is of interest to people with lactose tolerance issues. Goat milk is an important source of healthy fatty acids such as C18:1 c9 and C18:2 n–6. The percentage of C18:1 c9 decreased from colostrum to milk for all breeds, but the decrease in PL and TI (dairy breeds) was twofold compared to the decrease in GU and RE (meat breeds).

Suckling kid meat was also an important source of C18:1c9. Goat dairy breeds had higher percentages of *trans* MUFAs and most of the C18:1 isomers but lower amounts of total MUFAs. However, these dairy kids had meat with a lower percentage of CLA than meat kids. The meat of kids fed natural milk had higher amounts of CLA and BCFA and lower amounts of n–6 fatty acids than kids fed milk replacers. The presence of BCFAs on the meat of preruminant animals could be explained by the intake of BCFAs. Both milk and meat are a source of linoleic and alpha linolenic acids, which are essential fatty acids and healthy long-chain fatty acids, such as DHA, EPA and ARA. The effect of the rearing system on the fatty acid composition of milk and meat is clearly modulated by the breed. Therefore, investigations related to this topic should be afforded using more than one breed to avoid results open to misinterpretation.

Regarding the possibility of improving the fatty acid composition of suckling kid fat, several authors reported that the fatty acid profile in the fat of suckling kids usually reflects the fatty acid profile of the suckled meat [63,64].

Additionally, a lack of knowledge on the presence of branched chain fatty acids in both milk and meat of suckling kids was identified. Due to the rising importance of these fatty acids on human health, and the contribution to their flavor, more attention is warranted in future research.

**Author Contributions:** Conceptualization, B.P., M.J.A., M.d.G.C., A.A. and G.R.; Methodology, B.P., M.d.G.C., A.A. and G.R.; Data collection and acquisition, B.P., M.J.A., G.R., A.A. and M.d.G.C.; Writing—Original draft preparation, B.P., M.J.A. and G.R.; Writing—Review and editing, B.P., M.J.A., G.R., A.A. and M.d.G.C.; Funding acquisition, B.P., M.J.A. and A.A. All authors have read and agreed to the published version of the manuscript.

**Funding:** This research was funded by the Ministry of Economy and Competitiveness of Spain and FEDER, grant number RTA2012-23-C3 and the Research Group Funds of the Aragón Government (A14-17R SAGAS).

**Acknowledgments:** Appreciation is expressed to breeders and breeders' associations and the staff of CITA de Aragón for their help in data collection. We are particularly grateful to Noemi Castro and Aridany Suarez for their invaluable help with the samples.

**Conflicts of Interest:** The authors declare no conflict of interest.

## References

1. FAO. *FAOSTAT Livestock Primary*; Food and Agriculture Organization: Rome, Italy, 2017.
2. Yalcintan, H.; Akin, P.D.; Ozturk, N.; Ekiz, B.; Kocak, O.; Yilmaz, A. Carcass and meat quality traits of Saanen goat kids reared under natural and artificial systems and slaughtered at different ages. *Acta Vet. Brno* **2018**, *87*, 293–300. [CrossRef]
3. Aparnathi, K.; Mehta, B.; Velpula, S. Goat Milk in Human Nutrition and Health—A Review. *Int. J. Curr. Microbiol. Appl. Sci.* **2017**, *6*, 1781–1792. [CrossRef]

4. Castel, J.M.; Mena, Y.; Ruiz, F.A.; Gutiérrez, R. Situación y evolución de los sistemas de producción caprina en España. *Tierras Caprino* **2012**, *1*, 24–37.

5. Marichal, A.; Castro, N.; Capote, J.; Zamorano, N.; Arguello, A. Effects of live weight at slaughter (6, 10 and 25 kg) on kid carcass and meat quality. *Livest. Prod. Sci.* **2003**, *83*, 247–256. [CrossRef]

6. Argüello, A.; Castro, N.; Capote, J.; Solomon, M. Effects of diet and live weight at slaughter on kid meat quality. *Meat Sci.* **2005**, *70*, 173–179. [CrossRef]

7. Ripoll, G.; Alcalde, M.J.; Argüello, A.; Córdoba, M.G.; Panea, B. Effect of the rearing system on the color of four muscles of suckling kids. *Food Sci. Nutr.* **2019**, *7*, 1502–1511. [CrossRef]

8. Ripoll, G.; Alcalde, M.J.; Argüello, A.; Córdoba, M.G.; Panea, B. Consumer visual appraisal and shelf life of leg chops from suckling kids raised with natural milk or milk replacer. *J. Sci. Food Agric.* **2018**, *98*, 2651–2657. [CrossRef]

9. Ripoll, G.; Alcalde, M.J.; Argüello, A.; Panea, B. Web-based survey of consumer preferences for the visual appearance of meat from suckling kids. *Ital. J. Anim. Sci.* **2019**, *18*, 284–1293. [CrossRef]

10. Ripoll, G.; Córdoba, M.G.; Alcalde, M.J.; Martín, A.; Argüello, A.; Casquete, R.; Panea, B. Volatile organic compounds and consumer preference for meat from suckling goat kids raised with natural or replacers milk. *Ital. J. Anim. Sci.* **2019**, *18*, 1259–1270. [CrossRef]

11. Ribeiro, R.D.X.; Medeiros, A.N.; Oliveira, R.L.; de Araújo, G.G.L.; Queiroga, R.d.C.d.E.; Ribeiro, M.D.; Silva, T.M.; Bezerra, L.R.; Oliveira, R.L. Palm kernel cake from the biodiesel industry in goat kid diets. Part 2: Physicochemical composition, fatty acid profile and sensory attributes of meat. *Small Rumin. Res.* **2018**, *165*, 1–7. [CrossRef]

12. PAHO/WHO. Trans. fats free Americas. In *Conclusions and Recommendations*; PAHO/WHO: Washington, DC, USA, 2007; p. 5.

13. Alonso, L.; Fontecha, F.J.; Fraga, M.J.; Juárez, M.; Lozada, L. Fatty acid composition of caprine milk: Major, branched-chain, and trans fatty acids. *J. Dairy Sci.* **1999**, *82*, 878–884. [CrossRef]

14. Woo, A.; Lindsay, R. Concentrations of major free fatty acids and flavor development in Italian cheese varieties. *J. Dairy Sci.* **1984**, *67*, 960–968. [CrossRef]

15. Watkins, P.J.; Kearney, G.; Rose, G.; Allen, D.; Ball, A.J.; Pethick, D.W.; Warner, R.D. Effect of branched-chain fatty acids, 3-methylindole and 4-methylphenol on consumer sensory scores of grilled lamb meat. *Meat Sci.* **2014**, *96*, 1088–1094. [CrossRef] [PubMed]

16. Brennand, C.P.; Ha, J.K.; Lindsay, R.C. Aroma properties and thresholds of some branched-chaiin and other minor volatile fatty acids occurring in milk fat and meat lipids. *J. Sens. Stud.* **1989**, *4*, 105–120. [CrossRef]

17. Ran-Ressler, R.R.; Devapatla, S.; Lawrence, P.; Brenna, J.T. Branched chain fatty acids are constituents of the normal healthy newborn gastrointestinal tract. *Pediatric Res.* **2008**, *64*, 605. [CrossRef] [PubMed]

18. Yang, Z.; Liu, S.; Chen, X.; Chen, H.; Huang, M.; Zheng, J. Induction of apoptotic cell death and in vivo growth inhibition of human cancer cells by a saturated branched-chain fatty acid, 13-methyltetradecanoic acid. *Cancer Res.* **2000**, *60*, 505–509. [PubMed]

19. Duncan, W.R.H.; Garton, G.A. Differences in the proportions of branched-chain fatty acids in subcutaneous triacylglycerols of barley-fed ruminants. *Br. J. Nutr.* **2007**, *40*, 29–33. [CrossRef]

20. Serra, A.; Mele, M.; La Comba, F.; Conte, G.; Buccioni, A.; Secchiari, P. Conjugated Linoleic Acid (CLA) content of meat from three muscles of Massese suckling lambs slaughtered at different weights. *Meat Sci.* **2009**, *81*, 396–404. [CrossRef]

21. E.U. Directive 2010/63/EU of the European Parliament and of the Council of 22 September 2010 on the protection of animals used for scientific purposes. *Off. J. Eur. Union L* **2010**, *276*, 33–79.

22. E.U. Council Regulation (EC) No 1099/2009 of 24 September 2009 on the protection of animals at the time of killing. *Off. J. Eur. Union* **2009**, *303*, 1–30.

23. Molkentin, J.; Precht, D. Validation of a gas-chromatographic method for the determination of milk fat contents in mixed fats by butyric acid analysis. *Eur. J. Lipid Sci. Technol.* **2000**, *102*, 194–201. [CrossRef]

24. Folch, J.; Lees, M.; Stanley, G. A simple method for the isolation and purification of lipids from animal tissues. *J. Biol. Chem.* **1957**, *226*, 4977509.

25. Huerta-Leidenz, N.; Cross, H.; Lunt, D.; Pelton, L.; Savell, J.; Smith, S. Growth, carcass traits, and fatty acid profiles of adipose tissues from steers fed whole cottonseed. *J. Anim. Sci.* **1991**, *69*, 3665–3672. [CrossRef] [PubMed]

26. Sanchez-Macias, D.; Moreno-Indias, I.; Castro, N.; Morales-Delanuez, A.; Arguello, A. From goat colostrum to milk: Physical, chemical, and immune evolution from partum to 90 days postpartum. *J. Dairy Sci.* **2014**, *97*, 10–16. [CrossRef]

27. Sanchez-Macias, D.; Fresno, M.; Moreno-Indias, I.; Castro, N.; Morales-delaNuez, A.; Alvarez, S.; Arguello, A. Physicochemical analysis of full-fat, reduced-fat, and low-fat artisan-style goat cheese. *J. Dairy Sci.* **2010**, *93*, 3950–3956. [CrossRef]

28. Linzell, J.; Peaker, M. Changes in colostrum composition and in the permeability of the mammary epithelium at about the time of parturition in the goat. *J. Physiol.* **1974**, *243*, 129–151. [CrossRef]

29. Pecka-Kiełb, E.; Zachwieja, A.; Wojtas, E.; Zawadzki, W. Influence of nutrition on the quality of colostrum and milk of ruminants. *Mljekarstvo Časopis za Unaprjeđenje Proizvodnje i Prerade Mlijeka* **2018**, *68*, 169–181. [CrossRef]

30. Almeida, O.C.; Ferraz, M.V.C.; Susin, I.; Gentil, R.S.; Polizel, D.M.; Ferreira, E.M.; Barroso, J.P.R.; Pires, A.V. Plasma and milk fatty acid profiles in goats fed diets supplemented with oils from soybean, linseed or fish. *Small Rumin. Res.* **2019**, *170*, 125–130. [CrossRef]

31. Bobe, G.; Zimmerman, S.; Hammond, E.G.; Freeman, A.E.; Porter, P.A.; Luhman, C.M.; Beitz, D.C. Butter Composition and Texture from Cows with Different Milk Fatty Acid Compositions Fed Fish Oil or Roasted Soybeans. *J. Dairy Sci.* **2007**, *90*, 2596–2603. [CrossRef]

32. Hernandez-Castellano, L.E.; Arguello, A.; Almeida, A.M.; Castro, N.; Bendixen, E. Colostrum protein uptake in neonatal lambs examined by descriptive and quantitative liquid chromatography-tandem mass spectrometry. *J. Dairy Sci.* **2015**, *98*, 135–147. [CrossRef]

33. Palma, M.; Alves, S.P.; Hernandez-Castellano, L.E.; Capote, J.; Castro, N.; Arguello, A.; Matzapetakis, M.; Bessa, R.J.B.; de Almeida, A.M. Mammary gland and milk fatty acid composition of two dairy goat breeds under feed-restriction. *J. Dairy Res.* **2017**, *84*, 264–271. [CrossRef] [PubMed]

34. Moreno-Indias, I.; Sánchez-Macías, D.; Castro, N.; Morales-delaNuez, A.; Hernández-Castellano, L.E.; Capote, J.; Argüello, A. Chemical composition and immune status of dairy goat colostrum fractions during the first 10h after partum. *Small Rumin. Res.* **2012**, *103*, 220–224. [CrossRef]

35. Argüello, A.; Castro, N.; Álvarez, S.; Capote, J. Effects of the number of lactations and litter size on chemical composition and physical characteristics of goat colostrum. *Small Rumin. Res.* **2006**, *64*, 53–59. [CrossRef]

36. Romero, T.; Beltrán, M.C.; Rodríguez, M.; De Olives, A.M.; Molina, M.P. Short communication: Goat colostrum quality: Litter size and lactation number effects. *J. Dairy Res.* **2013**, *96*, 7526–7531. [CrossRef] [PubMed]

37. Capote, J.; Castro, N.; Caja, G.; Fernández, G.; Briggs, H.; Argüello, A. Effects of the frequency of milking and lactation stage on milk fractions and milk composition in Tinerfeña dairy goats. *Small Rumin. Res.* **2008**, *75*, 252–255. [CrossRef]

38. Delgado-Pertíñez, M.; Guzmán-Guerrero, J.L.; Caravaca, F.P.; Castel, J.M.; Ruiz, F.A.; González-Redondo, P.; Alcalde, M.J. Effect of artificial vs. natural rearing on milk yield, kid growth and cost in Payoya autochthonous dairy goats. *Small Rumin. Res.* **2009**, *84*, 108–115. [CrossRef]

39. Nudda, A.; Battacone, G.; Bee, G.; Boe, R.; Castanares, N.; Lovicu, M.; Pulina, G. Effect of linseed supplementation of the gestation and lactation diets of dairy ewes on the growth performance and the intramuscular fatty acid composition of their lambs. *Animal* **2015**, *9*, 800–809. [CrossRef]

40. Lou, X.; Li, J.; Zhang, X.; Wang, J.; Wang, C. Variations in fatty acid composition of Laoshan goat milk from partum to 135 days postpartum. *Anim. Sci. J.* **2018**, *89*, 1628–1638. [CrossRef]

41. Marziali, S.; Guerra, E.; Cerdán-Garcia, C.; Segura-Carretero, A.; Caboni, M.F.; Verardo, V. Effect of early lactation stage on goat colostrum: Assessment of lipid and oligosaccharide compounds. *Int. Dairy J.* **2018**, *77*, 65–72. [CrossRef]

42. LeDoux, M.; Rouzeau, A.; Bas, P.; Sauvant, D. Occurrence of trans-C18: 1 fatty acid isomers in goat milk: Effect of two dietary regimens. *J. Dairy Sci.* **2002**, *85*, 190–197. [CrossRef]

43. Andueza, D.; Rouel, J.; Chilliard, Y.; Leroux, C.; Ferlay, A. Prediction of the goat milk fatty acids by near infrared reflectance spectroscopy. *Eur. J. Lipid Sci. Technol.* **2013**, *115*, 612–620. [CrossRef]

44. Ha, J.K.; Lindsay, R. Release of volatile branched-chain and other fatty acids from ruminant milk fats by various lipases. *J. Dairy Sci.* **1993**, *76*, 677–690. [PubMed]

45. Haenlein, G. Goat milk in human nutrition. *Small Rumin. Res.* **2004**, *51*, 155–163. [CrossRef]

46. Massart-Leën, A.M.; De Pooter, H.; Decloedt, M.; Schamp, N. Composition and variability of the branched-chain fatty acid fraction in the milk of goats and cows. *Lipids* **1981**, *16*, 286–292. [CrossRef]

47. Vlaeminck, B.; Fievez, V.; Cabrita, A.R.J.; Fonseca, A.J.M.; Dewhurst, R.J. Factors affecting odd- and branched-chain fatty acids in milk: A review. *Anim. Feed Sci. Technol.* **2006**, *131*, 389–417. [CrossRef]

48. Albenzio, M.; Santillo, A.; Avondo, M.; Nudda, A.; Chessa, S.; Pirisi, A.; Banni, S. Nutritional properties of small ruminant food products and their role on human health. *Small Rumin. Res.* **2016**, *135*, 3–12. [CrossRef]

49. Calder, P.C. Fatty acids and inflammation: The cutting edge between food and pharma. *Eur. J. Pharmacol.* **2011**, *668*, S50–S58. [CrossRef]

50. Lordan, R.; Tsoupras, A.; Mitra, B.; Zabetakis, I. Dairy fats and cardiovascular disease: Do we really need to be concerned? *Foods* **2018**, *7*, 29. [CrossRef]

51. Van Soest, P. Function of the ruminant forestomach. In *Nutritional Ecology of the Ruminant*, 2nd ed.; Cornell University Press: Ithaca, NY, USA, 1994; pp. 230–252.

52. Valvo, M.A.; Lanza, M.; Bella, M.; Fasone, V.; Scerra, M.; Biondi, L.; Priolo, A. Effect of ewe feeding system (grass v. concentrate) on intramuscular fatty acids of lambs raised exclusively on maternal milk. *Anim. Sci.* **2005**, *81*, 431–436. [CrossRef]

53. Osorio, M.T.; Zumalacarregui, J.M.; Figueira, A.; Mateo, J. Fatty acid composition in subcutaneous, intermuscular and intramuscular fat deposits of suckling lamb meat: Effect of milk source. *Small Rumin. Res.* **2007**, *73*, 127–134. [CrossRef]

54. Wood, J.D.; Enser, M.; Fisher, A.V.; Nute, G.R.; Sheard, P.R.; Richardson, R.I.; Hughes, S.I.; Whittington, F.M. Fat deposition, fatty acid composition and meat quality: A review. *Meat Sci.* **2008**, *78*, 343–358. [CrossRef]

55. Horcada, A.; Ripoll, G.; Alcalde, M.J.; Sanudo, C.; Teixeira, A.; Panea, B. Fatty acid profile of three adipose depots in seven Spanish breeds of suckling kids. *Meat Sci.* **2012**, *92*, 89–96. [CrossRef] [PubMed]

56. Chilliard, Y.; Rouel, J.; Leroux, C. Goat's alpha-s1 casein genotype influences its milk fatty acid composition and delta-9 desaturation ratios. *Anim. Feed Sci. Technol.* **2006**, *131*, 474–487. [CrossRef]

57. Horcada, A.; Campo, M.D.M.; Polvillo, O.; Alcalde, M.J.; Cilla, I.; Sañudo, C. A comparative study of fatty acid profiles of fat in commercial Spanish suckling kids and lambs. *Span. J. Agric. Res.* **2014**, *12*, 427. [CrossRef]

58. Adeyemi, K.D.; Shittu, R.M.; Sabow, A.B.; Karim, R.; Sazili, A.Q. Myofibrillar protein, lipid and myoglobin oxidation, antioxidant profile, physicochemical and sensory properties of caprine *longissimus thoracis* during *postmortem* conditioning. *J. Food Process. Preserv.* **2017**, *41*, e13076. [CrossRef]

59. Binkoski, A.E.; Kris-Etherton, P.M.; Wilson, T.A.; Mountain, M.L.; Nicolosi, R.J. Balance of unsaturated fatty acids is important to a cholesterol-lowering diet: Comparison of mid-oleic sunflower oil and olive oil on cardiovascular disease risk factors. *J. Am. Diet. Assoc.* **2005**, *105*, 1080–1086. [CrossRef] [PubMed]

60. Pariza, M.W.; Park, Y.; Cook, M.E. The biologically active isomers of conjugated linoleic acid. *Prog. Lipid Res.* **2001**, *40*, 283–298. [CrossRef]

61. Marventano, S.; Kolacz, P.; Castellano, S.; Galvano, F.; Buscemi, S.; Mistretta, A.; Grosso, G. A review of recent evidence in human studies of n–3 and n–6 PUFA intake on cardiovascular disease, cancer, and depressive disorders: Does the ratio really matter? *Int. J. Food Sci. Nutr.* **2015**, *66*, 611–622. [CrossRef] [PubMed]

62. Scollan, N.; Hocquette, J.F.; Nuernberg, K.; Dannenberger, D.; Richardson, I.; Moloney, A. Innovations in beef production systems that enhance the nutritional and health value of beef lipids and their relationship with meat quality. *Meat Sci.* **2006**, *74*, 17–33. [CrossRef]

63. Manso, T.; Bodas, R.; Vieira, C.; Mantecon, A.R.; Castro, T. Feeding vegetable oils to lactating ewes modifies the fatty acid profile of suckling lambs. *Animal* **2011**, *5*, 1659–1667. [CrossRef]

64. Bas, P.; Morand-Fehr, P. Effect of nutritional factors on fatty acid composition of lamb fat deposits. *Livest. Prod. Sci.* **2000**, *64*, 61–79. [CrossRef]

65. Bañón, S.; Vila, R.; Price, A.; Ferrandini, E.; Garrido, M.D. Effects of goat milk or milk replacer diet on meat quality and fat composition of suckling goat kids. *Meat Sci.* **2006**, *72*, 216–221. [CrossRef] [PubMed]

66. Tsiplakou, E.; Papadomichelakis, G.; Sparaggis, D.; Sotirakoglou, K.; Georgiadou, M.; Zervas, G. The effect of maternal or artificial milk, age and sex on three muscles fatty acid profile of Damascus breed goat kids. *Livest. Sci.* **2016**, *188*, 142–152. [CrossRef]

67. De Palo, P.; Maggiolino, A.; Centoducati, N.; Tateo, A. Effects of different milk replacers on carcass traits, meat quality, meat color and fatty acids profile of dairy goat kids. *Small Rumin. Res.* **2015**, *131*, 6–11. [CrossRef]

68. Joy, M.; Ripoll, G.; Molino, F.; Dervishi, E.; Alvarez-Rodriguez, J. Influence of the type of forage supplied to ewes in pre- and post-partum periods on the meat fatty acids of suckling lambs. *Meat Sci.* **2012**, *90*, 775–782. [CrossRef] [PubMed]

69. Nudda, A.; Palmquist, D.L.; Battacone, G.; Fancellu, S.; Rassu, S.P.G.; Pulina, G. Relationships between the contents of vaccenic acid, CLA and n–3 fatty acids of goat milk and the muscle of their suckling kids. *Livest. Sci.* **2008**, *118*, 195–203. [CrossRef]

70. Sanz-Sampelayo, M.; Fernández, J.; Ramos, E.; Hermoso, R.; Extremera, F.G.; Boza, J. Effect of providing a polyunsaturated fatty acid-rich protected fat to lactating goats on growth and body composition of suckling goat kids. *Anim. Sci.* **2006**, *82*, 337–344. [CrossRef]

 *foods*

*Article*

# Influence of the Use of Milk Replacers and pH on the Texture Profiles of Raw and Cooked Meat of Suckling Kids

Guillermo Ripoll [1,2,*], María J. Alcalde [3], María G. Córdoba [4], Rocío Casquete [4], Anastasio Argüello [5], Santiago Ruiz-Moyano [4] and Begoña Panea [1,2]

[1]    Instituto Agroalimentario de Aragón, IA2, CITA-Universidad de Zaragoza, C/Miguel Servet, 177, 50013 Zaragoza, Spain; bpanea@aragon.es

[2]    Centro de Investigación y Tecnología Agroalimentaria de Aragón, Avda. Montañana, 930, 50059 Zaragoza, Spain

[3]    Department of Agroforestry Science. Universidad de Sevilla. Crta. Utrera, 41013 Sevilla, Spain; aldea@us.es

[4]    Instituto Universitario de Investigación de Recursos Agrarios (INURA), Nutrición y Bromatología, Escuela de Ingeniería Agrarias, Universidad de Extremadura, Avda. Adolfo Suarez s/n, 06007 Badajoz, Spain; mdeguia@unex.es (M.G.C.); rociocp@unex.es (R.C.); srmsh@unex.es (S.R.-M.)

[5]    Department of Animal Pathology, Animal Production and Science and Technology of Foods, Universidad de Las Palmas de Gran Canaria, 35416 Las Palmas, Spain; tacho@ulpgc.es

*    Correspondence: gripoll@aragon.es; Tel.: +34-976-716-452

Received: 31 October 2019; Accepted: 14 November 2019; Published: 19 November 2019

**Abstract:** The aim of this work was to study the texture profile of fresh and cooked *longissimus thoracis et lumborum* muscle from suckling kids raised with natural milk or milk replacers. Suckling male kids from eight goat breeds (Florida, FL; Cabra del Guadarrama, GU; Majorera, MA; Palmera, PL; Payoya, PY; Retinta, RE; Tinerfeña, TI; Verata, VE), all of single parturition, were raised with milk replacers (MR) or with natural milk from the dams (NM). The meat pH, Warner-Bratzler shear force, texture profile analysis and chemical composition were determined. Kids were clustered based on their pH by k-means clustering. The effect of the rearing system on the textural profile was strongly modulated by breed. The values of Warner-Bratzler shear force and hardness found in these breeds under both rearing systems were very low. Hence, the toughness of very light suckling kids should not be a determining factor in choosing a breed or rearing system. Nevertheless, the use of milk replacers increased the presence of meat with high pH, which modified the textural parameters, decreasing the shear force but increasing cohesiveness and adhesiveness. Consequently, depending on the commercial strategy of the farm, the election of the breed and rearing system must be considered together.

**Keywords:** rearing system; stress; DFD; TPA; hardness; toughness; shear force; Warner-Bratzler

---

## 1. Introduction

Approximately 4.7 million head of goats and kids were slaughtered within the European Union in 2017 [1]). Meat from goats is considered healthy because it is low in calories and fat [2]. However, Mediterranean goat farms are mainly focused on production of cheese and milk products because they have higher prices than cow milk [3–5]. When kid goats are reared with their dams, the availability of milk for cheese production is decreased, and the quality of milk may change. Although most of the incomes per goat on the dairy farm come from the sale of milk, 20% of the total income comes from the sale of kids [6]. These kids are weaned very early and reared with milk replacers. Milk replacers specifically formulated for kids result in good daily weight gain. The kids are mostly slaughtered at a very light carcass weight of 5–7 kg, and this meat is perceived by consumers to be of high quality [7]. However, some farmers believe that kids reared with milk replacers provide tougher meat [8] and are opposed to this practice. This belief could be explained by the fact that most of the kid meat with high pH comes from kids raised on milk replacers [9], which might induce tough meat. On the other hand, meat of kids reared with milk replacers was preferred by consumers based on its appearance. Additionally, the purchase intentions were greater for kids reared with milk replacers [10].

Meat sensory evaluation is determined from a complex interaction of sensory and physical processes during chewing, with tenderness being the most important [11]. Tenderness is the sensory variable that is most related to the overall appraisal [12]. Therefore, several instrumental methods have been developed to study the textural characteristics of meat. The most important are the Warner-Bratzler method [13], which provides a main variable based on the maximum force to shear the sample and is usually used with cooked meat; few studies have used the Warner-Bratzler method to assess raw meat. The texture profile analysis (TPA) is also a widely used method. This test provides a set of variables describing the rheological characteristics of meat and has been used in both raw and cooked meat. Both instrumental methods are often used as an approximation of sensory tenderness because they are easier and cheaper than sensory analysis. The use of raw meat is quick, but it is cooked meat that is consumed by people. TPA of raw meat predicted sensory tenderness better than the Warner-Bratzler method, but the Warner-Bratzler method was more correlated with the sensory tenderness of cooked meat than the TPA [14]. Therefore, it seems that the best options to analyze meat tenderness are the TPA for raw meat and the Warner-Bratzler method for cooked meat. There have been some studies about the Warner-Bratzler shear force of suckling kids [7,8,15–19], but there have been no such studies using TPA. Because pH and milk quality could affect kid meat texture and because information about TPA of suckling kid meat is scarce, the aim of this work was to study the texture profile of fresh and cooked meat from suckling kids raised with natural milk or milk replacers.

## 2. Materials and Methods

### 2.1. Animals

All procedures were conducted according to the guidelines of Directive 2010/63/EU on the protection of animals used for experimental and other scientific purposes [20]. Suckling male kids of eight goat breeds (Florida, FL; Cabra del Guadarrama, GU; Majorera, MA; Palmera, PL; Payoya, PY; Retinta, RE; Tinerfeña, TI; Verata, VE; US) were reared on two or three farms per breed in their respective local areas. Animals were all born from single parturition, and half were raised with milk replacers (MR), while the other half were raised with natural milk from the dams (NM). Kids in the MR rearing system were fed colostrum for the first 2 days and had free access to milk replacer 24 h a day, which was suckled from a teat connected to a unit for feeding a liquid diet. Commercial milk replacers were reconstituted at 17% (w/v) and given warm (40 °C). The main ingredients were skimmed milk (≈60%) and whey. The chemical composition of the milk replacers was as follows: total fat 25% ± 0.6, crude protein 24% ± 0.5, crude cellulose 0.1% ± 0.0, ash 7% ± 0.6, Ca 0.8% ± 0.1, Na 0.5% ± 0.2, P 0.7% ± 0.0, Fe 36 mg/kg ± 4.0, Cu 3 mg/kg ± 1.7, Zn 52 mg/kg ± 18.8, Mn 42 mg/kg ± 14.4, I 0.22 mg/kg ± 0.06, Se 0.1 mg/kg ± 0.06 and BHT 65 ppm ± 30. Kids in the NM rearing system were kept separated from their dams during the day while the dams grazed. At night, they were housed with their dams in a stable and suckled directly from dams with no additional feedstuff. Kids from both rearing systems had no access to concentrates, hay, forages or other supplements. The natural milk of goats at 30 d *postpartum* was collected in the morning, and the chemical composition of the milk was determined using a DMA2001 milk analyzer (Miris Inc., Uppsala, Sweden).

### 2.2. Carcass Sampling

The numbers of kids used are shown in Table 1. The 246 kids were slaughtered at a live weight of 8.47 kg ± 0.077 kg. Because the kids of the different breeds were raised in different places, to minimize the effect of the transport, they were slaughtered in a slaughterhouse close to each farm, hence, the duration of the transport from farm to slaughterhouse ranged from 30 to 60 min. The animals of different groups and farms were never mixed during transport or at the slaughterhouse. Standard commercial procedures according to the European normative for protection of animals at the time of killing [21] were followed. Head-only electrical stunning was applied (1.00 A) to kids, which were then exsanguinated and dressed. Thereafter, the hot carcasses, including head and kidneys, were weighed to achieve a hot carcass weight (HCW) of 4.97 kg ± 0.061 kg. Afterwards, the carcasses were hung by the Achilles tendon and chilled for 24 h at 4 °C.

**Table 1.** Means, standard error and *p*-value for breed effect on proximal composition of natural milk 30 d post birth in eight goat breeds.

| Breed | Protein, % | Fat, % | Lactose, % |
|---|---|---|---|
| Florida | 3.79 [b] | 5.11 [bc] | 4.05 |
| Cabra del Guadarrama | 3.35 [c] | 4.09 [e] | 4.10 |
| Majorera | 4.55 [a] | 4.78 [cd] | 4.11 |
| Palmera | 4.54 [a] | 5.40 [ab] | 4.16 |
| Payoya | 3.78 [b] | 4.29 [de] | 4.19 |
| Retinta | 3.79 [b] | 5.38 [ab] | 4.20 |
| Tinerfeña | 4.30 [a] | 4.58 [de] | 4.10 |
| Verata | 4.22 [ab] | 5.82 [a] | 4.25 |
| Standard error | 0.147 | 0.181 | 0.062 |
| Breed effect (*p-value*) | <0.001 | <0.001 | 0.33 |

Different superscripts indicate significant differences ($p < 0.05$).

After carcass chilling, the *longissimus thoracis et lumborum* muscle of both the left and right halves of the carcasses were extracted and sliced. The pH was measured on the left *longissimus thoracis* with a pH-meter equipped with a Crison 507 penetrating electrode (Crison Instruments S.A., Barcelona, Spain). Then, the left *longissimus thoracis* was vacuum packed and frozen at −20 °C until chemical composition analyses. The right *longissimus thoracis* was vacuum packed, aged for 3 days at 4 °C in darkness and frozen at −20 °C until Warner-Bratzler maximum stress determination. The *longissimus lumborum* muscle of both the left and right half carcasses were extracted, vacuum packed, aged for 3 days and frozen at −20 °C until TPA of raw and cooked meat, respectively.

### 2.3. Meat Chemical Composition

The moisture content of meat (Moist) was determined by dehydration at 100 °C to a constant weight by the ISO recommended methods (ISO, 1973). Crude protein (CP) was determined following the Kjeldahl method [22]. Intramuscular fat content (IMF) was quantified using the method of Bligh and Dyer [23]. Ash content was assessed by dividing the weight before and after ignition in a muffle furnace for 8 h [22] (Ash). Analyses were run in duplicate and expressed as the percentage of fresh meat. Non-protein nitrogen (NPN) was determined by the Nessler method using 4 g of sample after protein precipitation with 0.6 M perchloric acid, and amino acid nitrogen (AN) was determined from the 0.6 M perchloric acid protein precipitation fraction after peptide precipitation with 10% sulfosalicylic acid as described in Benito, et al. [24].

### 2.4. Meat Texture

An Instron machine model 5543 (Instron Limited, Cerdanyola, Spain) was used to determine the shear force of cooked *longissimus thoracis* (LT). Samples were thawed in tap water for 4 h until they reached an internal temperature of 16–19 °C. Then, the samples were heated in a water bath at 75 °C to an internal temperature of 70 °C. Temperature was controlled with a Testo 108-2 waterproof food thermometer with a Type T thermocouple (Instrumentos Testo S.A., Cabrils, Spain). Then, the steaks were cooled overnight at room temperature. Cross-sectioned meat blocks of 1 cm² and a

3 cm length were measured with a Mitutoyo digital caliper (Mitutoyo Co., Kawasaky, Japan) with a resolution of 0.01 mm. The samples were sheared perpendicularly to the long axis of the block using a Warner-Bratzler device with a cross-head speed of 2.5 mm s$^{-1}$. The maximum stress (load at maximum peak shear force per unit of cross-section, in N cm$^{-2}$) was determined.

Samples of *longissimus lumborum* (LL) were thawed in tap water for 4 h until they reached an internal temperature of 16–19 °C. Then, samples of the left LL were heated in a water bath at 75 °C to an internal temperature of 70 °C; samples of the right LL remained raw. TPA was performed at room temperature using a TA.XTA2i texture analyzer (Stable Micro Systems, Godalming, UK). One cylinder with a 1.5 cm height and 2 cm diameter was prepared from every sample. A double compression cycle test was performed at up to 50% compression of the original portion height with an aluminum cylinder probe with a 6 cm diameter. A time of 5 s was allowed to elapse between the two compression cycles. Force–time deformation curves were obtained with a 250 N load cell applied at a cross-head speed of 1 mm/s. The following parameters were quantified: hardness (maximum force of the first compression cycle required to compress the sample, N), adhesiveness (negative area under the abscissa after the first compression, N·s), springiness (ability of the sample to recover its original form after the deforming force was removed, cm), cohesiveness (extent to which the sample could be deformed prior to rupture, dimensionless), chewiness (work required to masticate a solid food before swallowing, J) and resilience (ability of a product to recover its original height, dimensionless).

*2.5. Statistical Analysis*

All statistics were calculated using the XLSTAT statistical package v.3.05 (Addinsoft, New York, NY, USA). Studied variables were analyzed using the ANCOVA procedure with the breed (B) and the rearing system (RS) as fixed effects and the hot carcass weight (HCW) as a covariate. Least square means were adjusted for an HCW of 4.965 kg, and differences were tested with the Bonferroni test at a 0.05 level of significance. Pearson's correlations between the raw studied variables and between the residuals of variables were calculated. Principal component analysis (PCA) was performed by projecting the pH, chemical composition and textural variables as active variables and the rearing system and breed as supplementary data to highlight the associations between the loadings of the variables [25]. There was used the varimax rotation and a biplot of variables and centroids were plotted. Kids were clustered together based on their pH by k-means clustering using Wilk's lambda as classification criterion. The statistical procedure tested the classification from 2 to 5 clusters to maximize the intergroup variability and minimize the intragroup variability. The number of clusters was selected to ensure significant pH differences among all clusters and to avoid clusters formed by 10 or fewer observations. The inter- and intra-class variabilities of the clusters were 80.8% and 19.2%. Then, ANCOVA was carried out for the pH and texture variables, with the pH cluster as a fixed effect and HCW as a covariate. A Duncan test was used to compare means, with a significance of $p < 0.05$. The independence between the rearing system and the pH clusters was tested with the $\chi^2$ test.

## 3. Results

*3.1. Chemical Composition of Natural Milk and Longissimus Thoracis Muscle*

The chemical composition of the natural milk of goats is shown in Table 1. Differences between breeds were found in protein and fat percentages ($p < 0.001$). Majorera, Palmera and Tinerfeña had the highest values of protein, while Cabra del Guadarrama had the lowest such value. In addition, Cabra del Guadarrama had the lowest value of fat, and Verata had the highest value. The lactose percentages ranged from 4.05 to 4.25 without differences among breeds ($p > 0.05$).

The chemical composition of longissimus thoracis muscle is shown in Table 2. The pH at 24 h ranged from 5.53 to 6.16. There was a significant interaction between breed and rearing system ($p < 0.001$). Most of the breeds had the same pH when the kids were reared in both rearing systems. However, the Payoya and Tinerfeña kids reared with milk replacers had greater pH than those kids reared with natural milk ($p < 0.05$). The rearing system did not affect the percentage of intramuscular fat (IMF), whereas breed did ($p < 0.001$). Guadarrama kids had the greatest values of IMF, and Florida, Majorera, Palmera, Payoya and Tinerfeña kids had the lowest ($p < 0.05$), while Retinta and Verata kids had intermediate values ($p < 0.05$). The protein percentage was affected by an interaction between breed and rearing system ($p < 0.001$). Therefore, kids of Guadarrama, Majorera and Payoya fed natural milk (NM) had a greater percentage of protein than kids fed milk replacers (MR) ($p < 0.05$). However, the kids of the other breeds had a similar percentage of protein in both rearing systems ($p > 0.05$).

**Table 2.** Value of pH at 24 h and chemical composition of *longissimus thoracis* muscle of kids reared with milk replacer (MR) or natural milk from their dams (NM).

| B [†] | RS | n | pH 24 h | Moist, % | IMF, % | CP, % | Ash, % | NPN, mg/g | AN, mg/g |
|---|---|---|---|---|---|---|---|---|---|
| FL | MR | 15 | 5.69 [de] | 78.03 [a] | 1.88 [def] | 18.30 [de] | 1.18 [ab] | 3.086 [bcde] | 0.822 [cde] |
|  | NM | 15 | 5.61 [efg] | 77.11 [ab] | 1.97 [def] | 18.68 [cd] | 1.18 [ab] | 3.491 [bcd] | 1.043 [bcd] |
| GU | MR | 15 | 5.66 [ef] | 76.88 [ab] | 4.29 [a] | 19.25 [cd] | 1.11 [abc] | 5.454 [a] | 0.236 [g] |
|  | NM | 16 | 5.67 [de] | 71.36 [f] | 5.16 [a] | 23.99 [ab] | 0.96 [e] | 4.699 [ab] | 0.290 [fg] |
| MA | MR | 16 | 5.81 [c] | 74.50 [d] | 1.78 [def] | 23.22 [b] | 1.08 [bcd] | 3.047 [bcde] | 1.619 [a] |
|  | NM | 16 | 5.86 [c] | 73.16 [e] | 0.89 [f] | 24.97 [a] | 1.10 [abc] | 1.702 [def] | 1.426 [ab] |
| PL | MR | 15 | 6.16 [a] | 75.31 [cd] | 1.72 [def] | 24.05 [ab] | 1.16 [ab] | 2.802 [bcde] | 0.826 [cde] |
|  | NM | 16 | 5.85 [c] | 74.07 [de] | 1.29 [ef] | 23.82 [ab] | 1.01 [de] | 1.743 [def] | 1.422 [ab] |
| PY | MR | 16 | 5.80 [c] | 76.88 [ab] | 1.63 [ef] | 16.89 [e] | 1.17 [ab] | 2.244 [cde] | 0.471 [efg] |
|  | NM | 14 | 5.78 [cd] | 76.90 [ab] | 1.11 [f] | 20.30 [c] | 1.12 [abc] | 2.695 [cde] | 0.635 [def] |
| RE | MR | 15 | 5.53 [g] | 78.08 [a] | 2.68 [bcd] | 19.51 [cd] | 1.10 [abc] | 1.466 [ef] | 0.790 [cde] |
|  | NM | 15 | 5.55 [fg] | 76.17 [bc] | 2.97 [bc] | 19.25 [cd] | 1.21 [a] | 2.210 [cdef] | 1.113 [bc] |
| TI | MR | 16 | 6.01 [b] | 75.06 [cd] | 1.32 [ef] | 23.81 [ab] | 1.08 [bcd] | 3.513 [bc] | 0.983 [cd] |
|  | NM | 16 | 5.88 [c] | 74.66 [d] | 1.45 [ef] | 23.56 [ab] | 1.04 [cde] | 2.602 [cde] | 1.754 [a] |
| VE | MR | 15 | 5.84 [c] | 76.99 [ab] | 3.16 [b] | 18.22 [de] | 0.96 [e] | 2.699 [cde] | 0.685 [de] |
|  | NM | 15 | 5.79 [cd] | 76.19 [bc] | 2.16 [cde] | 19.41 [cd] | 1.09 [bcd] | 0.476 [f] | 0.688 [de] |
|  | s.e. |  | 0.039 | 0.441 | 0.324 | 0.511 | 0.031 | 0.363 | 0.080 |
|  | B |  | <0.001 | <0.001 | 0.001 | <0.001 | 0.001 | <0.001 | <0.001 |
|  | RS |  | 0.006 | <0.001 | 0.26 | <0.001 | 0.47 | 0.002 | <0.001 |
|  | B*RS |  | <0.001 | <0.001 | 0.11 | <0.001 | <0.001 | <0.001 | <0.001 |

[†] B, Breed; RS, Rearing system; s.e., standard error; Moist, percentage of moisture on fresh basis; IMF, percentage of intramuscular fat on fresh basis; CP, percentage of crude protein on fresh basis; Ash, percentage of ashes on fresh basis; NPN, non-protein nitrogen in mg/g fresh meat; AN, amino acid nitrogen in mg AN/g fresh meat; FL, Florida; GU, del Guadarrama; MA, Majorera; PL, Palmera; PY, Payoya; RE, Retinta; TI, Tinerfeña; VE, Verata. Least square means has been adjusted for an HCW of 4.965 Kg. Different superscripts indicate significant differences ($p < 0.05$).

### 3.2. Meat Texture

The texture profile of raw longissimus thoracis muscle is shown in Table 3. The interaction between rearing system and breed was significant ($p < 0.001$) for every variable, but in general, the effect was more noticeable for breed than for rearing system. The Retinta breed was mostly affected by the rearing system, with a higher hardness in the milk replacer system than in the natural system, whereas adhesiveness, cohesiveness and resilience presented higher values for the natural milk system. In the Majorera breed, the rearing system affected only springiness and resilience, both of which were higher in the natural milk system than in the milk replacer system. In contrast, in the Verata breed, springiness was higher in the natural milk system than in the milk replacer system, with the rest of the variables being unaffected. Finally, in the Payoya breed, the only variable affected by the rearing system was chewiness, which was higher in the natural milk system than in the milk replacer system. The Florida, Cabra del Guadarrama, Palmera and Tinerfeña breeds were not affected at all ($p < 0.05$).

**Table 3.** Texture profile of raw *longissimus lumborum* muscle of suckling kids reared with milk replacer (MR) or natural milk from their dams (NM).

| B [†] | RS | Hardness (N) | Adhesiveness (-N·s) | Springiness (cm) | Cohesiveness (-) | Chewiness ($J·10^{-2}$) | Resilience (-) |
|---|---|---|---|---|---|---|---|
| FL | MR | 14.38 [bcd] | 0.26 [ab] | 0.83 [bcd] | 0.45 [ab] | 5.18 [cd] | 0.246 [ab] |
|    | NM | 14.97 [bcd] | 0.21 [c]  | 0.87 [abc] | 0.47 [ab] | 6.60 [bcd] | 0.247 [ab] |
| GU | MR | 10.46 [d]   | 0.09 [c]  | 0.90 [ab]  | 0.44 [ab] | 4.10 [d]   | 0.231 [abc] |
|    | NM | 11.97 [cd]  | 0.11 [c]  | 0.90 [ab]  | 0.47 [ab] | 5.08 [cd]  | 0.265 [a] |
| MA | MR | 11.90 [cd]  | 0.11 [c]  | 0.80 [cd]  | 0.44 [ab] | 4.14 [d]   | 0.195 [c] |
|    | NM | 14.21 [cd]  | 0.17 [c]  | 0.89 [ab]  | 0.47 [ab] | 5.96 [bcd] | 0.252 [ab] |
| PA | MR | 17.60 [bcd] | 0.37 [bc] | 0.93 [a]   | 0.46 [ab] | 7.44 [bc]  | 0.263 [a] |
|    | NM | 14.99 [bcd] | 0.21 [c]  | 0.89 [ab]  | 0.48 [a]  | 6.37 [bcd] | 0.266 [a] |
| PY | MR | 18.74 [bc]  | 0.29 [bc] | 0.86 [abc] | 0.43 [ab] | 6.80 [bcd] | 0.242 [ab] |
|    | NM | 33.41 [a]   | 0.34 [bc] | 0.84 [abcd]| 0.41 [b]  | 11.35 [a]  | 0.240 [ab] |
| RE | MR | 22.22 [b]   | 0.96 [a]  | 0.86 [abc] | 0.35 [c]  | 6.58 [bcd] | 0.120 [d] |
|    | NM | 13.81 [cd]  | 0.53 [b]  | 0.91 [ab]  | 0.48 [a]  | 5.77 [bcd] | 0.210 [bc] |
| TI | MR | 12.14 [cd]  | 0.14 [c]  | 0.90 [ab]  | 0.45 [ab] | 4.67 [cd]  | 0.237 [ab] |
|    | NM | 17.88 [bcd] | 0.26 [bc] | 0.89 [ab]  | 0.47 [ab] | 7.28 [bc]  | 0.246 [ab] |
| VE | MR | 16.72 [bcd] | 0.31 [bc] | 0.89 [ab]  | 0.46 [ab] | 8.74 [ab]  | 0.210 [bc] |
|    | NM | 14.45 [bcd] | 0.21 [c]  | 0.78 [d]   | 0.47 [ab] | 5.16 [ab]  | 0.236 [abc] |
|    | s.e. | 1.662     | 0.058     | 0.016      | 0.013     | 0.641      | 0.009 |
|    | B    | <0.001    | <0.001    | <0.001     | <0.001    | <0.001     | <0.001 |
|    | RS   | 0.086     | 0.044     | 0.773      | <0.001    | 0.023      | <0.001 |
|    | B*RS | <0.001    | <0.001    | <0.001     | <0.001    | <0.001     | <0.001 |

[†] B, Breed; RS, Rearing system; s.e., standard error; FL, Florida; GU, del Guadarrama; MA, Majorera; PL, Palmera; PY, Payoya; RE, Retinta; TI, Tinerfeña; VE, Verata. Least square means has been adjusted for an HCW of 4.965 Kg. Different superscripts indicate significant differences ($p < 0.05$).

Regarding the breed effect on the raw texture profile, the Retinta breed presented higher values for adhesiveness, and Verata had the highest values for chewiness, independent of the rearing system ($p < 0.05$). In addition, in the milk replacer system, Retinta presented the lowest values for cohesiveness and resilience, whereas Cabra del Guadarrama presented the lowest chewiness values, and Palmera presented the highest values for resilience. In the natural milk system, Payoya presented higher values for chewiness than the other breeds ($p < 0.05$). Chewiness was affected only by breed ($p < 0.001$).

The texture profile of cooked longissimus thoracis muscle is shown in Table 4. The interaction between rearing system and breed was significant ($p < 0.001$) for every variable except chewiness ($p > 0.05$). The rearing system affected the hardness, adhesiveness, springiness and cohesiveness only of Retinta kids. The use of milk replacers increased the two first variables and decreased the two latter variables ($p < 0.05$). There was an effect of breed but not rearing system on chewiness ($p < 0.001$). The resilience of Florida and Retinta was increased and decreased, respectively, by the use of milk replacers ($p < 0.05$). Cooking had a great effect, decreasing hardness and adhesiveness while decreasing cohesiveness. Chewiness and resilience were slightly affected by cooking, and springiness was not affected.

The Warner-Bratzler maximum stress is shown in Figure 1. The maximum stress was affected by the significant interaction ($p = 0.0002$) between the breed and rearing system. In addition, the covariate of HCW was also significant ($p = 0.004$). The maximum stress of Payoya and Retinta were affected by the rearing system ($p < 0.05$) but in opposite ways. Florida, Guadarrama, Payoya, Retinta and Verata had maximum stress values greater than 30 N cm−2, while Majorera, Palmera and Tinerfeña had lower values ($p < 0.05$).

**Table 4.** Texture profile of cooked *longissimus lumborum* muscle of suckling kids reared with milk replacer (MR) or natural milk from their dams (NM).

| B | RS [†] | Hardness (N) | Adhesiveness (-N·s) | Springiness (cm) | Cohesiveness (-) | Chewiness ($J·10^{-2}$) | Resilience (-) |
|---|---|---|---|---|---|---|---|
| FL | MR | 8.17 [de] | 0.002 [c] | 0.83 [a] | 0.77 [a] | 5.42 [cde] | 0.36 [ab] |
|  | NM | 9.08 [de] | 0.008 [c] | 0.85 [a] | 0.75 [abc] | 5.86 [cde] | 0.30 [c] |
| GU | MR | 8.27 [de] | 0.008 [c] | 0.86 [a] | 0.65 [d] | 4.67 [e] | 0.31 [bc] |
|  | NM | 8.51 [de] | 0.026 [c] | 0.82 [a] | 0.65 [d] | 4.53 [e] | 0.29 [c] |
| MA | MR | 6.92 [e] | 0.010 [c] | 0.87 [a] | 0.71 [abcd] | 4.36 [e] | 0.33 [bc] |
|  | NM | 7.77 [de] | 0.040 [bc] | 0.88 [a] | 0.73 [abcd] | 5.09 [de] | 0.34 [abc] |
| PA | MR | 8.13 [de] | 0.034 [bc] | 0.90 [a] | 0.70 [abcd] | 5.13 [cde] | 0.32 [bc] |
|  | NM | 8.77 [de] | 0.035 [bc] | 0.86 [a] | 0.68 [cd] | 5.23 [cde] | 0.32 [bc] |
| PY | MR | 11.72 [cd] | 0.012 [c] | 0.85 [a] | 0.76 [ab] | 7.55 [abc] | 0.39 [a] |
|  | NM | 10.38 [cde] | 0.012 [c] | 0.85 [a] | 0.74 [abc] | 6.68 [bcde] | 0.34 [abc] |
| RE | MR | 28.14 [a] | 0.271 [a] | 0.66 [b] | 0.52 [e] | 9.12 [a] | 0.14 [e] |
|  | NM | 17.53 [b] | 0.092 [b] | 0.80 [a] | 0.65 [d] | 8.31 [ab] | 0.23 [d] |
| TI | MR | 8.23 [de] | 0.016 [c] | 0.92 [a] | 0.70 [abcd] | 5.33 [cde] | 0.33 [bc] |
|  | NM | 9.08 [de] | 0.037 [bc] | 0.89 [a] | 0.69 [bcd] | 5.58 [cde] | 0.30 [c] |
| VE | MR | 15.01 [bc] | 0.039 [bc] | 0.82 [a] | 0.70 [abcd] | 8.51 [ab] | 0.28 [c] |
|  | NM | 11.96 [cd] | 0.012 [c] | 0.84 [a] | 0.77 [abcd] | 7.54 [abcd] | 0.32 [bc] |
|  | s.e. | 0.919 | 0.013 | 0.023 | 0.016 | 0.489 | 0.012 |
|  | B | <0.001 | <0.001 | <0.001 | <0.001 | <0.001 | <0.001 |
|  | RS | 0.002 | 0.011 | 0.365 | 0.042 | 0.518 | 0.859 |
|  | B*RS | <0.001 | <0.001 | 0.001 | <0.001 | 0.438 | <0.001 |

[†] B, Breed; RS, Rearing system; s.e., standard error; FL, Florida; GU, del Guadarrama; MA, Majorera; PL, Palmera; PY, Payoya; RE, Retinta; TI, Tinerfeña; VE, Verata. Least square means has been adjusted for an HCW of 4.965 Kg. Different superscripts indicate significant differences ($p < 0.05$).

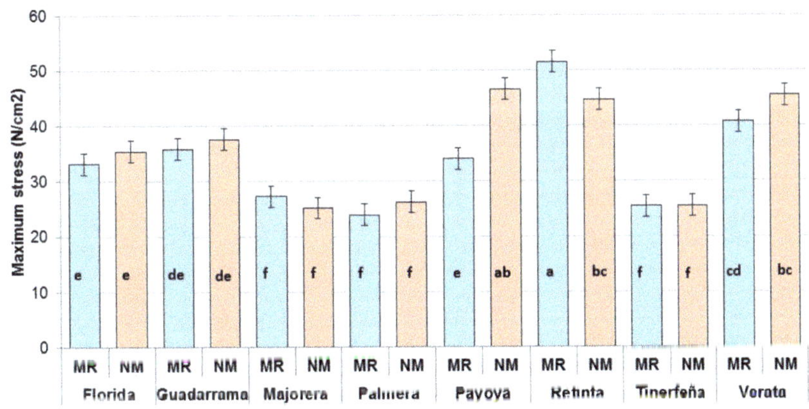


**Figure 1.** Warner-Bratzler maximum stress of *longissimus thoracis* muscle from kids reared with milk replacer (**MR**) or natural milk from their dams (**NM**). Different superscripts indicate significant differences (*p < 0.05*).

### 3.3. Principal Component Analysis and Correlations

There were many significant correlations among the studied variables. Warner-Bratzler maximum stress was significantly correlated with the TPA results of raw and cooked meat ($p < 0.001$), except for chewiness (raw) and resilience (cooked) ($p < 0.05$). However, most of the correlations became nonsignificant ($p > 0.05$) when correlation analysis was performed with residuals of variables. This general absence of correlations demonstrates that there was an important influence of carcass weight. Hence, the Warner-Bratzler maximum stress *was not* correlated with any of the variables ($p > 0.05$). Similarly, there was no correlation between the chemical composition of natural milk and muscle ($p > 0.05$) or between the chemical composition of muscle and the TPA results of cooked meat ($p > 0.05$). However, moisture was correlated with the cohesiveness (0.14; $p < 0.01$) and chewiness (0.14; $p < 0.01$) of raw meat, and protein percentage was correlated with resilience (0.18; $p = 0.006$). Raw meat TPA variables were correlated among themselves (from $-0.17$; $p = 0.008$ to 0.91; $p < 0.001$) similarly to cooked meat TPA variables (from 0.23; $p < 0.001$ to 0.90; $p < 0.001$). Correlations between the TPA variables of raw and cooked meat were small but significant (from 0.13; $p = 0.04$ to 0.27; $p < 0.001$).

A principal component analysis were made with the chemical composition, TPA on raw and cooked meat and Warner-Bratzler maximum stress. There were five principal components with eigenvalues higher than 1 explaining the 81.7% of variability. Figure 2 shows the bi-plot of the two first principal components, explaining the 55.2% of the variability.

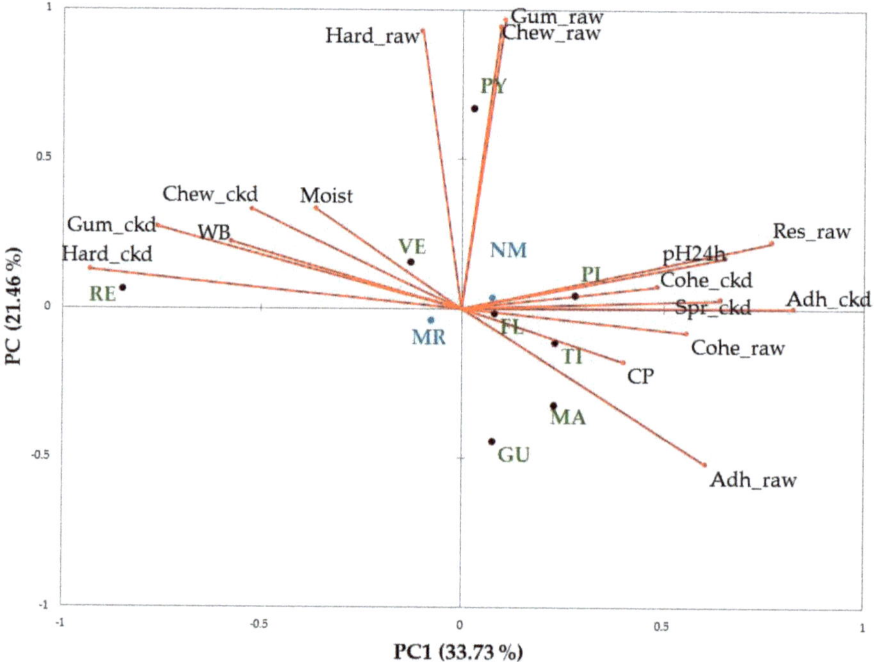


**Figure 2.** Bi-plot of the texture variables. MR, milk replacer; NM, natural milk; raw, variables determined on raw meat; ckd, variables determined on cooked meat; Hard, hardness; Adh, adhesiveness; Spr, springiness; Cohe, cohesiveness; Chew, chewiness; Res, resilience; WB, Warner-Bratzler maximum stress; Moist, moisture; CP, crude protein; FL, Florida; GU, del Guadarrama; MA, Majorera; PL, Palmera; PY, Payoya; RE, Retinta; TI, Tinerfeña; VE, Verata. The overall Kaiser-Mayer-Olkin score of the PCA was 0.73 (Bartlett's test of sphericity was significant, $p < 0.001$). The eigenvalues of PC 1 and PC2 are 5.6 and 3.2, respectively.

Resilience of cooked meat, springiness of raw meat, amino acidic nitrogen, non-protein nitrogen, ash and intramuscular fat were not included in the final PCA due to their Kaiser-Meyer-Olkin values. Three breeds (TI, PL and FL) were placed closer to the abscissas axis in the positive side, being related with cohesiveness on both raw and cooked meat, springiness and adhesiveness on cooked meat, resilience on raw meat and pH. The natural milk rearing system was also related with those variables but it had less importance than breeds. In the opposite side of the abscissas axis were placed moisture and chewiness, gumminess, hardness and Warner-Bratzler maximum stress on cooked meat. RE was the breed related with these variables while VE was placed close to the origin of coordinates. Chewiness, gumminess and hardness on raw meat were related positively to the second PC and Payoya was placed together with those variables. Finally, GU and MA were related with adhesiveness on raw meat. Although most of the studied variables were affected significantly by an interaction between rearing system and breed, according the PCA the effect of breed was more important than the effect of rearing system.

### 3.4. Effect of PH on Kids Meat Quality

Once the meat samples were clustered according their pH at 24 h, the percentage of samples within each cluster, average pH and percentage of kids in the MR group within each cluster were calculated and are shown in Table 5. There were significant differences in pH between clusters ($p < 0.001$). There was also a relationship between the rearing system and the pH cluster ($\chi^2 = 13.8$; $p = 0.001$). Therefore, the frequency of kids from both rearing systems was similar in the first and second clusters, but 76.7% of kids from cluster 3 with an average pH of 6.2 were in the MR group.

**Table 5.** Comparison of meat pH on chemical composition and texture of meat of the three pH clusters identified by the *k*-means algorithm. The percentage of kids per cluster is between brackets.

| Variables | CL1 (51.2%) | CL2 (36.6%) | CL3 (12.2%) | s.e. | Sig. |
|---|---|---|---|---|---|
| % of MR kids | 54.4 | 40.5 | 76.7 | | 0.001 |
| pH | 5.6 [c] | 5.8 [b] | 6.2 [a] | 0.020 | <0.001 |
| **Chemical composition** | | | | | |
| Moisture, % | 75.82 | 75.61 | 75.38 | 0.310 | 0.689 |
| Intramuscular fat, % | 2.59 [a] | 1.96 [b] | 2.12 [ab] | 0.196 | 0.015 |
| Crude protein, % | 20.42 [b] | 21.36 [b] | 22.34 [a] | 0.381 | 0.010 |
| Ash, % | 1.11 | 1.09 | 1.10 | 0.017 | 0.380 |
| NPN, mg/g | 3.08 [a] | 2.48 [b] | 2.89 [ab] | 0.223 | 0.041 |
| AN, mg/g | 0.91 | 0.97 | 0.85 | 0.065 | 0.397 |
| **TPA raw meat** | | | | | |
| Hardness, N | 15.47 | 16.15 | 17.94 | 1.023 | 0.390 |
| Adhesiveness,-N·s | 0.38 [a] | 0.20 [b] | 0.30 [ab] | 0.037 | <0.001 |
| Springiness, cm | 0.88 [b] | 0.86 [c] | 0.91 [a] | 0.009 | 0.001 |
| Cohesiveness | 0.43 [b] | 0.46 [a] | 0.46 [a] | 0.007 | 0.005 |
| Chewiness, $J\cdot10^{-2}$ | 5.86 [b] | 6.29 [b] | 7.60 [a] | 0.371 | 0.029 |
| Resilience | 0.21 [b] | 0.24 [a] | 0.26 [a] | 0.006 | <0.001 |
| **TPA cooked meat** | | | | | |
| Hardness, N | 13.79 [a] | 9.53 [b] | 9.13 [b] | 0.753 | <0.001 |
| Adhesiveness,-N·s | 0.08 [a] | 0.02 [h] | 0.02 [b] | 0.009 | <0.001 |
| Springiness, cm | 0.80 [b] | 0.86 [a] | 0.90 [a] | 0.012 | <0.001 |
| Cohesiveness | 0.66 [h] | 0.72 [a] | 0.72 [a] | 0.010 | <0.001 |
| Chewiness, $J\cdot10^{-2}$ | 6.10 | 5.99 | 6.14 | 0.296 | 0.466 |
| Resilience | 0.26 | 0.29 | 0.34 | 0.047 | 0.634 |
| **Warner-Bratzler** | | | | | |
| Max. Stress, $N/cm^2$ | 38.4 [a] | 34.0 [b] | 27.0 [c] | 1.284 | <0.001 |

NPN, non-protein nitrogen in mg/g fresh meat; AN, amino acid nitrogen in mg AN/g fresh meat; MR, milk replacer; s.e., standard error, Sig., signification of the effect. Different superscripts in the same row indicate significant differences ($p < 0.05$).

## 4. Discussion

On average, the goat's milk in this study had a higher content of fat and protein than those reported by other authors using comparable breeds [26–32]. However, the milk had similar lactose levels to those reported by several authors [26,28,30], but lower levels than those reported by others [29,31,33]. Both the fat and protein content are correlated with the energy of the diet, although in an opposite way. Undernutrition mainly due to grazing results in a decrease in protein and an increase in fat due to the mobilization of body fat [26]. However, the high protein content in this study demonstrates that the goats were fed adequately.

High pH values for kid meat are widespread in the literature, suggesting that goats are generally highly prone to stress [34,35]. The pH values found in the literature for kids were similar to those reported in the present study [7,9,36–41] with comparable farming systems and slaughter weights. Most of the reported values were in the range from 5.5 to 5.8, which is considered optimal for goat meat [42]. However, Tinerfeña raised with natural milk had a pH = 6.01, indicating preslaughtering stress [43]. While suckling lambs reared with natural milk or milk replacers had the same pH [44], kids are very sensitive to preslaughter stress (from transport, lairage, isolation, etc.) [39]. Young kids are more susceptible to emotional stress than old ones [40] because younger animals are still largely dependent on their dams [45]. Therefore, the higher frequency of kids fed milk replacers with high pH values could be explained because kids weaned early do not have enough skills to manage emotional stress [45]. The consequences of preslaughter stress are well known and are often responsible for DFD meats [43]. However, the meat of the group with high pH values did not show a modified moisture content. Moreover, the Warner-Bratzler shear force and the hardness of the cooked meat were lower in the meats with high pH than in the meat with low pH. Watanabe, et al. [46] reported that toughness increased from 5.5 to 5.8 with higher pH values. However, high pH values increased other textural parameters, such as chewiness and adhesiveness. Therefore, high pH kid meat is not tough but may be perceived as different by consumers. In addition to these different textural characteristics, high pH values are undesirable because the spoilage of meat increases when the pH is close to 7 [47].

The chemical composition of light suckling kid meat, especially intramuscular fat, was influenced mainly by the breed, but the rearing system had a slight influence. In agreement with these results, Zurita-Herrera, Delgado, Argüello, Camacho and Germano [18] and Argüello, Castro, Capote and Solomon [8] did not find differences in the chemical composition of LTL between the same rearing systems. It has been confirmed that the low amount of IMF is characteristic of suckling kids [48,49], because visceral fat deposits tend to be increased before intramuscular fat deposits in goats [50]. However, an exception was found in Cabra del Guadarrama with IMF higher than 4 %.

To the best of our knowledge, there have been no TPA studies of suckling kid meat, either raw or cooked. Nor has there been any TPA study of suckling lambs with comparable slaughter weights. Choi, et al. [51] reported that the meat of Australian lamb had higher hardness and adhesiveness, similar springiness, chewiness and cohesiveness and lower cohesiveness than those of kid meat. Önenç, et al. [52] also reported higher values of hardness but lower chewiness of lamb meat compared to suckling kid meat. Bañón, Vila, Price, Ferrandini and Garrido [36] did not use the TPA but studied the sensory characteristics of suckling kids. Hence, these researchers did not find differences in chewiness between rearing systems, but meat from kids fed milk replacers was tenderer than that from natural milk-fed kids. Comparing more meats to suckling kid meat, duck cooked breast had higher hardness, lower chewiness and springiness and similar cohesiveness and resilience as kid meat [53]. Both raw and cooked chicken breast had higher hardness and lower chewiness than kid meat [54]. Romero de Ávila, et al. [55] performed a TPA of cooked hams, which had higher hardness and lower adhesiveness, cohesiveness and springiness than suckling kid meat. However, different variations in the TPA parameters, such as the compression ratio and speed, and the dimensions of the samples make the comparison of results challenging [56]. Therefore, Wee, et al. [57] measured the texture profile of 59 foods and found significant correlations between the chemical composition and the textural properties of food. These authors reported that carbohydrate content decreases hardness.

Adhesiveness was the variable most influenced by chemical composition, being increased by humidity and decreased by protein and fat contents. Regarding the effect of cooking, Ruiz de Huidobro, Miguel, Blazquez and Onega [14] reported that hardness, chewiness and springiness of beef increase with cooking in disagreement with our results.

While TPA data are scarce, there is more information about the application of the Warner-Bratzler method to suckling kids and lambs. The literature often compares the use of natural milk and milk replacers to raise kids of just one breed, so conclusions about the influence of milk replacers on meat quality are misleading. Hence, it has been reported that meat toughness is not affected by the rearing system when suckling lambs and kids are slaughtered at very low live weight [8,44]. This is likely because collagen content and solubility are more affected by the age [47] than by the rearing system [8,18,36]. However, we found that some breeds, such as Payoya and Retinta, were affected by the rearing system but were affected in contrasting ways. Unfortunately, as far as we know, there are no similar studies focused on these breeds. Zurita-Herrera, Delgado, Argüello, Camacho and Germano [18] reported low Warner-Bratzler shear force at 1 d *postmortem* on kids of Murciano-Granadina fed milk replacers. Values of approximately 30 N have been reported for very light suckling kids, such as *capretto* and *cabrito* at 1–2 d *postmortem* [18,19,43], in agreement with the results of our study. Other authors reported higher values than 30 N for kid meat [4,5,58] in meat aged from 1 d to 3 d. These values are also lower than those reported for other meats [59–61]. Additionally, these values are lower than the values of extremely tender beef [62,63]. Shackelford, et al. [64] reported that meat having Warner-Bratzler shear force values higher than 54 N would be assessed as tough by consumers. However, the transition from tough to tender occurred between 42 N and 48 N [63]. In our study, cooked meat had slightly less hardness than raw meat. Cooking softens the connective tissue but toughs the myofibrils. Therefore, the meat would be tougher or tenderer depending on the temperature and cooking time [47]. Machlik and Draudt [65] studied the influence of cooking time and temperature in very small cylinders of beef. These authors concluded that heating meat at 71°C decreased the toughness during the first 9 min of cooking. The samples of the kids were also small, and the samples reached the temperature endpoint quickly. Thus, toughness diminished due to cleavage of the peptide bonds and mature crosslinks [47].

## 5. Conclusions

The effect of rearing system on the textural profile was strongly modulated by breed. The values of Warner-Bratzler shear force and hardness found in these breeds under both rearing systems were very low. Hence, the toughness of very light suckling kids should not be a determining factor in choosing a breed or rearing system. Nevertheless, the use of milk replacers increased the pH of meat, which modified the textural parameters, decreasing the shear force but increasing cohesiveness, adhesiveness and cohesiveness. Consequently, depending on the commercial strategy of the farm, the election of the breed and rearing system must be considered together.

**Author Contributions:** Conceptualization, B.P., M.J.A. and G.R.; Methodology, B.P., M.J.A., R.C., M.G.C., S.R.-M. and G.R.; Data collection and acquisition, B.P., M.J.A., G.R., A.A., R.C., M.G.C. and S.R.-M.; Writing-Original Draft Preparation, B.P., M.J.A. and G.R.; Writing-Review & Editing, B.P., M.J.A., G.R., A.A., R.C., M.G.C. and S.R.-M.; Funding Acquisition, B.P., M.J.A. and A.A.

**Funding:** This research was funded by Instituto Nacional de Investigación y Tecnología Agraria y Alimentaria (RTA2012-23-C3), Research Group Funds of the Aragón Government (A14-17R SAGAS) and CYTED (116RT0503).

**Acknowledgments:** Appreciation is expressed to breeders and the staff of CITA de Aragón for their help in data collection.

**Conflicts of Interest:** The authors declare no conflict of interest. The founding sponsors had no role in the design of the study; in the collection, analyses, or interpretation of data; in the writing of the manuscript, and in the decision to publish the results.

## References


1. Eurostat. *Agricultural Statistics*; Statistical Office of the European Communities: Luxemburg, 2019.
2. Ribeiro, R.D.X.; Medeiros, A.N.; Oliveira, R.L.; de Araújo, G.G.L.; Queiroga, R.d.C.d.E.; Ribeiro, M.D.; Silva, T.M.; Bezerra, L.R.; Oliveira, R.L. Palm kernel cake from the biodiesel industry in goat kid diets. Part 2: Physicochemical composition, fatty acid profile and sensory attributes of meat. *Small Rumin. Res.* **2018**, *165*, 1–7. [CrossRef]
3. Castel, J.M.; Mena, Y.; Delgado-Pertíñez, M.; Camúñez, J.; Basulto, J.; Caravaca, F.; Guzmán-Guerrero, J.L.; Alcalde, M.J. Characterization of semi-extensive goat production systems in southern spain. *Small Rumin. Res.* **2003**, *47*, 133–143. [CrossRef]
4. Yalcintan, H.; Akin, P.D.; Ozturk, N.; Ekiz, B.; Kocak, O.; Yilmaz, A. Carcass and meat quality traits of saanen goat kids reared under natural and artificial systems and slaughtered at different ages. *Acta Vet. Brno* **2018**, *87*, 293–300. [CrossRef]
5. Yalcintan, H.; Ekiz, B.; Ozcan, M. Comparison of meat quality characteristics and fatty acid composition of finished goat kids from indigenous and dairy breeds. *Trop. Anim. Health Prod.* **2018**, *50*, 1261–1269. [CrossRef] [PubMed]
6. Castel, J.M.; Mena, Y.; Ruiz, F.A.; Gutiérrez, R. Situación y evolución de los sistemas de producción caprina en españa. *Tierras Caprino* **2012**, *1*, 24–37.
7. Marichal, A.; Castro, N.; Capote, J.; Zamorano, N.; Arguello, A. Effects of live weight at slaughter (6, 10 and 25 kg) on kid carcass and meat quality. *Livest. Prod. Sci.* **2003**, *83*, 247–256. [CrossRef]
8. Argüello, A.; Castro, N.; Capote, J.; Solomon, M. Effects of diet and live weight at slaughter on kid meat quality. *Meat Sci.* **2005**, *70*, 173–179.
9. Ripoll, G.; Alcalde, M.J.; Argüello, A.; Córdoba, M.G.; Panea, B. Effect of the rearing system on the color of four muscles of suckling kids. *Food Sci. Nutr.* **2019**, *7*, 1502–1511. [CrossRef]
10. Ripoll, G.; Alcalde, M.J.; Argüello, A.; Córdoba, M.G.; Panea, B. Consumer visual appraisal and shelf life of leg chops from suckling kids raised with natural milk or milk replacer. *J. Sci. Food Agric.* **2018**, *98*, 2651–2657. [CrossRef]
11. Caine, W.R.; Aalhus, J.L.; Best, D.R.; Dugan, M.E.R.; Jeremiah, L.E. Relationship of texture profile analysis and warner-bratzler shear force with sensory characteristics of beef rib steaks. *Meat Sci.* **2003**, *64*, 333–339. [CrossRef]
12. Panea, B.; Carrasco, S.; Ripoll, G.; Joy, M. Diversification of feeding systems for light lambs: Sensory characteristics and chemical composition of meat | diversificación de los sistemas de producción de corderos ligeros: Características sensoriales y composición química de la carne. *Span. J. Agric. Res.* **2011**, *9*, 74–85. [CrossRef]
13. Bratzler, L.J. *Measuring the Tenderness of Meat by Means of a Mechanical Shear*; Kansas State College of Agriculture and Applied Science: Manhattan, KS, USA, 1932.
14. Ruiz de Huidobro, F.; Miguel, E.; Blazquez, B.; Onega, E. A comparison between two methods (warner-bratzler and texture profile analysis) for testing either raw meat or cooked meat. *Meat Sci.* **2005**, *69*, 527–536. [CrossRef] [PubMed]
15. Borgogno, M.; Corazzin, M.; Saccà, E.; Bovolenta, S.; Piasentier, E. Influence of familiarity with goat meat on liking and preference for capretto and chevon. *Meat Sci.* **2015**, *106*, 69–77. [CrossRef] [PubMed]
16. Bonvillani, A.; Peña, F.; de Gea, G.; Gómez, G.; Petryna, A.; Perea, J. Carcass characteristics of criollo cordobés kid goats under an extensive management system: Effects of gender and liveweight at slaughter. *Meat Sci.* **2010**, *86*, 651–659. [CrossRef] [PubMed]
17. Ozcan, M.; Yalcintan, H.; Tölü, C.; Ekiz, B.; Yilmaz, A.; Savaş, T. Carcass and meat quality of gokceada goat kids reared under extensive and semi-intensive production systems. *Meat Sci.* **2014**, *96*, 496–502. [CrossRef] [PubMed]
18. Zurita-Herrera, P.; Delgado, J.V.; Argüello, A.; Camacho, M.E.; Germano, R. Effects of three management systems on meat quality of dairy breed goat kids. *J. Appl. Anim. Res.* **2013**, *41*, 173–182. [CrossRef]
19. De Palo, P.; Maggiolino, A.; Centoducati, N.; Tateo, A. Effects of different milk replacers on carcass traits, meat quality, meat color and fatty acids profile of dairy goat kids. *Small Rumin. Res.* **2015**, *131*, 6–11. [CrossRef]
20. E.U. Directive 2010/63/eu of the european parliament and of the council of 22 september 2010 on the protection of animals used for scientific purposes. *Off. J. Eur. Un.* **2010**, L276/233–L276/279.

21. E.U. Council regulation (ec) no 1099/2009 of 24 september 2009 on the protection of animals at the time of killing. *Off. J. Eur. Un.* **2009**, L303/301–L303/330.

22. AOAC. *Official Methods of Analysis. Association of Official Analytical Chemist*, 17th ed.; Horwitz, W., Latimer, G., Eds.; Association of Analytical Communities: Arlington, VA, USA, 2000.

23. Bligh, E.G.; Dyer, W.J. A rapid method of total lipid extraction and purification. *Can. J. Biochem. Physiol.* **1959**, *37*, 911–917. [CrossRef]

24. Benito, M.J.; Núñez, F.; Córdoba, M.G.; Martín, A.; Córdoba, J.J. Generation of non-protein nitrogen and volatile compounds by penicillium chrysogenum pg222 activity on pork myofibrillar proteins. *Food Microbiol.* **2005**, *22*, 513–519. [CrossRef]

25. Gagaoua, M.; Picard, B.; Monteils, V. Associations among animal, carcass, muscle characteristics, and fresh meat color traits in charolais cattle. *Meat Sci.* **2018**, *140*, 145–156. [CrossRef] [PubMed]

26. Zervas, G.; Tsiplakou, E. The effect of feeding systems on the characteristics of products from small ruminants. *Small Rumin Res* **2011**, *101*, 140–149. [CrossRef]

27. Sanchez-Macias, D.; Moreno-Indias, I.; Castro, N.; Morales-Delanuez, A.; Arguello, A. From goat colostrum to milk: Physical, chemical, and immune evolution from partum to 90 days postpartum. *J. Dairy Sci.* **2014**, *97*, 10–16. [CrossRef]

28. Kusza, S.; Cziszter, L.T.; Ilie, D.E.; Sauer, M.; Padeanu, I.; Gavojdian, D. Kompetitive allele specific pcr (kasp) genotyping of 48 polymorphisms at different caprine loci in french alpine and saanen goat breeds and their association with milk composition. *PeerJ* **2018**, *6*, e4416-10. [CrossRef]

29. Delgado-Pertíñez, M.; Guzmán-Guerrero, J.L.; Mena, Y.; Castel, J.M.; González-Redondo, P.; Caravaca, F.P. Influence of kid rearing systems on milk yield, kid growth and cost of florida dairy goats. *Small Rumin. Res.* **2009**, *81*, 105–111.

30. Hernandez-Castellano, L.E.; Almeida, A.M.; Renaut, J.; Arguello, A.; Castro, N. A proteomics study of colostrum and milk from the two major small ruminant dairy breeds from the canary islands: A bovine milk comparison perspective. *J. Dairy Res.* **2016**, *83*, 366–374. [CrossRef]

31. Torres, A.; Hernandez-Castellano, L.E.; Morales-delaNuez, A.; Sanchez-Macias, D.; Moreno-Indias, I.; Castro, N.; Capote, J.; Arguello, A. Short-term effects of milking frequency on milk yield, milk composition, somatic cell count and milk protein profile in dairy goats. *J. Dairy Res.* **2014**, *81*, 275–279. [CrossRef]

32. Rota, A.M.; Rodríguez, P.; Rojas, A.; Martin, L.; Tovar, J. Evolution of the cuantitative and quality characteistics in the goat milk (verata breed) through the lactation. *Arch. Zootec.* **1993**, *42*, 137–146.

33. Margatho, G.; Rodríguez-Estévez, V.; Medeiros, L.; Simões, J. Seasonal variation of serrana goat milk contents in mountain grazing system for cheese manufacture. *Rev. Med. Vet.* **2018**, *169*, 157–165.

34. Webb, E.C.; Casey, N.H.; Simela, L. Goat meat quality. *Small Rumin. Res.* **2005**, *60*, 153–166. [CrossRef]

35. Casey, N.H.; Webb, E.C. Managing goat production for meat quality. *Small Rumin. Res.* **2010**, *89*, 218–224. [CrossRef]

36. Bañón, S.; Vila, R.; Price, A.; Ferrandini, E.; Garrido, M.D. Effects of goat milk or milk replacer diet on meat quality and fat composition of suckling goat kids. *Meat Sci.* **2006**, *72*, 216–221.

37. Santos, V.A.C.; Silva, A.O.; Cardoso, J.V.F.; Silvestre, A.J.D.; Silva, S.R.; Martins, C.; Azevedo, J.M.T. Genotype and sex effects on carcass and meat quality of suckling kids protected by the pgi "cabrito de barroso". *Meat Sci.* **2007**, *75*, 725–736. [CrossRef] [PubMed]

38. Peña, F.; Bonvillani, A.; Freire, B.; Juárez, M.; Perea, J.; Gómez, G. Effects of genotype and slaughter weight on the meat quality of criollo cordobes and anglonubian kids produced under extensive feeding conditions. *Meat Sci.* **2009**, *83*, 417–422.

39. Ripoll, G.; Alcalde, M.J.; Horcada, A.; Panea, B. Suckling kid breed and slaughter weight discrimination using muscle colour and visible reflectance. *Meat Sci.* **2011**, *87*, 151–156. [CrossRef]

40. Sañudo, C.; Campo, M.M.; Muela, E.; Olleta, J.L.; Delfa, R.; Jiménez-Badillo, R.; Alcalde, M.J.; Horcada, A.; Oliveira, I.; Cilla, I. Carcass characteristics and instrumental meat quality of suckling kids and lambs. *Span. J. Agric. Res.* **2012**, *10*, 690.

41. Teixeira, A.; Jimenez-Badillo, M.R.; Rodriguez, S. Effect of sex and carcass weight on carcass traits and meat quality in goat kids of cabrito transmontano. *Span. J. Agric. Res.* **2011**, *9*, 753–760. [CrossRef]

42. Herold, P.; Snell, H.; Tawfik, E.S. Growth, carcass and meat quality parameters of purebred and crossbred goat kids in extensive pasture. *Arch. Anim. Breed.* **2007**, *50*, 186–196. [CrossRef]

43. Dhanda, J.S.; Taylor, D.G.; Murray, P.J. Part 1. Growth, carcass and meat quality parameters of male goats: Effects of genotype and liveweight at slaughter. *Small Rumin. Res.* **2003**, *50*, 57–66. [CrossRef]

44. Osorio, M.T.; Zumalacárregui, J.M.; Cabeza, E.A.; Figueira, A.; Mateo, J. Effect of rearing system on some meat quality traits and volatile compounds of suckling lamb meat. *Small Rumin. Res.* **2008**, *78*, 1–12. [CrossRef]

45. Napolitano, F.; Marino, V.; De Rosa, G.; Capparelli, R.; Bordi, A. Influence of artificial rearing on behavioral and immune response of lambs. *Appl. Anim. Behav. Sci.* **1995**, *45*, 245–253. [CrossRef]

46. Watanabe, A.; Daly, C.; Devine, C. The effects of the ultimate ph of meat on tenderness changes during ageing. *Meat Sci.* **1996**, *42*, 67–78. [CrossRef]

47. Lawrie, R.A. *Meat Science*, 3th ed.; Acribia, S.A., Ed.; Elsevier: Zaragoza, Spain, 1998.

48. Ripoll, G.; Alcalde, M.J.; Horcada, A.; Campo, M.M.; Sanudo, C.; Teixeira, A.; Panea, B. Effect of slaughter weight and breed on instrumental and sensory meat quality of suckling kids. *Meat Sci.* **2012**, *92*, 62–70. [CrossRef]

49. Vacca, G.; Dettori, M.; Paschino, P.; Manca, F.; Puggioni, O.; Pazzola, M. Productive traits and carcass characteristics of sarda suckling kids. *Large Anim. Rev.* **2014**, *20*, 169–173.

50. Banskalieva, V.; Sahlu, T.; Goetsch, A.L. Fatty acid composition of goat muscles and fat depots: A review. *Small Rumin. Res.* **2000**, *37*, 255–268. [CrossRef]

51. Choi, M.J.; Abduzukhurov, T.; Park, D.H.; Kim, E.J.; Hong, G.P. Effects of deep freezing temperature for long-term storage on quality characteristics and freshness of lamb meat. *Korean J. Food Sci. Anim. Resour.* **2018**, *38*, 959–969. [CrossRef]

52. Önenç, S.; Özdoğan, M.; Aktümsek, A.; Taşkın, T. Meat quality and fatty acid composition of chios male lambs fed under traditional and intensive conditions. *Emir. J. Food Agric.* **2015**, *27*, 636.

53. Khan, M.A.; Ali, S.; Yang, H.; Kamboh, A.A.; Ahmad, Z.; Tume, R.K.; Zhou, G. Improvement of color, texture and food safety of ready-to-eat high pressure-heat treated duck breast. *Food Chem.* **2019**, *277*, 646–654. [CrossRef]

54. Rabeler, F.; Feyissa, A.H. Modelling the transport phenomena and texture changes of chicken breast meat during the roasting in a convective oven. *J. Food Eng.* **2018**, *237*, 60–68. [CrossRef]

55. Romero de Ávila, M.D.; Cambero, M.I.; Ordóñez, J.A.; de la Hoz, L.; Herrero, A.M. Rheological behaviour of commercial cooked meat products evaluated by tensile test and texture profile analysis (tpa). *Meat Sci.* **2014**, *98*, 310–315.

56. Novaković, S.; Tomašević, I. *A Comparison between Warner-Bratzler Shear Force Measurement and Texture Profile Analysis of Meat and Meat Products: A Review*; IOP Publishing: Bristol, UK, 2017.

57. Wee, M.S.M.; Goh, A.T.; Stieger, M.; Forde, C.G. Correlation of instrumental texture properties from textural profile analysis (tpa) with eating behaviours and macronutrient composition for a wide range of solid foods. *Food Funct.* **2018**, *9*, 5301–5312. [CrossRef] [PubMed]

58. Rotondi, P.; Colonna, M.A.; Marsico, G.; Giannico, F.; Ragni, M.; Facciolongo, A.M. Dietary supplementation with oregano and linseed in garganica suckling kids: Effects on growth performances and meat quality. *Pak. J. Zool.* **2018**, *50*. [CrossRef]

59. Babiker, S.; El Khider, I.; Shafie, S. Chemical composition and quality attributes of goat meat and lamb. *Meat Sci.* **1990**, *28*, 273–277. [CrossRef]

60. Ripoll, G.; Alberti, P.; Casasus, I.; Blanco, M. Instrumental meat quality of veal calves reared under three management systems and color evolution of meat stored in three packaging systems. *Meat Sci.* **2013**, *93*, 336–343. [CrossRef]

61. Calvo, J.H.; Iguácel, L.P.; Kirinus, J.K.; Serrano, M.; Ripoll, G.; Casasús, I.; Joy, M.; Pérez-Velasco, L.; Sarto, P.; Alberti, P.; et al. A new single nucleotide polymorphism in the calpastatin (cast) gene associated with beef tenderness. *Meat Sci.* **2014**, *96*, 775–782. [CrossRef]

62. Huffman, K.L.; Miller, M.F.; Hoover, L.C.; Wu, C.K.; Brittin, H.C.; Ramsey, C.B. Effect of beef tenderness on consumer satisfaction with steaks consumed in the home and restaurant. *J. Anim. Sci.* **1996**, *74*, 91–97. [CrossRef]

63. Miller, M.F.; Carr, M.A.; Ramsey, C.B.; Crockett, K.L.; Hoover, L.C. Consumer thresholds for establishing the value of beef tenderness. *J. Anim. Sci.* **2001**, *79*, 3062–3068. [CrossRef]

64. Shackelford, S.D.; Morgan, J.B.; Cross, H.R.; Savell, J.W. Identificación of threshold levels for warner-bratzler shear force in beef top loin steaks. *J. Muscle Foods* **1991**, *2*, 289–296. [CrossRef]
65. Machlik, S.M.; Draudt, H.N. The effect of heating time and temperature on the shear of beef semitendinosus muscle a. *J. Food Sci.* **1963**, *28*, 711–718. [CrossRef]

 *foods*

*Article*

# Value-Added Carp Products: Multi-Class Evaluation of Crisp Grass Carp by Machine Learning-Based Analysis of Blood Indexes

**Bing Fu [1], Gen Kaneko [2], Jun Xie [1], Zhifei Li [1], Jingjing Tian [1], Wangbao Gong [1], Kai Zhang [1], Yun Xia [1], Ermeng Yu [1,*] and Guangjun Wang [1]**

[1] Pearl River Fisheries Research Institute, CAFS, Guangzhou 510380, China; fub@prfri.ac.cn (B.F.); xj@prfri.ac.cn (J.X.); lzf@prfri.ac.cn (Z.L.); tianjj@prfri.ac.cn (J.T.); gwb@prfri.ac.cn (W.G.); zk@prfri.ac.cn (K.Z.); xy@prfri.ac.cn (Y.X.); gjwang@prfri.ac.cn (G.W.)

[2] School of Arts & Sciences, University of Houston-Victoria, Victoria, TX 77901, USA; KanekoG@uhv.edu

* Correspondence: yem@prfri.ac.cn

Received: 10 October 2020; Accepted: 4 November 2020; Published: 6 November 2020

**Abstract:** Crisp grass carp products from China are becoming more prevalent in the worldwide fish market because muscle hardness is the primary desirable characteristic for consumer satisfaction of fish fillet products. Unfortunately, current instrumental methods to evaluate muscle hardness are expensive, time-consuming, and wasteful. This study sought to develop classification models for differentiating the muscle hardness of crisp grass carp on the basis of blood analysis. Out of the total 264 grass carp samples, 12 outliers from crisp grass carp group were removed based on muscle hardness (<9 N), and the remaining 252 samples were used for the analysis of seven blood indexes including hydrogen peroxide ($H_2O_2$), glucose 6-phosphate dehydrogenase (G6PD), malondialdehyde (MDA), glutathione (GSH/GSSH), red blood cells (RBC), platelet count (PLT), and lymphocytes (LY). Furthermore, six machine learning models were applied to predict the muscle hardness of grass carp based on the training (152) and testing (100) datasets obtained from the blood analysis: random forest (RF), naïve Bayes (NB), gradient boosting decision tree (GBDT), support vector machine (SVM), partial least squares regression (PLSR), and artificial neural network (ANN). The RF model exhibited the best prediction performance with a classification accuracy of 100%, specificity of 93.08%, and sensitivity of 100% for discriminating crisp grass carp muscle hardness, followed by the NB model (93.75% accuracy, 91.83% specificity, and 94% sensitivity), whereas the ANN model had the lowest prediction performance (85.42% accuracy, 81.05% specificity, and 85% sensitivity). These machine learning methods provided objective, cheap, fast, and reliable classification for in vivo crisp grass carp and also prove useful for muscle quality evaluation of other freshwater fish.

**Keywords:** meat quality; muscle hardness; classification model; random forest

## 1. Introduction

Fish are a valuable source of high-quality animal protein throughout the world, with its annual consumption outpacing population growth between 1961 and 2016 [1]. Grass carp (*Ctenopharyngodon idella*), the largest freshwater fish species, has a global production of about six million tons [1]. Crisp grass carp (*Ctenopharyngodon idellus* C. et V) is one of the most representative varieties of grass carp that shows improved textural characteristics (hardness, chewiness, springiness, etc.) after being fed solely with whole faba bean (*Vicia faba* L.) for 90 to 120 days [2,3]. Crisp grass carp has been deemed a value-added product and is protected as a "China Geographical Indication Product". The fillets of crisp grass carp are exported to various countries in Southeast Asia and Latin America as well as Hong Kong [4].

Hardness is the most prominent quality indicator of crisp grass carp and is directly related to the consumer's acceptability [3,4]. As mentioned above, faba bean feeding for 90–120 days is used to improve textual characteristics of crisp grass carp. However, according to aquaculture experiences, approximately 5% of the treated fish still exhibit low muscle hardness similar to ordinary grass carp after 120 d of faba bean feeding, which financially affects producers, regulatory agencies, and consumers. To prevent this, it is necessary to assess the muscle hardness of crisp grass carp products over different culture periods. Sensory evaluation has been the primary method for the evaluation of the muscle hardness of crisp grass carp [5], but this method is subjective and is greatly influenced by the experience of the evaluator [6]. Yang et al. [4] proposed an alternative method for evaluating the muscle hardness of crisp grass carp via instrumental texture analysis, but its widespread application is limited because of high equipment costs and long preparation and analysis times. As such, it is necessary to develop objective, cheap, fast, and reliable in vivo analytical methods for analyzing muscle hardness of crisp grass carp.

Machine learning techniques have emerged as a potential in vivo analytical tool. Machine learning identifies patterns in large datasets and aids in predicting outcomes based on various algorithms [7], which have been applied to classify aquatic animals. For example, support vector machine (SVM) can differentiate between organically and conventionally farmed salmon with an accuracy of 98.2% based on hyperspectral imaging and computer vision [8] and can obtain 82% accuracy using skin images [9]. Additionally, multi-class SVM achieved a high accuracy (97.77%) in classifying six freshwater fish species using skin color and texture [10]. The artificial neural network (ANN) achieved 91.86% accuracy in the automated identification of fish species when combined with machine learning algorithms [11]. In living cattle, muscle quality was evaluated by machine learning based on blood analysis, and the random forest (RF) model distinguished organic cattle with a classification accuracy close to 90% [12]. However, machine learning combined with blood analysis has yet to be applied to predict fish quality.

Therefore, to develop objective and reliable in vivo analytical methods for analyzing the muscle hardness of crisp grass carp, taking advantage of machine learning techniques, the present study first evaluated the quality (muscle hardness) of crisp grass carp. Samples selected based on muscle hardness were used for the analysis of seven blood indexes. Six machine learning models were applied to predict muscle hardness of grass carp based on the training and testing datasets obtained from the blood analysis. The performance of the machine learning methods in classifying muscle hardness was evaluated in view of accuracy, sensitivity, specificity, and the area under the receiver operating characteristic curve (AUC). This work will establish objective, cheap, fast, and reliable in vivo analytical methods for evaluating freshwater fish quality.

## 2. Materials and Methods

### 2.1. Experimental Fish and Sample Collection

The feeding trial of grass carp was conducted at the Pearl River Fisheries Research Institute (Guangdong, China). A total of 540 fish (512.12 ± 10.67 g) were randomly distributed into 6 experimental tanks (tank size: 4 × 4 × 1.5 m) comprising a crisp grass carp group and ordinary grass carp groups (per group in triplicate). The crisp grass carp and ordinary grass carp were fed solely with faba bean and a commercial diet, respectively, for 120 days (d). Thirty-three individuals were sampled from both the crisp grass carp group and ordinary grass carp on 30, 60, 90, and 120 days, and total 264 fishes were used. Two-milliliter blood samples were drawn for blood cell analysis from the caudal vein using a sterile heparinized syringe and immediately transferred to tubes containing ethylenediaminetetraacetic acid (EDTA) that prevents blood from clotting. The 2 mL whole blood was stored at 4 °C for 3 h followed by centrifugation at 3500× $g$ for 10 min. The separated serum was stored at −80 °C for biochemistry analysis.

For muscle sampling, the fish were firstly anesthetized with tricaine methanesulfonate (MS-222). Each fish was killed, and the scale, skin, and red muscle were removed. For the texture determination and sensory evaluation, the dorsal white muscle ($2 \times 2 \times 1$ cm) was sampled at the junction of the dorsal fin and the lateral line scales from the right and left sides of the fish, respectively.

The experimental protocols used in the present study were approved by the Animal Ethics Committee of the Guangdong Provincial Zoological Society, China (permit number GSZ-AW012).

### 2.2. Muscle Hardness Measurement and Sensory Evaluation

As hardness is one of the key texture indicators for crisp grass carp muscle, we mainly measured muscle hardness using a Universal TA texture analyzer (Tengba instrument company, Shanghai, China) in a double compression Texture Profile Analysis (TPA) test. Each sample was treated using a flattened cylindrical probe (3.5 cm diameter) moving at $1 \text{ mm} \cdot \text{s}^{-1}$ to compress the tissue to 25% of its original height at room temperature. TPA was performed at least three times for each fillet.

Sensory evaluation of crisp grass carp was performed referring to the procedure of Yang et al. [4]. Texture properties were assessed by a panel of 5 trained experts (male, ages 35–50) using a five-class scale rating test (first level—minimal hardness; second level—moderate hardness; third level—normal hardness; fourth level—high hardness; fifth level—maximal hardness). Prior to the sensory evaluation, the muscle samples were cut into small chunks ($2 \times 2 \times 2$ cm), steamed over boiling water for 15 min, and cooled down to room temperature. Before evaluating each sample, the panelists rinsed their mouths five times with water to prevent interference from the previous samples. The final results of the sensory evaluation required a minimum of three identical ratings to be included.

### 2.3. Outlier Samples Removal

In general, crisp grass carp is characterized by muscle hardness greater than 1000 g after being fed faba bean for a couple of months. To improve the accuracy of the established models, muscle hardness was analyzed for all samples, with outlier samples removed based on a muscle hardness boxplot. In the boxplot, points below $Q1 - \alpha IQR$ and above $Q3 + \alpha IQR$ are considered as hardness outliers, where IQR is the interval quartile range, Q1 and Q3 indicate the first and third quartiles, respectively, and $\alpha$ is defined as 1.5 [13].

### 2.4. Measurement of Blood Indexes

Red blood cells (RBC), platelet count (PLT), and lymphocyte (LY) from whole blood samples were measured using a hematology analyzer Mek-7222K (Nihon Kohden, Tokyo, Japan) according to the manufacturer's instructions. The serum samples were used for measuring hydrogen peroxide ($H_2O_2$) (Kit No. A064-1), glucose-6-phosphate dehydrogenase (G6PD) (Kit No. M015), glutathione (GSH/GSSH) (Kit No. A006-1-1), and malondialdehyde (MDA) (Kit No. A003-1-2) using detection kits from Nanjing Jiancheng Bioengineering Institute (Nanjing, China).

### 2.5. Description of the Algorithms

To construct machine learning-based classification models, the datasets from the blood indexes were divided into two datasets—training and testing datasets. The training dataset was formed from 60% of the total samples used for calculating the classifiers, with the remaining 40% used for the testing dataset to validate the constructed models. Data processing was done using the Python3 sklearn package, and the algorithms used in this study were executed under default settings [14].

We first used unsupervised principal component analysis (PCA) to visualize the natural data distribution in a reduced dimensional space, which also allowed us to verify the relationship between the variables in the multidimensional space [15,16]. After PCA, the association patterns of variables can be clearly described [17].

Following the unsupervised PCA, we applied six supervised learning methods to build models that predict the hardness of crisp grass carp muscle from blood parameters. The models include two linear methods, linear Support Vector Machine (SVM) and Naïve Bayes (NB), and four non-linear models, Gradient Boosting Decision Tree (GBDT), Artificial Neural Network (ANN), non-linear SVM, and Random Forest (RF). Supervised algorithms require labeled training data to generate reasonable classifications for new data, whereas unsupervised algorithms do not.

Gradient boosting decision tree (GBDT) is an algorithm that consists of multiple decision trees, in which the final conclusion is derived from all of the decision trees [18]. The base learner of GBDT is the categorical regression tree (CART), which is a binary tree-based machine learning algorithm that can handle both regression and classification problems.

Artificial neural networks (ANNs) are adaptive non-linear decision-making tools inspired by the structure of the human brain [19]. ANNs consist of a number of nodes connected to each other, which mimic neurons in the human brain, which receive signals from the input links. Each input link (corresponding to a synapse) has an assigned weight that corresponds to synaptic efficiency. ANNs are typically trained by back-propagation consisting of at least three layers: input, output, and the hidden layer that connect the two layers.

Support vector machine (SVM) is a supervised model generally used for sample classification and regression [20]. This algorithm conducts non-linear transformation of the data to fit them into a K-dimensional hyperplane (K > original dimension). The SVM shows an excellent generalization ability when a specialized learning procedure is applied [21].

Partial least squares regression (PLSR) is a standard multi-linear regression model. This model is able to find linear relationships between observable variables and predicted variables [22]. This method is particularly useful when the data suffers from the multicollinearity [23] because it can reduce the number of observable variables and extracts a number of components like PCA.

Naïve Bayes (NB) is a simple algorithm that requires a small amount of data for training because it can be trained very efficiently by supervised learning [24]. The theoretical base of this algorithm is the Bayes theorem, in which each variable is treated as an independent variable.

Random forest (RF) can be used for either classification or regression through the construction of many decision trees [25]. The RF method performs a bootstrap sample from the training dataset and makes a decision tree using each of them. The final prediction is made by the set of trees [26].

*2.6. Classification Performance, Statistical Analysis, and Calculations*

In this paper, the results were validated by a tenfold cross-validation procedure. For this purpose, four indicators were calculated: accuracy, sensitivity, specificity, and the area under receiver operating characteristic curve (AUC). We calculated both micro- and macro-averages of the performance metrics as well as the confusion matrices for each model to present their predictive capabilities.

The performance measures are defined as follows. TP, TN, FP, and FN stand for true positives, true negatives, false positives, and false negatives, respectively.

Accuracy refers to the average number of samples properly categorized.

$$\text{Classification accuracy} = \frac{(\text{TN} + \text{TP})}{(\text{FN} + \text{TP} + \text{FP} + \text{TN})} \times 100\%$$

Sensitivity is the ability to correctly classify samples (i.e., the fraction of target samples correctly classified as target samples).

$$\text{Classification Sensitivity} = \frac{\text{TP}}{(\text{FN} + \text{TP})} \times 100\%$$

Specificity is the fraction of non-target samples correctly classified as non-target samples.

$$\text{Classification Specificity} = \frac{\text{TN}}{(\text{TN} + \text{FP})} \times 100\%$$

Continuous variables are presented as the mean ± standard deviation. Student's t-test and Duncan's test were used for statistical analysis. A *p* value of less than 0.05 was considered to be statistically significant.

## 3. Results and Discussion

### 3.1. Removal of Outlier Samples

Boxplots can be used to detect and eliminate outliers from a dataset [27]. This is an important step in machine learning-based analysis because outliers can misdirect the training process and produce a less accurate model [28]. As muscle hardness is the most obvious texture feature of crisp grass carp and increases with the faba bean feeding time [2], outliers were eliminated on the basis of muscle hardness. In general, there were a higher number of outliers in crisp grass carp than ordinary grass carp (Figure 1A). The boxplot revealed 12 crisp grass carp outliers (out of 132; 9.09%) exceeding the interquartile range by ± 1.5 times. The number of outliers varied across the feeding periods, where four outliers were found at 30 day (W1), three found at 60 and 120 day (W2 and W4, respectively), and two found at 90 day (W3).

The sensory evaluation results, which are known to be consistent with those of instrumental texture analyses in crisp grass carp evaluation [4], can be seen in Figure 1B. All the ordinary grass carp samples (132 fishes) were evaluated as level 1. For the crisp grass carp group, at 30 day, 29 samples were evaluated as level 2; at 60 day, 30 samples were evaluated as level 3; at 90 day, 31 samples were evaluated as level 4; at 120 day, 30 samples were evaluated as level 5; the remaining 12 samples were evaluated as level 1. These evaluation results were consistent with the result of the boxplot analysis (Figure 1A). Upon eliminating the outliers, 252 observations were retained for further analysis.

### 3.2. Blood Indexes Analysis

Our previous study found that faba bean suppresses the immune and antioxidant responses of grass carp [2,29]. To include the effects in the predictive models, seven blood indexes including blood red cells (BRC), platelet counts (PLT), lymphocyte (LY), hydrogen peroxide ($H_2O_2$), 6-phosphate dehydrogenase (G6PD), glutathione (GSH/GSSH), and malondialdehyde (MDA) were selected for this study. All values from 252 samples (120 crisp grass carp and 132 ordinary grass carp) are shown in the violin plots (Figure 2), in which some differences were observed depending on the culture periods and treatment. $H_2O_2$ levels were not significantly different between both groups on 30 and 60 day, but were higher in crisp grass carp than ordinary grass carp at 90 and 120 day (Figure 2A). The levels of G6PD, GSH/GSSH, and LY of crisp grass carp were significantly lower than those of ordinary grass carp throughout the culture period (Figure 2B,C,G). MDA markedly increased between 30 and 120 day in crisp grass carp and was notably higher than ordinary grass carp at 60, 90, and 120 day (Figure 2D). Compared to ordinary grass carp, the RBCs of crisp grass carp were significantly higher at 30 and 60 day but were lower at 90 and 120 day (Figure 2E). The PLT values between the two groups were also significantly different during the entire culturing period (Figure 2F). These differences were used to establish the classification models.

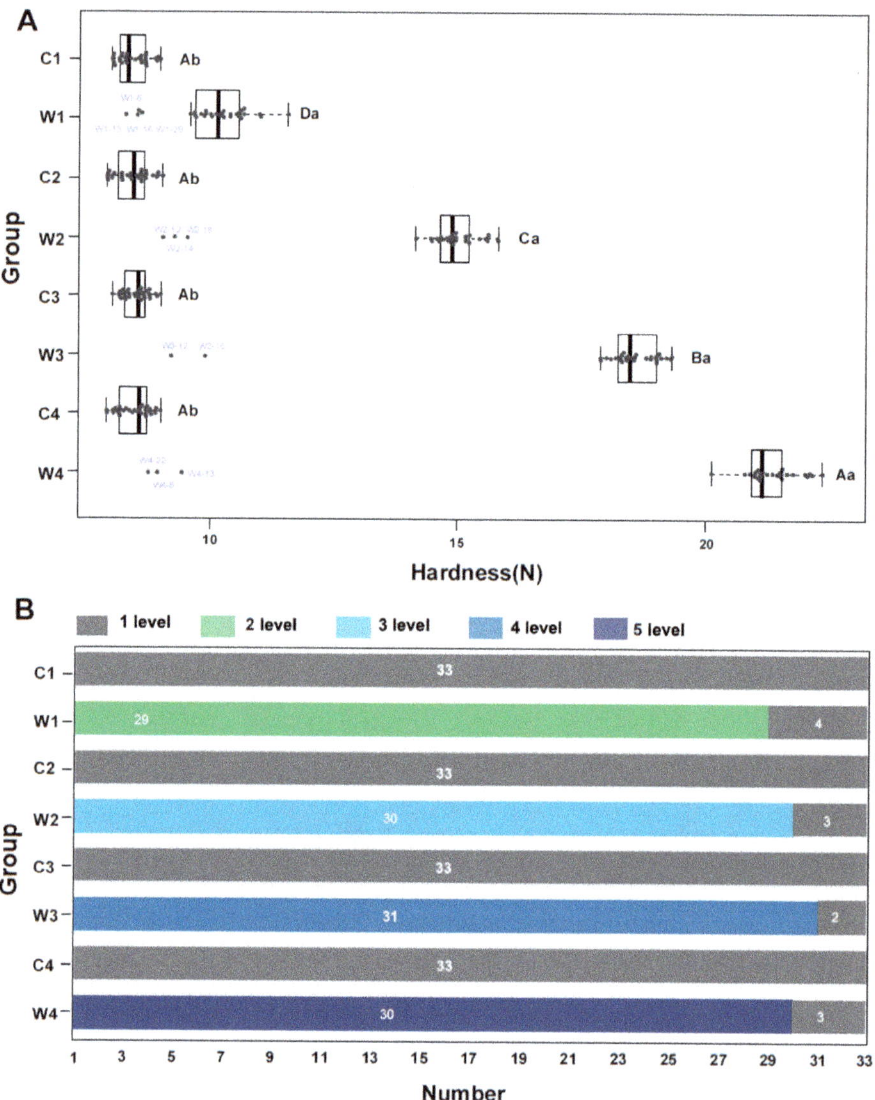

**Figure 1.** Removal of outliers using muscle hardness. C1, C2, C3, and C4 represent the ordinary grass carp from 30, 60, 90, and 120 day, respectively. W1, W2, W3, and W4 represent the crisp grass carp from 30, 60, 90, and 120 day, respectively. (**A**) Boxplot of muscle hardness of crisp grass carp and ordinary grass carp during different farming stages. Boxes represent the interquartile range, and whiskers delineate ± 1.5 times the interquartile range beyond the box boundaries. Points falling outside of the whiskers were considered outliers in this study. Statistical analyses were performed using Student's *t*-test and Duncan's test. Different lowercase letters represent significant differences in muscle hardness ($p < 0.05$). Different uppercase letters represent significant differences in muscle hardness ($p < 0.05$). (**B**) Sensory texture evaluation of crisp grass carp and ordinary grass carp during different farming stages. Samples from 33 fish were evaluated for each group.

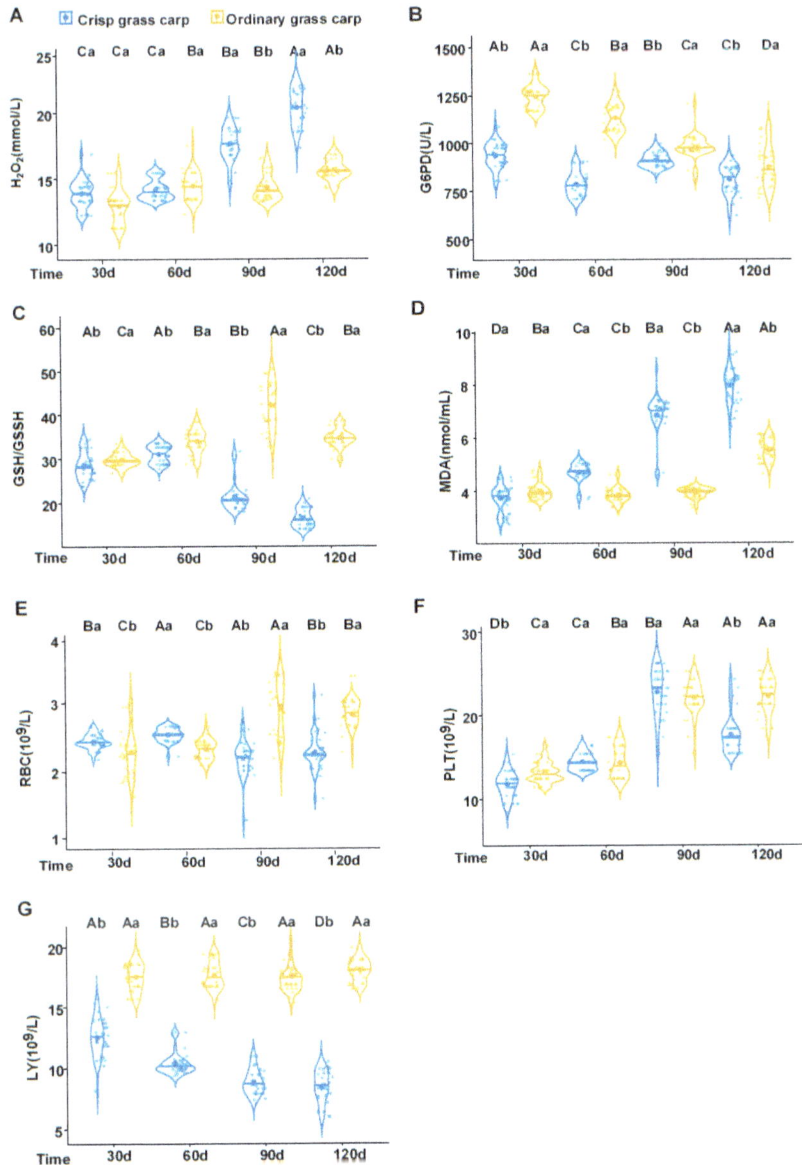

**Figure 2.** Violin plots of seven blood indexes in ordinary and crisp grass carp samples. The medians are represented by a horizontal blue or yellow line. The whiskers span the entire range. (**A**) $H_2O_2$. (**B**) G6PD. (**C**) GSH/GSSH. (**D**) MDA. (**E**) RBC. (**F**) PLT. (**G**) LY. Statistical analyses were performed using Student's *t*-test and Duncan's test. Different lowercase letters represent significant differences in blood indexes. Different uppercase letters represent significant differences in blood indexes ($p < 0.05$).

### 3.3. Natural Clustering Based on PCA Analysis

The unsupervised PCA was used for the exploratory data analysis. The PCA model was built using the blood index data from 252 grass carp samples (120 crisp grass carp and 132 ordinary grass carp). Most of the variance in the data could be visualized in the first a few principal components (PCs). PC1 and PC2 represented 50.64% and 24.77% of the variance from the original data, respectively (Figure 3A). From the PCA plots, samples originating from W1, C1, and C2 strongly overlapped and showed negative scores in PC2. There was a slight separation between the centroids of the W1 and W2 samples. W3 samples slightly overlapped W4 samples and had positive scores in the PC1. There were marked differences between the centroids of W1, W2, W3, and W4 samples, especially between W2 and W3, which indicated that long-term feeding with faba bean exerted an obvious effect on blood indexes. In contrast, C3 and C4 samples slightly overlapped in the negative side of the PC2. The orientation of the variables in the PC2–PC1 plane is observed in Figure 3B. PC1 was strongly influenced by positive contributions from GSH/GSSH and LY and by negative contributions from MDA and $H_2O_2$. The dominant variables in PC2 included PLT and RBC.

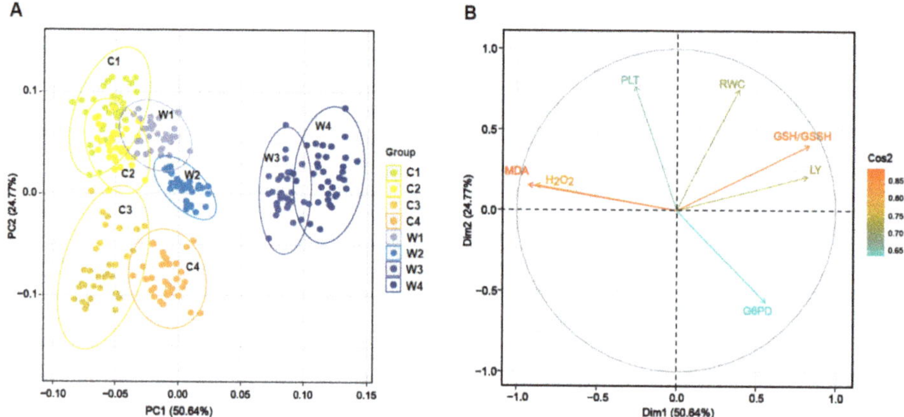

**Figure 3.** Natural clustering of crisp grass carp and ordinary grass carp. (**A**) Scatter plot of PCA scores from ordinary grass carp and crisp grass carp at different culturing stages. C1, C2, C3, and C4 represent the ordinary grass carp from 30, 60, 90, and 120 day, respectively. W1, W2, W3, and W4 represent the crisp grass carp from 30, 60, 90, and 120 day, respectively. (**B**) Loading plot for the original variables in the first two principal components (PCs).

The application of PCA allowed for a natural grouping of the 252 of grass carp samples, with a slight tendency of some samples to group more favorably. However, this approach could not systematically separate samples due to the overlap. Therefore, we subsequently applied several supervised methods classify crisp grass carp samples with different textures based on their blood indexes.

### 3.4. Classification and Comparing Classification Performance

Six machine learning techniques, GBDT, ANN, SVM, PLSR, NB, and RF, were applied to predict muscle hardness levels. The training and testing datasets were formed by 60% (152) and 40% (100) of the total samples, respectively.

In our analysis, the area under the receiver operating characteristic curve (ROC) was obtained for all models to aid in evaluations. The ROC represents a model's success across varying discrimination thresholds, with the AUC representing the overall probability of correct classification. The shape of the ROC curve also provides insight into the model's success [30]. The AUC of a binary class prediction could not be considered since our investigation involved a five-class prediction. Thus, micro- and macro-averages were used to obtain the ROC curves [31]. For the testing set, the AUCs of the micro- and macro-averages, respectively, were as follows: GBDT, 0.98 and 0.97 (Figure 4A); ANN, 0.92 and 0.90 (Figure 4B); SVM, 0.99 and 1 (Figure 4C); PLSR, 0.98 and 1 (Figure 4D); NB, 1 and 1 (Figure 4E); and RF, 1 and 1 (Figure 4F).

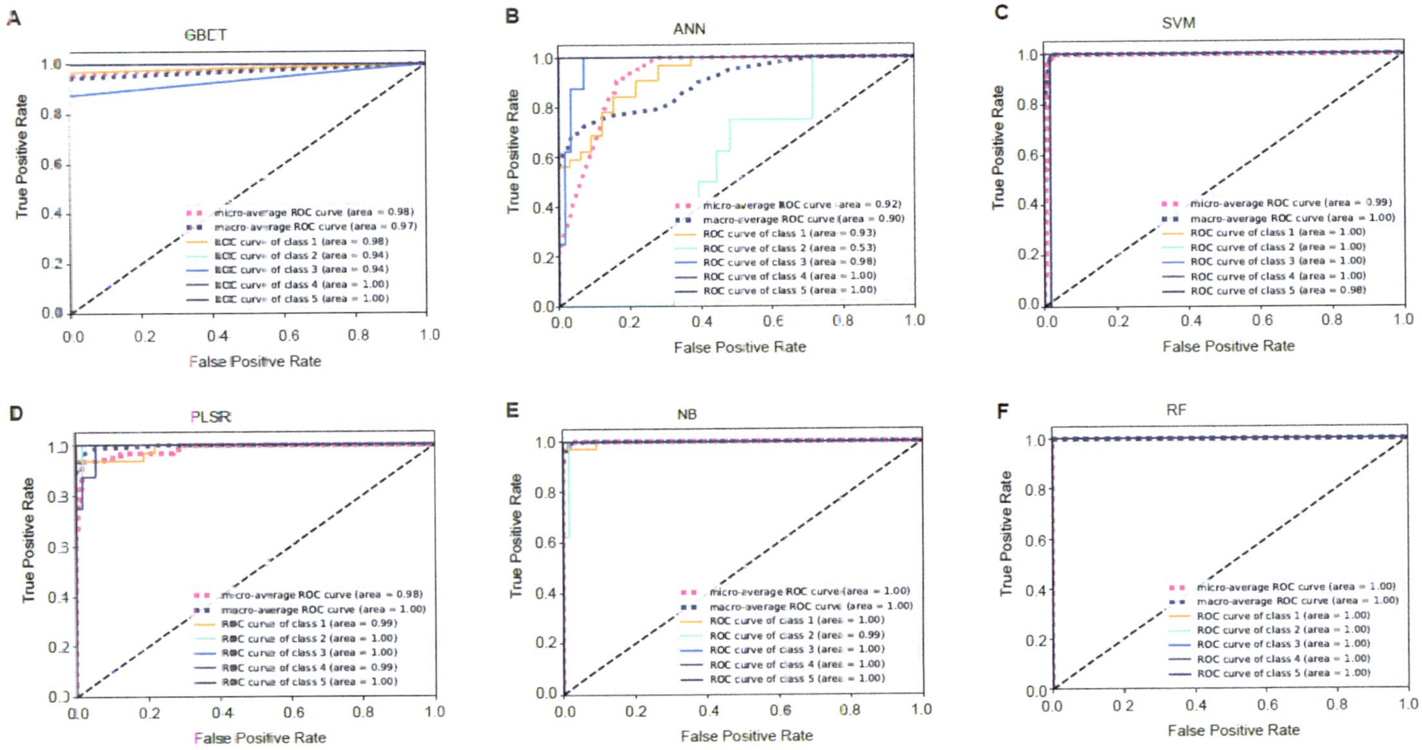

**Figure 4.** Receiver operating characteristic (ROC) curves for the six machine learning models. (**A**) Gradient boosting decision tree model. (**B**) Artificial neural network model. (**C**) Support vector machine model. (**D**) Partial least squares regression model. (**E**) Naïve Bayes model. (**F**) Random forest model.

A confusion matrix (Tables 1 and 2) was used to compare the discrimination performances of the six prediction models. The prediction accuracy of RF was the highest (96.01%) followed by NB (94.0%), PLSR (92.00%), GBDT (92.00%), SVM (91.00%), and ANN (89.00%). It is interesting to underline that the RF model exhibited the best prediction performance, with a classification accuracy of 100% for discriminating the crisp grass carp sampled across different culturing stages, while other models could not properly classify all of the target/authentic samples. The NB model incorrectly classified three observations from crisp grass carp with the third level hardness. GBDT and PLSR models correctly classified all but four crisp grass carp samples. ANN and SVM displayed lower classification performance with accuracies of 85.42% and 87.50%, respectively. RF demonstrated excellent average sensitivity (100%) and average specificity (93.08%) in classification of crisp grass carp samples.

**Table 1.** Confusion matrix of GBDT, ANN, and SVM classification for crisp grass carp. (**a**) GBDT. Accuracy of all samples = 92.00%; Accuracy of crisp grass carp samples = 91.67%. (**b**) ANN. Accuracy of all samples = 89.00%; Accuracy of crisp grass carp samples = 85.42%. (**c**) SVM. Accuracy of all samples = 91.00%; Accuracy of crisp grass carp samples = 87.50%.

| (a) GBDT model | | | | | | | | |
|---|---|---|---|---|---|---|---|---|
| Target | Level 1 | Level 2 | Level 3 | Level 4 | Level 5 | Sensitivity | ASC | ASA |
| level 1 | 48 | 3 | 0 | 1 | 0 | 92.31% | - | |
| level 2 | 0 | 12 | 0 | 0 | 0 | 100% | | |
| level 3 | 3 | 1 | 8 | 0 | 0 | 66.67% | 92.00% | 91.80% |
| level 4 | 0 | 0 | 0 | 12 | 0 | 100% | | |
| level 5 | 0 | 0 | 0 | 0 | 12 | 100% | | |
| Specificity | 94.12% | 75.00% | 100% | 90.91% | 100% | - | - | - |
| APC | - | 91.485 | | | | - | - | - |
| APA | | 92.01% | | | | - | - | - |

| (b) ANN model | | | | | | | | |
|---|---|---|---|---|---|---|---|---|
| Target | Level 1 | Level 2 | Level 3 | Level 4 | Level 5 | Sensitivity | ASC | ASA |
| level 1 | 48 | 3 | 0 | 1 | 0 | 92.31% | - | |
| level 2 | 1 | 9 | 2 | 0 | 0 | 75.00% | | |
| level 3 | 0 | 2 | 10 | 0 | 0 | 83.33% | 85.00% | 86.79% |
| level 4 | 0 | 0 | 0 | 10 | 2 | 83.33% | | |
| level 5 | 0 | 0 | 0 | 0 | 12 | 100% | | |
| Specificity | 97.95% | 64.26% | 83.33% | 90.91% | 85.71% | - | - | - |
| APC | - | 81.05% | | | | - | - | - |
| APA | | 84.43% | | | | - | - | - |

| (c) SVM model | | | | | | | | |
|---|---|---|---|---|---|---|---|---|
| Target | Level 1 | Level 2 | Level 3 | Level 4 | Level 5 | Sensitivity | ASC | ASA |
| level 1 | 49 | 2 | 0 | 1 | 0 | 94.23% | - | |
| level 2 | 0 | 8 | 3 | 0 | 0 | 66.67% | | |
| level 3 | 0 | 3 | 8 | 0 | 0 | 66.67% | 83.00% | 85.51% |
| level 4 | 0 | 0 | 0 | 12 | 0 | 100% | | |
| level 5 | 0 | 0 | 0 | 0 | 12 | 100% | | |
| Specificity | 100% | 61.54% | 72.73% | 92.31% | 100% | - | - | - |
| APC | - | 81.65% | | | | - | - | - |
| APA | | 85.32% | | | | - | - | - |

Note: ASC—average sensitivity of crisp grass carp samples; ASA—average sensitivity of all testing samples; APC—average specificity of crisp grass carp samples; APA—average specificity of all testing samples. Accuracy is the proportion of properly predicted samples to the total sample. Sensitivity and specificity are calculated as described in the Materials and Method. Green color denotes correct classification, red color denotes false classification.

**Table 2.** Confusion matrix of PLSR, NB, and RF classification for crisp grass carp. (**a**) PLSR. Accuracy of all samples = 92.00%; Accuracy of crisp grass carp samples = 91.67%. (**b**) NB. Accuracy of all samples = 94.00%; Accuracy of crisp grass carp samples = 93.75%. (**c**) RF. Accuracy of all samples = 96.00%; Accuracy of crisp grass carp samples = 100%.

| (a) PLSR model | | | | | | | | |
|---|---|---|---|---|---|---|---|---|
| **Target** | **Level 1** | **Level 2** | **Level 3** | **Level 4** | **Level 5** | **Sensitivity** | **ASC** | **ASA** |
| level 1 | 48 | 3 | 0 | 1 | 0 | 92.31% | - | |
| level 2 | 0 | 12 | | 0 | 0 | 100% | | |
| level 3 | 3 | 0 | 9 | 0 | 0 | 75.00% | 92.00% | 91.79% |
| level 4 | 0 | 0 | 0 | 11 | 1 | 91.66% | | |
| level 5 | 0 | 0 | 0 | 0 | 12 | 100% | | |
| Specificity | 94.12% | 80.00% | 100% | 91.67% | 92.31% | - | - | - |
| APC | - | 91.00% | | | | - | - | - |
| APA | | 91.62% | | | | - | - | - |

| (b) NB model | | | | | | | | |
|---|---|---|---|---|---|---|---|---|
| **Target** | **Level 1** | **Level 2** | **Level 3** | **Level 4** | **Level 5** | **Sensitivity** | **ASC** | **ASA** |
| level 1 | 49 | 2 | 0 | 1 | 0 | 94.23% | - | |
| level 2 | 0 | 12 | 0 | 0 | 0 | 100% | | |
| level 3 | 1 | 2 | 9 | 0 | 0 | 75.00% | 94.00% | 93.85% |
| level 4 | 0 | 0 | 0 | 12 | 0 | 100% | | |
| level 5 | 0 | 0 | 0 | 0 | 12 | 100% | | |
| Specificity | 98.00% | 75.00% | 100% | 92.31% | 100% | - | - | - |
| APC | - | 91.83% | | | | - | - | - |
| APA | | 93.06% | | | | - | - | - |

| (c) RF model | | | | | | | | |
|---|---|---|---|---|---|---|---|---|
| **Target** | **Level 1** | **Level 2** | **Level 3** | **Level 4** | **Level 5** | **Sensitivity** | **ASC** | **ASA** |
| level 1 | 48 | 3 | 0 | 1 | 0 | 92.30% | - | |
| level 2 | 0 | 12 | 0 | 0 | 0 | 100% | | |
| level 3 | 0 | 0 | 12 | 0 | 0 | 100% | 100% | 98.46% |
| level 4 | 0 | 0 | 0 | 12 | 0 | 100% | | |
| level 5 | 0 | 0 | 0 | 0 | 12 | 100% | | |
| Specificity | 100% | 80.00% | 100% | 92.31% | 100% | - | - | - |
| APC | - | 93.08% | | | | - | - | - |
| APA | | 94.46% | | | | - | - | - |

Note: Refer to Table 1 for abbreviations.

In sum, the RF, NB, PLSR, and GBDT models all presented excellent accuracy (>90%), being capable of separating crisp grass carp samples into classes. The RF model was superior to other models as it allowed the discrimination with an accuracy of 100%. However, the present study has several limitations that need to be addressed. First, we only used data from our experimental facility, and the results should be validated using samples from other culture ponds or conditions. Additionally, increasing the number of samples will enhance the reliability of the machine learning models. Both of these limitations are currently being addressed in our research group.

## 4. Conclusions

This study established six machine learning-based approaches for the classification of muscle hardness in different crisp grass carp samples based on seven blood indexes ($H_2O_2$, G6PD, GSH/GSSH, MDA, RBC, PLT, and LY). The results showed that the RF model has the highest classification accuracy of 100%, followed by the NB model (93.75% accuracy), whereas the ANN model was the least accurate

(85.42%). This approach provides an objective, cheap, fast, and reliable method that could help producers and consumers in evaluating the quality of in vivo crisp grass carp. Moreover, this system can be easily applied to evaluating the muscle quality of other freshwater fishes in vivo.

**Author Contributions:** Conceptualization, E.Y.; methodology, G.K.; software, G.K.; validation, J.X. and Z.L.; formal analysis, J.T. and W.G.; investigation, B.F.; resources, Y.X.; data curation, K.Z.; writing—original draft preparation, B.F.; writing—review and editing, E.Y., G.K., G.W.; supervision, J.X.; project administration, G.W.; funding acquisition, G.W. All authors have read and agreed to the published version of the manuscript.

**Funding:** This study was funded by the National Key R&D Program of China (2019YFD0900303), and Modern Agro-industry Technology Research System (No.CARS-45-21).

**Conflicts of Interest:** The authors declare no conflict of interest.

## References


1. FAO. *The State of World Fisheries and Aquaculture 2018-Meeting the Sustainable Development Goals*; Licence: CC BY-NC-SA 3.0 IGO; FAO: Rome, Italy, 2018.
2. Yu, E.; Fu, B.; Wang, G.; Li, Z.; Ye, D.; Jiang, Y.; Ji, H.; Wang, X.; Yu, D.; Ehsan, H.; et al. Proteomic and metabolomic basis for improved textural quality in crisp grass carp (*Ctenopharyngodon idellus C.* et V) fed with a natural dietary pro-oxidant. *Food Chem.* **2020**, *325*, 126906. [CrossRef] [PubMed]
3. Yu, E.M.; Zhang, H.F.; Li, Z.F.; Wang, G.J.; Wu, H.K.; Xie, J.; Yu, D.G.; Xia, Y.; Zhang, K.; Gong, W.B. Proteomic signature of muscle fibre hyperplasia in response to faba bean intake in grass carp. *Sci. Rep.* **2017**, *7*, e45950. [CrossRef] [PubMed]
4. Yang, S.; Li, L.; Qi, B.; Wu, Y.; Hu, X.; Lin, W.; Hao, S.; Huang, H. Quality evaluation of crisp grass carp (*Ctenopharyngodon idellus C.* ET V) based on instrumental texture analysis and cluster analysis. *Food Anal. Method.* **2015**, *8*, 2107–2114. [CrossRef]
5. Administration of Quality and Technology Supervision of Guangdong Province. *Zhongshan Crisp Grass Carp*; Standards Press of China: Beijing, China, 2010.
6. Koç, H.; Vinyard, C.J.; Essick, G.K.; Foegeding, E.A. Food oral processing: Conversion of food structure to textural perception. *Annu. Rev. Food Sci.* **2013**, *4*, 237–266. [CrossRef] [PubMed]
7. Du, C.J.; Sun, D.W. Learning techniques used in computer vision for food quality evaluation: A review. *J. Food Eng.* **2006**, *72*, 39–55. [CrossRef]
8. Xu, J.L.; Riccioli, C.; Sun, D.W. Comparison of hyperspectral imaging and computer vision for automatic differentiation of organically and conventionally farmed salmon. *J. Food Eng.* **2017**, *196*, 170–182. [CrossRef]
9. Saberioon, M.; Císař, P.; Labbé, L.; Souček, P.; Pelissier, P.; Kerneis, T. Comparative performance analysis of support vector machine, random forest, logistic regression and k-nearest neighbours in rainbow trout (*Oncorhynchus mykiss*) classification using image-based features. *Sensors* **2018**, *18*, 1027. [CrossRef]
10. Hu, J.; Li, D.; Duan, Q.; Han, Y.; Chen, G.; Si, X. Fish species classification by color, texture and multi-class support vector machine using computer vision. *Comput. Electron. Agric.* **2012**, *88*, 133–140. [CrossRef]
11. Hernández-Serna, A.; Jiménez-Segura, L.F. Automatic identification of species with neural networks. *PeerJ* **2014**, *2*, e563.
12. Rodríguez-Bermúdez, R.; Herrero-Latorre, C.; López-Alonso, M.; Losada, D.E.; Iglesias, R.; Miranda, M. Organic cattle products: Authenticating production origin by analysis of serum mineral content. *Food Chem.* **2018**, *264*, 210–217. [CrossRef]
13. James, G.; Witten, D.; Hastie, T.; Tibshirani, R. *An Introduction to Statistical Learning: With Applications in R*; Springer: New York, NY, USA, 2013.
14. Pedregosa, F.; Varoquaux, G.; Gramfort, A.; Michel, V.; Thirion, B.; Grisel, O. Scikit-learn: Machine learning in Python. *J. Mach. Learn. Res.* **2011**, *12*, 2825–2830.
15. Bro, R.; Smilde, A.K. Principal component analysis. *Anal. Methods* **2014**, *6*, 2812–2831. [CrossRef]
16. Varmuza, K.; Filzmoser, P. *Introduction to Multivariate Statistical Analysis in Chemometrics*; CRC Press: Boca Raton, FL, USA, 2009.
17. Hastie, T.; Tibshirani, R.; Friedman, J. *The Elements of Statistical Learning: Data Mining, Inference, and Prediction*; Springer: New York, NY, USA, 2009.
18. Sun, R.; Wang, G.; Zhang, W.; Hsu, L.T.; Ochieng, W.Y. A gradient boosting decision tree based GPS signal reception classification algorithm. *Appl. Soft Comput.* **2020**, *86*, 105942. [CrossRef]

19. Dwivedi, A.K. Artificial neural network model for effective cancer classification using microarray gene expression data. *Neural Comput. Appl.* **2018**, *29*, 1545–1554. [CrossRef]
20. Cortes, C.; Vapnik, V. Support-vector networks. *Mach. Learn.* **1995**, *20*, 273–297. [CrossRef]
21. Fung, G.M.; Mangasarian, O.L. Multicategory proximal support vector machine classifiers. *Mach. Learn.* **2005**, *59*, 77–97. [CrossRef]
22. Lozano, J.; Santos, J.P.; Arroyo, T.; Aznar, M.; Cabellos, J.M.; Gil, M.; del Carmen Horrillo, M. Correlating e-nose responses to wine sensorial descriptors and gas chromatography–mass spectrometry profiles using partial least squares regression analysis. *Sens. Actuators B Chem.* **2007**, *127*, 267–276. [CrossRef]
23. Abdi, H. Partial least squares regression and projection on latent structure regression (PLS Regression). *Comput. Stat.* **2010**, *2*, 97–106. [CrossRef]
24. Liu, B.; Blasch, E.; Chen, Y.; Shen, D.; Chen, G. Scalable sentiment classification for big data analysis using naive bayes classifier. In Proceedings of the 2013 IEEE International Conference on Big Data, Silicon Valley, CA, USA, 6–9 October 2013; pp. 99–104.
25. Breiman, L. Random forests. *Mach. Learn.* **2001**, *45*, 5–32. [CrossRef]
26. Casella, G.; Fienberg, S.; Olkin, I. *An Introduction to Statistical Learning. Design*; Springer: New York, NY, USA, 2006; Volume 102.
27. Schwertman, N.C.; Owens, M.A.; Adnan, R. A simple more general boxplot method for identifying outliers. *Comput. Stat. Data Anal.* **2004**, *47*, 165–174. [CrossRef]
28. Sim, C.H.; Gan, F.F.; Chang, T.C. Outlier labeling with boxplot procedures. *J. Am. Stat. Assoc.* **2005**, *100*, 642–652. [CrossRef]
29. Ma, L.L.; Kaneko, G.; Wang, X.J.; Xie, J.; Tian, J.J.; Zhang, K.; Wang, G.J.; Yu, D.G.; Li, Z.F.; Gong, W.B.; et al. Effects of four faba bean extracts on growth parameters, textural quality, oxidative responses, and gut characteristics in grass carp. *Aquaculture* **2020**, *516*, 734620. [CrossRef]
30. Mandrekar, J.N. Receiver operating characteristic curve in diagnostic test assessment. *J. Thorac. Oncol.* **2010**, *5*, 1315–1316. [CrossRef]
31. Xu, K.; Lam, M.; Pang, J.; Gao, X.; Band, C.; Mathur, P. Multimodal machine learning for automated ICD coding. In Proceedings of the Machine Learning for Healthcare Conference, Ann Arbor, MI, USA, 8–10 August 2019; pp. 197–215.

**Publisher's Note:** MDPI stays neutral with regard to jurisdictional claims in published maps and institutional affiliations.

MDPI

St. Alban-Anlage 66

4052 Basel

Switzerland

Tel. +41 61 683 77 34

Fax +41 61 302 89 18

www.mdpi.com

*Foods* Editorial Office

E-mail: foods@mdpi.com

www.mdpi.com/journal/foods